Climate Change and United States Forests

ADVANCES IN GLOBAL CHANGE RESEARCH

VOLUME 57

Editor-in-Chief

Martin Beniston, *University of Geneva, Switzerland*

Editorial Advisory Board

B. Allen-Diaz, *Department ESPM-Ecosystem Sciences, University of California, Berkeley, CA, U.S.A.*
R.S. Bradley, *Department of Geosciences, University of Massachusetts, Amherst, MA, U.S.A.*
W. Cramer, *Department of Global Change and Natural Systems, Potsdam Institute for Climate Impact Research, Potsdam, Germany.*
H.F. Diaz, *Climate Diagnostics Center, Oceanic and Atmospheric Research, NOAA, Boulder, CO, U.S.A,*
S. Erkman, *Institute for Communication and Analysis of Science and Technology–ICAST, Geneva, Switzerland*
R. Garcia Herrera, *Faculated de Fisicas, Universidad Complutense, Madrid, Spain*
M. Lal, *Center for Atmospheric Sciences, Indian Institute of Technology, New Delhi, India.*
U. Luterbacher, *The Graduate Institute of International Studies, University of Geneva, Geneva, Switzerland.*
I. Noble, *CRC for Greenhouse Accounting and Research School of Biological Science, Australian National University, Canberra, Australia.*
L. Tessier, *Institut Mediterranéen d'Ecologie et Paléoécologie, Marseille, France.*
F. Toth, *International Institute for Applied Systems Analysis Laxenburg, Austria.*
M.M. Verstraete, *Institute for Environment and Sustainability, Ec Joint Research Centre, Ispra (VA), Italy.*

For further volumes:
http://www.springer.com/series/5588

David L. Peterson • James M. Vose
Toral Patel-Weynand
Editors

Climate Change and United States Forests

Springer

Editors
David L. Peterson
U.S. Forest Service
Pacific Northwest Research Station
Seattle, WA, USA

James M. Vose
U.S. Forest Service
Southern Research Station
Raleigh, NC, USA

Toral Patel-Weynand
U.S. Forest Service
Research and Development
Arlington, VA, USA

ISSN 1574-0919
ISBN 978-94-007-7514-5   ISBN 978-94-007-7515-2 (eBook)
DOI 10.1007/978-94-007-7515-2
Springer Dordrecht Heidelberg New York London

Library of Congress Control Number: 2013957377

© Springer Science+Business Media Dordrecht (outside the USA) 2014
This work is subject to copyright. All rights are reserved by the Publisher, whether the whole or part of the material is concerned, specifically the rights of translation, reprinting, reuse of illustrations, recitation, broadcasting, reproduction on microfilms or in any other physical way, and transmission or information storage and retrieval, electronic adaptation, computer software, or by similar or dissimilar methodology now known or hereafter developed. Exempted from this legal reservation are brief excerpts in connection with reviews or scholarly analysis or material supplied specifically for the purpose of being entered and executed on a computer system, for exclusive use by the purchaser of the work. Duplication of this publication or parts thereof is permitted only under the provisions of the Copyright Law of the Publisher's location, in its current version, and permission for use must always be obtained from Springer. Permissions for use may be obtained through RightsLink at the Copyright Clearance Center. Violations are liable to prosecution under the respective Copyright Law.
The use of general descriptive names, registered names, trademarks, service marks, etc. in this publication does not imply, even in the absence of a specific statement, that such names are exempt from the relevant protective laws and regulations and therefore free for general use.
While the advice and information in this book are believed to be true and accurate at the date of publication, neither the authors nor the editors nor the publisher can accept any legal responsibility for any errors or omissions that may be made. The publisher makes no warranty, express or implied, with respect to the material contained herein.

Cover caption: ©vovan/2003–2013 Shutterstock, Inc.

Printed on acid-free paper

Springer is part of Springer Science+Business Media (www.springer.com)

The remainder of the twenty-first century will pose significant challenges for natural resource managers as they respond to rapid and unexpected changes. We have spent most of our careers studying how forest ecosystems work and how to keep them working so they can continue to provide a myriad of benefits to society. We hope that our efforts in producing this book will help ensure the sustainability of forests in the face of climate change for future generations. We dedicate this book to Gordon Weynand for his unwavering support and to the next generation, future stewards of the global environment: Aadya, Aaron, Christian, Christina, Nate, and Zak.

Foreword

Managing forest ecosystems has always been about dealing with change and providing for the future. *Climate Change and United States Forests* shows how changes in the climate are causing pervasive and far-reaching changes in forest ecosystems. The book helps us, the benefactors of services provided by forest ecosystems, connect the scientific dots and better understand the big patterns. Hopefully, these insights will drive our thinking and actions as we confront recent and future changes in our forests.

The authors have methodically surveyed the scientific literature for a wide range of climatic effects, organizing them into regional projections for the future and calling for flexibility and nuance in management and policy action. Their exploration of a large and growing body of science gives us a clear, intricate, and balanced picture of both challenges and opportunities, unencumbered by ideological advocacy and policy prescriptions. What follows is not a doomsday prophecy supported by selections from the literature, but a clear-eyed synthesis of observations and insights from climate and forest science. It offers both alarm and hope, challenging us to address the overarching asynchrony of a climate changing faster than some forest systems and species can adapt. Whether, when, and how to intervene with proactive adaptation are ultimately society's decisions, but the assessment that follows can help assure that these choices are well-informed.

The picture here is not of a new scientific terrace where we can stop and catch our breath, but of systems in motion, where we must use scientific inquiry and management experience to provide signals of pattern shifts, new configurations, and emerging issues. This is the science of the unsettled, where decision makers and citizens must learn to refuel in flight.

This is a summary for those who take and learn from actions. It raises hope and provides examples of taking action to deal with the changing climate. It encourages us to move forward with actions that help us understand and deal more effectively with the complexities of different climatic effects in different systems. While the myriad uncertainties about climate change are obsessively debated and amplified by theorists, policy scholars, wonks, ideologues, and advocates, it is refreshing that science can be interpreted from the perspective of proactive adaptation and learning. We have the tools—silviculture, genetics, fire and fuels management, engineering, hydrology, forest products, forest economics—that can be modified or used in different combinations to help steer adaptive processes. We may need to deploy these tools not just to manipulate systems but also for learning, because as the book fully discloses, there is a lot we do not yet know. It is heartening to know that resource managers and institutions are already taking action, establishing new partnerships, and innovating strategies and techniques that will allow us to adapt to a cadence of change that will accelerate.

The authors emphasize that actions must be based on new realities. Most obvious but perhaps underappreciated is the fact that the climate is indeed changing. These changes, which have become increasingly supported by observation by scientists, managers, and citizens, have profound effects on forest resources and our abilities to use, appreciate, and manage them. And they call on us to reexamine and challenge assumptions about stationarity and heretofore predictable and recurring cycles that may underlie some of our forest management practices. These changes must be put in temporal and spatial perspective and understood in the context of what we already know about how forests grow and change. We know a lot about how forests vary geographically, how they respond to multiple stresses, and the roles they play in various biogeochemical cycles. The book challenges us to integrate these new findings into the design of actions and measures of success as the forest around us moves and changes.

We are now encouraged to build "climate smartness" into how we undertake forest management and how we assess issues of policy and social expectations. The changing climate should not be a stand-alone issue, but rather a property of all resource discussions and decisions. By describing climate change effects in a risk management framework, the authors have provided a structure to guide integration. All decisions involve weighing the tradeoffs among benefits, costs, opportunities, and risks, and it is easier to blend climatic effects on forests into ongoing decision processes when they are expressed in the common language of risk assessment and risk management. Climate change is a component of broader risk-based thinking in which all elements of forest enterprise are integrated—vegetation protection and management, roads and access, harvesting and products.

The book is a vivid reminder that we humans influence forests through (1) the overarching and increasingly evident role of global climate change, and (2) multiple, direct influences of an expanding population on urban development, fire management, commerce, water use, and other resource-dependent activities. Where these two factors collide in systems already attuned to patterns of natural disturbance, we are seeing changes we have not had to deal with before. Deforestation of the last two centuries may have been acute and visible, but the solution was inherently simple yet massive: protect, reforest, manage. Today's issues created by climate change intersecting with the intricacies of an existing forest are more complex and less amenable to blunt national policy prescriptions.

The overall impact of the book may be to change the way we think and talk about forests. The authors assure us that despite the effects of a changing climate, all is not chaos. But we do need to "get up and move around a little" to get new blood flowing to our collective brain. As climate changes play out into different regional weather patterns and responses, we cannot assume that forests will stay as they are or where they are today. We need to reexamine what we expect from forests and what ecosystem services they can provide. Forests and their changing provision of services are not limited to the traditionally defined rural sector that produces forest products; forests are also vital elements of infrastructure in urban and agricultural systems. Changes are underway not only for the forest landscape, but also for the forest in the landscape, wherever it may occur.

We can now address multiple, interacting sources of stress and disturbance and the rapid changes they create as they combine, and not limit our thinking to just fire, insects, disease, air pollution, invasive species, and human development as separate influences. Paying more attention to extreme conditions and events will allow us to understand how their patterns differ over time and space and their influence in the life cycles of forest systems. It is critical that we closely follow regenerative pathways after these events, no longer assuming that the system will be "reset" predictably to some familiar forest condition. Rather than looking at effects on one species at a time, we need to monitor and understand changes in entire forest systems, positioned at the intersection of cycles for water, carbon, and other vital ecosystem functions.

An emerging imperative from the book concerns our most basic approaches to creating and using knowledge in a rapidly changing world. The results described here serve as both warning and inspiration to develop better ways to convert existing and emerging knowledge into proactive decisions about tomorrow. We need to become better at interpreting trends, describing alternative futures and designing forest management actions that are robust and flexible to a wider range of future conditions. Research and development are vital for finding our way forward through a changing climate. Without the integration of advancing forest and climate science, a broad picture of our nation's changing forests is not possible. Syntheses such as this help us identify gaps, reset our focus, and remind us that resource management is itself a learning device.

But this book also shows that the best science for adaptation will be conceived in strong adaptive research-management partnerships. We must find new ways for scientists, managers, and citizens to work together to pool their observational powers and intelligence to continually reexamine the realities of forest systems. With so many changes underway and more to come, we cannot afford to follow the linear, sequential model of science-based management in which actions are contingent on research providing the absolute certainty that never quite arrives. New relationships between science and management, as demonstrated in examples here, will streamline our learning and integrate emerging lessons from experimental science, experiential learning, and traditional forms of knowledge.

Ultimately, adaptation is not an action or a set of actions. It is a way of gaining and using knowledge, of creating and preserving options, and of cultivating new institutions that are agile, open, and flexible enough to provide enduring values in dynamic forest ecosystems. *Climate Change and United States Forests* can be viewed as a call for new strategies and institutional arrangements to address adaptation in this larger sense, as we venture from forests as they are today to forests of the future, shaped by the interacting forces of climatic, demographic, and economic change.

Climate Change Advisor David Cleaves
U.S. Forest Service
Washington, DC

Preface

Climate Change and United States Forests assesses the current condition and likely future condition of forest resources in the United States relative to climatic variability and change. Derived from a report that provides technical input to the 2014 U.S. Global Change Research Program National Climate Assessment, it serves as a framework for managing forest resources in the United States in the context of climate change. A complete synthesis of all of the effects of climatic variability and change on forest resources in the United States would require a multi-volume effort, especially given the enormous scientific literature on climate change over the past 20 years. Therefore, we focus on topics that have the greatest potential to alter the structure and function of forest ecosystems, and therefore ecosystem services, by the end of the twenty-first century.

Part I provides an environmental context for assessing the effects of climate change on forest resources. First, recent changes in environmental stressors, including climatic (e.g., temperature, droughts) and other biophysical phenomena (e.g., wildfire, insects), are summarized in Chap. 1. Then, state-of-science projections for future climate are presented for parameters relevant to forest ecosystems (Chap. 2).

Part II provides a wide-ranging assessment of vulnerability of forest ecosystems and ecosystem services to climate change. Biogeochemical cycling (including carbon), hydrology, and forest dynamics, which are strongly affected by climate and are expected to change significantly in some regions of the United States, are the focus of Chap. 3. We anticipate that altered disturbance regimes and stressors will have the biggest effects on forest ecosystems, causing long-term and in some cases permanent changes in forest conditions. Chap. 4 documents the effects of ecological disturbance and examines projected future disturbance regimes. Forest values and the socioeconomic context for human-forest interactions in the United States, ranging from rural to urban environments, are discussed in Chap. 5. Chapters 3, 4 and 5 cannot capture the enormous variability in biogeographic phenomena across U.S. forests; therefore, Part II concludes with a series of short summaries of climate change effects, issues, and adaptation for eight regions of the United States (Chap. 6).

Part III describes social and management responses to climate change in U.S. forests. Current status and trends in forest carbon, effects of carbon management, and carbon mitigation strategies are summarized in Chap. 7. Current and projected greenhouse gas emissions make climate change inevitable, so it is imperative that we prepare forest ecosystems and land management organizations for a permanently warmer climate. We are fortunate that principles of climate change adaptation are well established and that tools and resources to facilitate this management transition are available (Chap. 8). Risk assessment is regarded as a foundation for the 2013 National Climate Assessment and the Fifth Assessment of the Intergovernmental Panel on Climate Change (IPCC) (expected in 2014). Part III concludes with a framework for risk assessment, including case studies, to provide a structured approach for projecting future changes in resource conditions and ecosystem services (Chap. 9).

Finally, Part IV describes how sustainable forest management, the paradigm that guides activities on most public and private lands in the United States, can provide an overarching structure for mitigation of and adaptation to climate change (Chap. 10).

Because of the complexity of forest ecosystems, it is often difficult to conclude whether recently observed trends or changes in ecological phenomena are the direct result of human-caused climate change, climatic variability, or other factors. Regardless of the cause, forest ecosystems in the United States at the end of the twenty-first century will differ from those of today as a result of changing climate. Surprises are likely—some forests may change faster than we expect, some forests may be more tolerant of a warmer climate than we expect, or a new non-native insect may be a "game changer" by quickly killing large areas of native forest species. Because the current trajectory of greenhouse gas emissions implies at least one to three centuries of higher temperatures, preparing for future changes in forest ecosystems is imperative.

Climate change science must quickly move from the academic realm to the applied world of resource management. Land managers in the United States are faced with a landscape that has been greatly altered, with some 90 % of the nation's forest having been harvested in the nineteenth and twentieth centuries. Urban areas are encroaching on wildlands. Private forest land is becoming increasingly fragmented and is expected to decrease in the future. Non-native flora often comprises more than 10 % of the vegetation in a given location. Although production forestry is still important in some regions, especially the southeastern United States, restoration is dominant in other regions. Because restoration must now occur in a warmer climate, we can no longer use static images of the past (e.g., historic range of variability) as targets for future conditions. We must provide land managers with the expertise, scientific principles, and techniques for transitioning forest ecosystems into a warmer, more variable climate.

Because our charge was to provide input to the U.S. National Climate Assessment, we have a provincial focus on United States forests and have not considered the broader geographic realm of North America and other continents. However, we anticipate that this book will contribute to ongoing efforts to synthesize information at continental and global scales (e.g., the Fifth IPCC Assessment). In terms of on-the-ground management of forest resources, vulnerability assessments and adaptation strategies are most useful at the regional to sub-regional scales, and we hope that recent collaborative efforts described in this book will propagate across all landscapes in the United States.

We are optimistic about the future of forest resources in the United States, assuming that a strong commitment to monitor and respond to climate change is institutionalized within land management agencies and other organizations. Failure to do so may preclude future options for ensuring the long-term productivity and functionality of forest ecosystems. How will future generations judge the resource stewardship of our generation?

Seattle, WA, USA	David L. Peterson
Raleigh, NC, USA	James M. Vose
Arlington, VA, USA	Toral Patel-Weynand

Acknowledgments

We were fortunate to work with an outstanding team of authors who provided succinct, timely contributions to this book. We are humbled by their expertise in climate change science and their ability to articulate complex issues so they can be understood by a broad readership. We are grateful to land managers in the U.S. Forest Service and other organizations for feedback and ideas that contributed to the content of the book. They have developed many of the adaptation concepts and options discussed here, and we are impressed at how they can quickly respond to challenging climate change scenarios with constructive management approaches. We are also grateful to participants in a national stakeholders workshop held in Atlanta, Georgia, in July, 2011, who provided input that helped frame the subject matter content and management options in this book, ensuring relevance for decision makers and resource managers.

We thank the following individuals for thoughtful reviews of material in the book: Craig Allen, Dan Binkley, Barry Bollenbacher, Paul Bolstad, Marilyn Buford, Catherine Dowd, Daniel Fagre, Bruce Goines, Margaret Griep, Sue Hagle, Edie Sonne Hall, Jeffrey Hicke, Randy Johnson, Evan Mercer, David Meriwether, James Morrison, Charles Sams, Amy Snover, and David Theobald. Ellen Eberhardt assisted with formatting, checking references, and tracking many details of the editorial process. Patricia Loesche carefully edited the entire manuscript and provided many helpful suggestions on writing quality. Robert Norheim created and revised several maps. Nancy Walters provided skillful facilitation of the national stakeholders workshop mentioned above. Takeesha Moerland-Torpey of Springer, Environmental Science, provided helpful guidance throughout the preparation of the book manuscript. Financial support was provided by the U.S. Forest Service Pacific Northwest Research Station (Peterson), Southern Research Station (Vose), and national research and development office (Patel-Weynand).

Contents

Part I Seeking the Climate Change Signal

1. **Recent Changes in Climate and Forest Ecosystems** 3
   David L. Peterson and Kailey W. Marcinkowski

2. **Projected Changes in Future Climate** 13
   Chelcy F. Miniat and David L. Peterson

Part II Effects of Climatic Variability and Change

3. **Forest Processes** 25
   Michael G. Ryan, James M. Vose, Paul J. Hanson,
   Louis R. Iverson, Chelcy F. Miniat, Charles H. Luce,
   Lawrence E. Band, Steven L. Klein, Don McKenzie,
   and David N. Wear

4. **Disturbance Regimes and Stressors** 55
   Matthew P. Ayres, Jeffrey A. Hicke, Becky K. Kerns,
   Don McKenzie, Jeremy S. Littell, Lawrence E. Band,
   Charles H. Luce, Aaron S. Weed, and Crystal L. Raymond

5. **Climate Change and Forest Values** 93
   David N. Wear, Linda A. Joyce, Brett J. Butler,
   Cassandra Johnson Gaither, David J. Nowak,
   and Susan I. Stewart

6. **Regional Highlights of Climate Change** 113
   David L. Peterson, Jane M. Wolken, Teresa N. Hollingsworth,
   Christian P. Giardina, Jeremy S. Littell, Linda A. Joyce,
   Christopher W. Swanston, Stephen D. Handler,
   Lindsey E. Rustad, and Steven G. McNulty

Part III Responding to Climate Change

7   Managing Carbon 151
Kenneth E. Skog, Duncan C. McKinley, Richard A. Birdsey,
Sarah J. Hines, Christopher W. Woodall,
Elizabeth D. Reinhardt, and James M. Vose

8   Adapting to Climate Change 183
Constance I. Millar, Christopher W. Swanston,
and David L. Peterson

9   Risk Assessment 223
Dennis S. Ojima, Louis R. Iverson, Brent L. Sohngen,
James M. Vose, Christopher W. Woodall, Grant M. Domke,
David L. Peterson, Jeremy S. Littell, Stephen N. Matthews,
Anantha M. Prasad, Matthew P. Peters, Gary W. Yohe,
and Megan M. Friggens

Part IV Scientific Issues and Priorities

10   Research and Assessment in the Twenty-First Century 247
Toral Patel-Weynand, David L. Peterson, and James M. Vose

Index 253

Contributors

Matthew P. Ayres Department of Biological Sciences, Dartmouth College, Hanover, NH, USA

Lawrence E. Band Department of Geography, University of North Carolina, Chapel Hill, NC, USA

Richard A. Birdsey Northern Research Station, U.S. Forest Service, Newtown Square, PA, USA

Brett J. Butler Northern Research Station, U.S. Forest Service, Amherst, MA, USA

Grant M. Domke Northern Research Station, U.S. Forest Service, St. Paul, MN, USA

Megan M. Friggens Rocky Mountain Research Station, U.S. Forest Service, Albuquerque, NM, USA

Christian P. Giardina Pacific Southwest Research Station, U.S. Forest Service, Hilo, HI, USA

Stephen D. Handler Northern Research Station, U.S. Forest Service, Houghton, MI, USA

Paul J. Hanson Environmental Sciences Division, Oak Ridge National Laboratory, Oak Ridge, TN, USA

Jeffrey A. Hicke Department of Geography, University of Idaho, Moscow, ID, USA

Sarah J. Hines Rocky Mountain Research Station, U.S. Forest Service, Fort Collins, CO, USA

Teresa N. Hollingsworth Pacific Northwest Research Station, U.S. Forest Service, Fairbanks, AK, USA

Louis R. Iverson Northern Research Station, U.S. Forest Service, Delaware, OH, USA

Cassandra Johnson Gaither Southern Research Station, U.S. Forest Service, Athens, GA, USA

Linda A. Joyce Rocky Mountain Research Station, U.S. Forest Service, Fort Collins, CO, USA

Becky K. Kerns Pacific Northwest Research Station, U.S. Forest Service, Corvallis, OR, USA

Steven L. Klein National Health and Environmental Effects Research Laboratory, Western Ecology Division, U.S. Environmental Protection Agency, Corvallis, OR, USA

Jeremy S. Littell Alaska Climate Science Center, U.S. Geological Survey, Anchorage, AK, USA

Charles H. Luce Rocky Mountain Research Station, U.S. Forest Service, Boise, ID, USA

Kailey W. Marcinkowski Northern Institute of Applied Climate Science, Michigan Technological University, Houghton, MI, USA

Stephen N. Matthews Northern Research Station, U.S. Forest Service, Delaware, OH, USA

Don McKenzie Pacific Northwest Research Station, U.S. Forest Service, Seattle, WA, USA

Duncan C. McKinley National Research and Development, U.S. Forest Service, Washington, DC, USA

Steven G. McNulty Southern Research Station, U.S. Forest Service, Raleigh, NC, USA

Constance I. Millar Pacific Southwest Research Station, U.S. Forest Service, Albany, CA, USA

Chelcy F. Miniat Southern Research Station, U.S. Forest Service, Otto, NC, USA

David J. Nowak Northern Research Station, U.S. Forest Service, Syracuse, NY, USA

Dennis S. Ojima Natural Resource Ecology Laboratory, Colorado State University, Fort Collins, CO, USA

Toral Patel-Weynand National Research and Development, U.S. Forest Service, Arlington, VA, USA

Matthew P. Peters Northern Research Station, U.S. Forest Service, Delaware, OH, USA

David L. Peterson Pacific Northwest Research Station, U.S. Forest Service, Seattle, WA, USA

Anantha M. Prasad Northern Research Station, U.S. Forest Service, Delaware, OH, USA

Crystal L. Raymond Seattle City Light, Seattle, WA, USA

Elizabeth D. Reinhardt Fire and Aviation Management, U.S. Forest Service, Washington, DC, USA

Lindsey E. Rustad Northern Research Station, U.S. Forest Service, Durham, NH, USA

Michael G. Ryan Natural Resource Ecology Laboratory, Colorado State University, Fort Collins, CO, USA

Kenneth E. Skog Forest Products Laboratory, U.S. Forest Service, Madison, WI, USA

Brent L. Sohngen Department of Agricultural, Environmental, and Development Economics, Ohio State University, Columbus, OH, USA

Susan I. Stewart Northern Research Station, U.S. Forest Service, Evanston, IL, USA

Christopher W. Swanston Northern Research Station, U.S. Forest Service, Houghton, MI, USA

James M. Vose Southern Research Station, U.S. Forest Service, Raleigh, NC, USA

David N. Wear Southern Research Station, U.S. Forest Service, Raleigh, NC, USA

Aaron S. Weed Department of Biological Sciences, Dartmouth College, Hanover, NH, USA

Jane M. Wolken School of Natural Resources and Agricultural Sciences, University of Alaska, Fairbanks, AK, USA

Christopher W. Woodall Northern Research Station, U.S. Forest Service, St. Paul, MN, USA

Gary W. Yohe Economics Department, Wesleyan University, Middletown, CT, USA

List of Figures

Fig. 1.1 Cumulative mortality area (ha) from 1997 to 2010 for the western conterminous United States and British Columbia for trees killed by bark beetles. Data are adjusted for underestimation (calculated by comparison with classified imagery) 6

Fig. 1.2 Annual area burned by wildfire on federal lands in the 11 large western states in the conterminous United States, since 1916, including an indication of warm and cool phases of the Pacific Decadal Oscillation (*PDO*)............... 7

Fig. 2.1 Multi-model mean annual differences in temperature between three future periods compared to 1971–2000, from 15 GCMs using two emission scenarios (*A2* and *B1*). The A2 scenario is for higher emissions than for B1 (see text). For most interior states, models project a 1.4–1.9 °C temperature increase, rising to 2.5–3.6 °C for 2051–2071, and to greater than 4.2 °C for 2071–2099, depending on the emission scenario 15

Fig. 2.2 Spatial distribution of the mean change in the annual number of days with a maximum temperature above 35 °C (a), and in the annual number of consecutive days with a maximum temperature above 35 °C (b) between 1971 and 2000 and 2041–2070. Models project that much of the southeastern and southwestern United States will experience more days with maximum temperature above 35 °C, and more consecutive days above that temperature. Results are for the high (A2) emission scenario only, from the North American Regional Climate Change Assessment Program multi-model means (n = 9 GCMs).................................... 17

Fig. 2.3   Mean percentage of annual differences in U.S.
            precipitation between three future periods relative to a
            1971–2000 reference period. The Northeast, northern
            Midwest and Pacific Northwest are projected to have
            slightly more precipitation, and the Southwest is
            projected to have 2–12 % less precipitation, depending
            on the emission scenario, location, and time period.
            Means are for 15 GCMs.. 18

Fig. 2.4   Trend in the Palmer Drought Severity Index (*PDSI*) per
            decade for (a) observed data and the mean of the (b)
            first half and (c) second half of the twenty-first century.
            The PDSI is projected to decrease by 0.5–1 unit per
            decade for the period 2050–2096. For the PDSI, −1.9 to
            1.9 is near normal, −2 to −2.9 is moderate drought, −3
            to −3.9 is severe drought and less than −4 is extreme
            drought. Projections are made by HadCM3 with the A2
            emission scenario... 19

Fig. 2.5   The projected average annual proportion of the global
            land surface in drought each month shows drought
            increasing over the current century. Drought is defined
            as extreme, severe, or moderate, which represents 1, 5
            and 20 %, respectively, of the land surface in drought
            under present-day conditions. Results from the three
            simulations are from HadCM3 with the A2 emission scenario 20

Fig. 3.1   Risk analysis diagram for the forest C cycle. Western
            forests are considered inherently limited by water
            demands that exceed precipitation supplies during
            substantial portions of the year. Xeric Eastern forests
            include those growing on shallow or coarse-textured
            soils or those present at the prairie-forest transition zone
            that experience water deficits in some years. Mesic
            Eastern forests experience severe water deficits only in
            occasional years and for relatively brief periods 28

Fig. 3.2	Maps of current and potential future suitable habitat for sugar maple in the United States show potential northward movement of habitat by 2100. In addition to showing the range of sugar maple in Little (1971), the map includes the current inventory estimate of abundance from U.S. Forest Service Forest Inventory and Analysis (FIA-current) sampling and the modeled current distribution (RF-current). Model projections for future climate are: (1) low emission scenario (B1) using the average of three global climate models (GCM3 Avg lo), (2) low emission scenario (B1) using the National Center for Atmospheric Research Parallel Climate Model (PCM lo), (3) high emission scenario (A1F1) using the average of three global climate models (GCM3 Avg hi), (4) high emission scenario (A1F1) using the HadleyCM3 model (Hadley hi)	37
Fig. 3.3	Maps of current and potential future suitable habitat for U.S. Forest Service forest types (named according to dominant species) in the eastern United States show potential northward movement of forest types by 2100. The map includes the current inventory estimate of abundance from U.S. Forest Service Forest Inventory and Analysis (FIA-current) sampling and modeled current distribution (RF-current). Model projections for future climate are: (1) low emission scenario (B1) using the average of three global climate models (GCM3 Avg lo), (2) low emission scenario (B1) using the National Center for Atmospheric Research Parallel Climate Model (PCM lo), (3) high emission scenario (A1F1) using the average of three global climate models (GCM3 Avg hi), (4) high emission scenario (A1F1) using the HadleyCM3 model (Hadley hi)	38
Fig. 3.4	Maps of Upper Merced River watershed, Yosemite Valley, California, showing areas with differences in transpiration in (a) warmest vs. coldest simulation years, (b) wet vs. average precipitation year, and (c) dry vs. average precipitation year. The largest decreases in transpiration between years are shown in *red*; increases between years are shown in *green*	40
Fig. 3.5	Linkages between ecosystem services and human well-being	43
Fig. 4.1	General pathways by which atmospheric changes associated with increasing greenhouse gases can influence forest disturbance from insects and pathogens. CO_2 carbon dioxide, CH_4 methane	59

Fig. 4.2	Conceptual model of stress complexes in mixed conifer forests of the southern Sierra Nevada and southern California. The effects of insects and fire disturbance regimes (*red box*) and of fire exclusion are exacerbated by higher temperature. Stand-replacing fires and drought-induced mortality both contribute to species changes and invasive species	79
Fig. 4.3	Mortality of white spruce from bark beetle attack on the Kenai Peninsula, Alaska	79
Fig. 4.4	Conceptual model of stress complexes in the interior and coastal forests of Alaska. Rapid increases in the severity of disturbance regimes (insects and fire) are triggered by a warmer climate. Stand-replacing fires, massive mortality from insects, and permafrost degradation contribute to species changes and conversion to deciduous life forms	80
Fig. 4.5	Interactions between wildfire and hurricanes are synergistic in the southern United States. Figure depicts a longleaf pine/saw palmetto flatwoods stand on the Atlantic coastal plain, 2.5 years after a hurricane and with a previous history of prescribed fire	81
Fig. 4.6	Conceptual model of stress complexes in the interior and coastal forests of the Southeast. Increases in the severity of hurricanes are triggered by global warming as sea level rises. Warmer and drier climate in uplands leads to longer periods with flammable fuels. Changes in fire and hydrologic regimes, and responses to them, lead to species change and altered C dynamics	82
Fig. 5.1	Forest ownership in the United States, 2006	95
Fig. 5.2	Distribution of public and private forest ownership in the United States	95
Fig. 5.3	Family forest ownerships in the United States by size of forest holdings, 2006	96
Fig. 5.4	Economic dependence in the United States	104
Fig. 5.5	Reservations with significant timberland resources. Numbers 1 through 41 have over 4,000 ha of commercial timberland per reservation. Numbers 41 through 83 have less area in timberland, but what they have is economically viable	106
Fig. 6.1	Regions of the United States as defined by the U.S. Global Change Research Program National Climate Assessment	114
Fig. 6.2	In 2004, Alaska's largest wildfire season on record, the Boundary Fire, burned 217,000 ha of forest in interior Alaska	116

Fig. 6.3	In Hawaii's high-elevation forests (shown here) and in forests across the Pacific, projected warming and drying will increase invasive plants such as fire-prone grasses, resulting in novel fire regimes and conversion of native forests to exotic grasslands. For areas already affected in this way, climate change will increase the frequency and in some cases intensity of wildfire	119
Fig. 6.4	The effectiveness of fuel treatments is seen in this portion of the 2011 Wallow Fire near Alpine, Arizona. High-intensity crown fire was common in this area, but forest that had been thinned and had surface fuels removed experienced lower fire intensity, and structures in the residential area were protected	126
Fig. 6.5	Ecoregions in the Midwest, according to Bailey (1995)	130
Fig. 6.6	Suitable habitat for forest vegetation in New England is expected to shift with changes in climate (year 2100) associated with different emissions scenarios	133
Fig. 6.7	Climate change (year 2100) is expected to affect bird species richness more intensely in some areas of the northeastern United States than in others	135
Fig. 6.8	Percentage change in water supply stress owing to climate change, as defined by the water supply stress index (*WaSSI*) for 2050 using the CSIROMK2 B2 climate scenario. WaSSI is calculated by dividing water demand by supply, where higher values indicate higher stress on watersheds and water systems	139
Fig. 7.1	Growing stock carbon change is affected by growth, mortality, and removals, along with timberland area, 1953–2007	154
Fig. 7.2	Aboveground live biomass in forests	154
Fig. 7.3	Aboveground live forest carbon change	155
Fig. 7.4	Forest sector and non-forest sector greenhouse gas emissions and stock changes that are influenced by forest management	156

Fig. 7.5   Carbon (C) balance from two hypothetical management
projects with different initial ecosystem C stocks and
growth rates. Cumulative C stocks in forest, C removed
from forest for use in wood projects (long [L]- and
short-lived [S]), substitution, and biomass energy are
shown on land that (a) has been replanted or afforested,
or (b) has an established forest with high C stocks. The
dotted line represents the trajectory of forest C stocks if
no harvest occurred. Actual C pathways vary by project.
Carbon stocks for trees, litter, and soils are net C stocks
only. The scenario is harvested in x-year intervals,
which in the United States could be as short as 15 years
or longer than 100 years. This diagram assumes that all
harvested biomass will be used and does not account
for logging emissions. Carbon is sequestered by (1)
increasing the average ecosystem C stock (tree biomass)
by afforestation, or (2) accounting for C stored in wood
products in use and in landfills, as well as preventing
the release of fossil fuel C through product substitution
or biomass energy. The product-substitution effect is
assumed to be 2:1 on average. Biomass is assumed to
be a 1:1 substitute for fossil fuels in terms of C, but this
is not likely for many wood-to-energy options. This
represents a theoretical maximum C benefit for these
forest products and management practices. Carbon
"debt" is any period of time at which the composition
of forest products and remaining forest C stocks after
harvest is lower than estimated C stocks under a
no-harvest scenario 168

Fig. 8.1   Conceptual diagram of educational and training efforts
leading to increased complexity of adaptation planning
and activities. These elements are integrated but need
not be taken consecutively. Distance learning can be
incorporated into all activities...................................... 184

Fig. 8.2   A continuum of adaptation options to address needs at
appropriate scales, and examples of each (*shaded boxes*) 193

Fig. 8.3   Four dimensions of action outlined by the U.S. Forest
Service roadmap for responding to climate change.................. 197

Fig. 8.4   The U.S. Forest Service Eastern Region approach to
climate change response works from ecoregional scales
down to the stand scale by moving information to action
through partnerships, science, and communication 209

List of Figures

Fig. 8.5 The Climate Change Response Framework uses an adaptive management approach to help land managers understand the potential effects of climate change on forest ecosystems and integrate climate change considerations into management 210

Fig. 9.1 A conceptual risk framework used to help identify risks associated with climate change and prioritize management decisions 225

Fig. 9.2 Ratio of precipitation (P) to potential evapotranspiration (PET) for forests in the continental United States. PET was calculated using the Hamon (1961) model 228

Fig. 9.3 Changes in risk (*arrows* indicate transition from current risk to future risk) as a consequence of increased drought frequency and severity. P/PET $< 1 = \bigcirc$; P/PET $> 1 = \square$; P/PET $= 1 = \bigstar$; *green* = ecological flows, *blue* = ecosystem service flows. Risks are higher for ecological flows than for ecosystem service flows because some of the risks to the latter can be offset by engineering and conservation 229

Fig. 9.4 Climate change risk matrix for forest ecosystem carbon (C) pools in the United States, in which climate change may cause C pools to move in a positive (sink = net annual sequestration) or negative (source = net annual emission) direction. Likelihood of change in C stocks is based on the coefficient of variation across the national Forest Inventory and Assessment plot network (x-axis). Size of C stocks is based on the U.S. National Greenhouse Gas Inventory (y-axis). Societal response (e.g., immediate adaptive response or periodic monitoring) to climate change events depends on the size and relative likelihood of change in stocks. The dead wood pool, a relatively small stock, exhibits increasingly high variability across the landscape and therefore may be affected by climate change and disturbance events such as wildfire. In contrast, the forest floor is a relatively small C stock, and has low variability. Potential climate change effects are not incorporated in the matrix, because they represent many complex feedbacks both between C stocks (e.g., live aboveground biomass transitioning to the dead wood pool) and the atmosphere (e.g., forest floor decay) 231

Fig. 9.5	Percentage of increase (relative to 1950–2003) in median area burned for western United States ecoprovinces for a 1 °C temperature increase. Color intensity is proportional to the magnitude of the projected increase in area burned	234
Fig. 9.6	Risk matrix of potential change in suitable habitat for three tree species in northern Wisconsin that are expected to either lose habitat (*black ash*), gain habitat (*white oak*), or become a potential new migrant because of newly appearing habitat (*yellow poplar*).	237
Fig. 9.7	Risk of the effects of climate change on the northern cardinal and mourning warbler, expressed as a combination of likelihood of habitat change (x-axis) and magnitude of adaptability (y-axis). Values are rescaled from calculations that used the approach in SAVS	240

List of Tables

Table 4.1	Common invasive plant species and environmental impacts for forests and woodlands in the United States............	67
Table 6.1	Potential climate change related risks, and confidence in projections..	120
Table 7.1	Net annual changes in carbon (C) stocks in forest and harvested wood pools, 1990–2009	152
Table 7.2	Carbon (C) stocks in forest and harvested wood pools, 1990–2010	152
Table 7.3	Mitigation strategies, timing of impacts, uncertainty in attaining carbon (C) effects, co-benefits and tradeoffs	160
Table 7.4	Programs that influence carbon mitigation	172
Table 7.5	Tools and processes to inform forest management	174
Table 8.1	Factors that affect the relevance of information for assessing vulnerability to climate change of large, intermediate, and small spatial scales...............................	188
Table 8.2	Factors that affect the relevance of information for assessing vulnerability to climate change of large, intermediate, and small time scales	189
Table 8.3	Climate adaptation guides relevant to the forest sector	190
Table 8.4	Climate change adaptation strategies under broad adaptation options..	194
Table 8.5	Resources that can assist climate change adaptation in forest ecosystems	195

Table 8.6 Performance scorecard used by the U.S. Forest Service for annual review of progress and compliance, and to identify deficit areas in implementation of the national roadmap for responding to climate change 199

Table 9.1 Likelihood and magnitude of increased wildfire risk for fire regimes in forests of the western United States, based on a temperature increase of 2 °C 236

Part I
Seeking the Climate Change Signal

Chapter 1
Recent Changes in Climate and Forest Ecosystems

David L. Peterson and Kailey W. Marcinkowski

1.1 Atmospheric Environment

At the time this chapter is being written (January 2013), the United States has just experienced a drought that was unprecedented in the climatic record for its overall magnitude, spatial extent, and persistence (Karl et al. 2012), including being the warmest year since 1895, the beginning of formal measurements (NCDC n.d.). The standardized temperature anomaly for spring and summer of 2012 was a 1 in 1,600-year event for maximum temperature and a 1 in 450-year event for minimum temperature (Karl et al. 2012). July 2012 recorded the highest monthly mean temperature ever measured, and there were many individual-month records and near-records for various states. Globally, 2012 had the warmest summer on record (June through August), and in the United States only 2011 and 1936 had warmer summers (NCDC n.d.). In 2012, crop yields were reduced on nearly 80 % of U.S. agricultural lands (NIDIS n.d.), reducing the rate of national economic growth (Wiseman 2012). In late October, Hurricane Sandy swept northward from the Caribbean along the eastern coastal region, killing over 250 people, displacing tens of thousands from their homes, disrupting energy supplies, and causing $65 billion in damage. Spanning 1,800 km in diameter, Sandy was the largest Atlantic hurricane and the second costliest hurricane on record (Sullivan and Doan 2012; Wikipedia n.d.).

D.L. Peterson (✉)
Pacific Northwest Research Station, U.S. Forest Service, Seattle, WA, USA
e-mail: peterson@fs.fed.us

K.W. Marcinkowski
Northern Institute of Applied Climate Science, Michigan Technological University, Houghton, MI, USA
e-mail: kfmarcin@mtu.edu

The weather patterns of 2012 in the United States represent extreme conditions that may be associated with a well-documented, long-term warming trend. Between 1948 and 2010, mean temperature increase in North America was 0.2 °C per decade (Isaac and van Wijngaarden 2012), and 7 of the 10 warmest years on record occurred since 1990 (USEPA n.d.(a)). Over the last 50 years, annual mean temperature in Alaska has increased twice as much as the rest of the United States (Karl et al. 2009). Minimum temperature in U.S. urban areas has increased about 25 % faster than in rural areas (Mishra et al. 2012), and minimum temperatures have been increasing faster than maximum temperatures (Mishra and Lettenmaier 2011).

Over the last 40 years, the mean duration of dry episodes has increased in the eastern and southwestern United States (Groisman and Knight 2008). In the West and Southwest, droughts have increased in duration, severity, and frequency as a result of increased temperature (Andreadis and Lettenmaier 2006). In the Southeast, the area of moderate to severe drought has increased by 14 % in summer and 12 % in spring (Karl et al. 2009). In the Southwest, droughts have been more severe since 2000 than in the twentieth century (Breshears et al. 2005), and the western United States has experienced continuous drought at some location since 1999 (MacDonald 2007).

Recent increases in temperature and drought are consistent with the projected effects of elevated ambient carbon dioxide (CO_2), which reached 399 ppm at the long-term monitoring station on Mauna Loa, Hawaii, in June 2013 (Tans and Keeling n.d.). This level represents an increase of 80 ppm since 1960 and approximately 110 ppm since 1850 when fossil fuel combustion began contributing to atmospheric CO_2. The growth rate for global average CO_2 for 2000–2006 was higher than at any time since measurements began at Mauna Loa (1959); the annual rate of increase for total human-caused CO_2 emissions in the 2000s was nearly 3 %, compared to 0.7 % in the 1990s. The annual rate of fossil fuel CO_2 emissions increased from 1.3 % in the 1990s to over 3 % in the 2000s (Tans and Keeling n.d.); fossil fuel and cement emissions were 35 % higher in 2006 than in 1990 (Canadell et al. 2007). In the United States, CO_2 emissions have increased by 12 % since 1990 (USEPA n.d.(b)).

1.2 Trends and Extreme Events in Forest Ecosystems

One of the biggest changes in U.S. forested ecosystems in recent decades has been a decrease in the quantity and persistence of snow (Grundstein et al. 2010). Since the 1920s, snowfall has been declining in the West and the mid-Atlantic coast, and more recently in the Northeast (Kunkel et al. 2009). The ratio of snowfall to precipitation has decreased greatly in the Northwest because of lower snowfall (Feng and Hu 2007). Snowpack in the northern portion of the U.S. Rocky Mountains has decreased significantly since the 1980s (Pederson et al. 2011). Snow water equivalent (SWE) throughout the West has been lower since 1980 than during the rest of the twentieth century (McCabe and Wolock 2010), and winter precipitation in the northwestern United States has decreased since 1950 (Mote al. 2005). In the Colorado Rockies, snowmelt now occurs 2–3 weeks earlier than in 1978, and April 1 SWE and maximum SWE have respectively declined 4.1 and 3.6 cm per decade (Clow 2010). Glaciers, which are an iconic component of Western mountains and an important source of water in many locations, have been receding for the past century. Loss of ice has been especially rapid since around 1980 (Hodge et al. 1998; Josberger et al. 2007), with the largest losses of ice mass occurring in Alaska and the Northwest (Granshaw and Fountain 2006).

Concurrent with changes in temperature and the physical environment of forest ecosystems has been an apparent increase in the extent of ecological disturbances. Insect outbreaks have been especially prominent, extending over more land area than any other disturbance (see Chap. 4). Insect-caused mortality causes rapid changes in forest structure, productivity, and hydrology, and provides opportunities for tree regeneration and establishment of invasive species. Current epidemics of mountain pine beetle (*Dendroctonus ponderosae* Hopkins) and other beetles in the western United States have increased rapidly over the past decade (Meddens et al. 2012) (Fig. 1.1), mostly in lodgepole pine (*Pinus contorta* var. *latifolia* Engelm. ex S. Watson) forests. Over 4 million ha of forest have been killed in the United States, and another 8 million ha in British Columbia. Although most mortality has occurred in older stands that are physiologically stressed, higher temperatures have stimulated the reproductive cycle of beetles and reduced winter beetle mortality, allowing for rapid population increases (see Chap. 4). For the first time, higher temperature has also allowed beetles to attack high-elevation species such as whitebark pine (*Pinus albicaulis* Engelm.) (Gibson et al. 2008; Millar et al. 2012). The largest spruce beetle (*D. rufipennis* Kirby) epidemic ever observed in North America occurred in southern Alaska in the 1990s (Hayes and Lundquist 2009). Pinyon ips (*Ips confusus* LeConte) and other twig beetles have contributed to mortality in drought-stressed pinyon pine (*Pinus edulis* Engelm.) in the Southwest (Shaw et al. 2005), and southern pine beetle (*D. frontalis* Zimmermann) has caused extensive mortality in seven Southern states (Nowak 2004).

After insect outbreaks, wildfire is the second most important ecological disturbance in U.S. forests in terms of area affected and is strongly influenced by climate (see Chaps. 4 and 9). Annual area burned and duration of fire season in the West have increased since the 1980s, including several individual fires larger than 200,000 ha since 2000. This trend has in some cases been attributed to climate change (Westerling et al. 2006), although, a longer term perspective indicates that annual area burned in the past 20 years is not much different than in the early twentieth century (Fig. 1.2), especially for fires larger than 100,000 ha (Morgan et al. 2008). The extent of area burned is correlated with alternating multi-decadal periods of warm and cool climate (associated with phases of the Pacific Decadal Oscillation [PDO]), and it is reasonable to assume that if future climate looks like a warm-phase PDO (or if warm PDOs become more persistent or extreme), then more area will burn. The effect of climate on wildfire is clear and quantifiable (Littell et al. 2009), although fire severity (typically expressed as magnitude of tree mortality) is often modulated by fuel quantities (Miller et al. 2012). We may be entering a cool

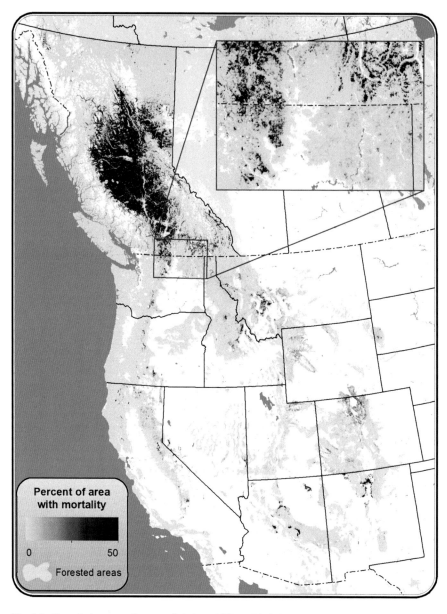

Fig. 1.1 Cumulative mortality area (ha) from 1997 to 2010 for the western conterminous United States and British Columbia for trees killed by bark beetles. Data are adjusted for underestimation (calculated by comparison with classified imagery) (From Meddens et al. (2012), with permission)

phase of the PDO, and if the extent of area burned in the West continues to remain high, then we can confidently infer that the increasing temperature trend associated with climate change is indeed affecting wildfire. A long-term trend of increased area burned, especially if severe, could maintain young forest age classes across

1 Recent Changes in Climate and Forest Ecosystems   7

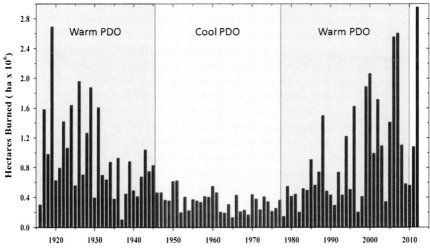

Fig. 1.2 Annual area burned by wildfire on federal lands in the 11 large western states in the conterminous United States, since 1916, including an indication of warm and cool phases of the Pacific Decadal Oscillation (*PDO*) (Based on data from U.S. Forest Service, modified from Littell et al. (2009), with permission)

large landscapes, lead to significant changes in the distribution and abundance of forest species in some locations (see Chap. 4), and cause rapid changes in carbon (C) dynamics (see Chap. 7).

An increase in extreme climatic and biophysical events will be the most important effect of climate change on forest ecosystems in future decades. Analogous to the concept of punctuated equilibrium in evolutionary biology (Gould and Eldredge 1977), we expect that rare but extreme climate-related events will cause faster, more pervasive effects than a gradual increase in temperature over time. For example, large wildfires essentially "clear the slate" across a particular landscape, and postfire biophysical conditions plus climate set the course for species composition, productivity, animal habitat, and many other ecosystem properties over decades to centuries. As a result, ecosystem structure and function can shift rapidly, especially if a big change in species composition occurs. This may already be occurring following large, severe wildfires in the Southwest (from forest and woodland to nonforest) (C. Allen, personal communication) and Alaska (from conifer-dominated forest to hardwood-dominated forest) (Wolken et al. 2011) (see Chap. 6). These shifts in vegetation physiognomy may not be reversible in a permanently warmer climate with more droughts and altered disturbance regimes. Large-scale, "clear the slate" disturbances are not limited to fire. Increased hurricane and storm intensity (Walsh and Ryan 2000), sea level rise, and severe and prolonged drought could have similar effects. Moreover, interactions of multiple disturbances and stressors may result in new combinations of species and ecosystem conditions for which we have no precedent in historical or paleoecological records (Williams and Jackson 2007) (see Chap. 4).

1.3 Resilience of Ecosystems and Institutions

Rapid shifts in climate and disturbance may strain both the resilience of forest ecosystems and the capacity of social systems and management institutions (Moser and Luers 2008) (see Chaps. 5 and 8). During the 1980s and 1990s, public land management in the United States shifted from an emphasis on resource extraction (e.g., timber harvest) to management for multiple resource values. Ecological restoration has become a dominant paradigm on federal forest land, focused on establishment of native species, older forests, and diverse habitat across large landscapes. Restoration targets are often based on "reference conditions" which may in turn be based on "historic range of variation" (HRV) for species and forest structure. In novel climates of the future, static concepts like HRV, plant associations, and potential vegetation types will probably be ineffective in attaining reference conditions. Rather, it will be more effective to recognize that most forests are dynamic, non-equilibrium systems, and to manage them to retain desired functions and processes (e.g., productivity, C retention, hydrologic flow). This approach would shift the management focus from *restoring* systems to *building resilience* in systems.

Current environmental policies and regulations in the United States, most of which were developed before climate change was recognized as an influence on forest resources, may not be flexible enough to accommodate rapidly changing climate and disturbance regimes (Peterson et al. 2011). For example, the capacity (budget and personnel) of agencies to suppress wildfires is already being stretched. Since 2000, the U.S. Forest Service has typically spent $1–2 billion per year on fire suppression, and nearly 50 % of the total agency budget is currently allocated to fire management. If annual area burned doubles during the twenty-first century, a conservative projection in most modeling studies (e.g., McKenzie et al. 2004), then balancing fire suppression versus other management functions will be a critical policy issue in the absence of a large increase in agency budget. Although fuel treatments reduce fire severity and build landscape resilience to wildfire, it will be difficult to offset increasing fire suppression costs without greatly increasing budgets for proactive, strategically placed treatments.

In December 2012, the United Nations Conference of the Parties extended the Kyoto Protocol, a pact that curbs greenhouse gas emissions from industrialized nations but covers only about 15 % of global C output. Failure to make meaningful reductions in emissions ensures that atmospheric CO_2 will continue to increase unabated in the absence of a technological fix or dramatically altered policies in countries that use most of the fossil fuels. In fact, current global emissions appear to exceed the high-end A2 emission scenario (see Chap. 2). This level guarantees that temperature will continue to rise for the next several decades, a trend that will be very difficult to reverse.

Concurrent with increasing temperature, forest ecosystems appear to be nearing important thresholds for the effects of climate change (Fagre et al. 2009). Although physical changes (e.g., temperature, hydrology) are better documented than biological changes, we anticipate that documentation of climate-related changes in U.S. forests will increase in future decades. Modifications of U.S. forest ecosystems will be superimposed on landscapes that have already been greatly altered by timber harvest and other land uses, as well as by the presence of 312 million people (and increasing) and their need for ecosystem services (see Chap. 3). Along with uncertainty in the extent and magnitude of extreme events in a warmer climate, we can expect surprises in how these events will affect the structure and functionality of forest ecosystems at large spatial scales. Responding to known challenges and unanticipated surprises will require shifting the focus of research and management from individual species and forest stands to landscapes that cover millions of hectares (see Chap. 10), while updating policy and regulations to facilitate this shift.

References

Andreadis, K. M., & Lettenmaier, D. P. (2006). Trends in 20th century drought over the continental United States. *Geophysical Research Letters, 33*, L10403. doi:10.1029/2006GL025711.

Breshears, D. B., Cobb, N. S., Rich, P. M., et al. (2005). Regional vegetation die-off in response to global-change-type drought. *Proceedings of the National Academy of Sciences, USA, 102*, 15144–14148.

Canadell, J. G., Le Quéré, C., Raupach, M. R., et al. (2007). Contributions to accelerating atmospheric CO_2 growth from economic activity, carbon intensity, and efficiency of natural sinks. *Proceedings of the National Academy of Sciences, USA, 104*, 18866–18870.

Clow, D. W. (2010). Changes in the timing of snowmelt and streamflow in Colorado: A response to recent warming. *Journal of Climate, 23*, 2293–2306.

Fagre, D. B., Charles, C. W., Allen, C. D., et al. (2009). *Thresholds of climate change in ecosystems: A report by the U.S. Climate Change Science Program and the Subcommittee on Global Change Research* (70pp). Reston: U.S. Department of the Interior, Geological Survey.

Feng, S., & Hu, Q. (2007). Changes in winter snowfall/precipitation ratio in the contiguous United States. *Journal of Geophysical Research, 112*, D15109. doi:10.1029/2007JD008397.

Gibson, K., Skov, K., Kegley, S. et al. (2008). *Mountain pine beetle impacts in high-elevation five-needle pines: Current trends and challenges* (R1-08-020, 32pp). Missoula: U.S. Department of Agriculture, Forest Service, Forest Health Protection.

Gould, S. J., & Eldredge, N. (1977). Punctuated equilibria: The tempo and mode of evolution reconsidered. *Paleobiology, 3*, 115–151.

Granshaw, F. D., & Fountain, A. G. (2006). Glacier change (1958–1998) in the North Cascades National Park Complex, Washington, USA. *Journal of Glaciology, 52*, 251–256.

Groisman, P. Y., & Knight, R. W. (2008). Prolonged dry episodes over the conterminous United States: New tendencies emerging during the last 40 years. *Journal of Climate, 21*, 1850–1862.

Grundstein, A., & Mote, T. L. (2010). Trends in average snow depth across the western United States. *Physical Geography, 31*, 172–185.

Hayes, J. L., & Lundquist, J. E. (compilers). (2009). *The western bark beetle research group: A unique collaboration with forest health protection—Proceedings of a symposium at the 2007 Society of American Foresters conference* (Gen. Tech. Rep. PNW-GTR-784, 134pp). Portland: U.S. Department of Agriculture, Forest Service, Pacific Northwest Research Station.

Hodge, S. M., Trabant, D. C., Krimmel, R. M., et al. (1998). Climate variations and changes in mass of three glaciers in western North America. *Journal of Climate, 11*, 2161–2179.

Isaac, V., & van Wijngaarden, W. A. (2012). Surface water vapor pressure and temperature trends in North America during 1948–2010. *American Meteorological Society, 25*, 3599–3609.

Josberger, E. G., Bidlake, W. R., March, R. S., & Kennedy, B. W. (2007). Glacier mass-balance fluctuations in the Pacific Northwest and Alaska, USA. *Annals of Glaciology, 46*, 291–296.

Karl, T. R., Melillo, J. M., Peterson, T. C., & Hassol, S. J. (Eds.). (2009). *Global climate change impacts in the United States. A report of the U.S. Global Change Research Program* (192pp). Cambridge: Cambridge University Press.

Karl, T. R., Gleason, B. E., Menne, M. J., et al. (2012). U.S. temperature and drought: Recent anomalies and trends. *EOS, Transactions of the American Geophysical Union, 93*(47), 473.

Kunkel, K. E., Palecki, M. A., Ensor, L., et al. (2009). Trends in twentieth-century U.S. extreme snowfall seasons. *Journal of Climate, 22*, 6204–6216.

Littell, J. S., McKenzie, D., Peterson, D. L., & Westerling, A. L. (2009). Climate and wildfire area burned in the western U.S. ecoprovinces, 1916–2003. *Ecological Applications, 19*, 1003–1021.

MacDonald, G. M. (2007). Severe and sustained drought in southern California and the West: Present conditions and insights from the past on causes and impacts. *Quaternary International, 173*, 87–100.

McCabe, G. J., & Wolock, D. M. (2010). Long-term variability in Northern hemisphere snow cover and associations with warmer winters. *Climatic Change, 99*, 141–153.

McKenzie, D., Gedalof, Z., Peterson, D. L., & Mote, P. (2004). Climatic change, wildfire and conservation. *Conservation Biology, 18*, 890–902.

Meddens, A. J. H., Hicke, J. A., & Ferguson, C. A. (2012). Spatiotemporal patterns of observed bark beetle-caused tree mortality in British Columbia and the western United States. *Ecological Applications, 22*, 1876–1891.

Millar, C. L., Westfall, R. D., Delany, D. L., et al. (2012). Forest mortality in high-elevation whitebark pine (*Pinus albicaulis*) forests of eastern California, USA: Influence of environmental context, bark beetles, climatic water deficit, and warming. *Canadian Journal of Forest Research, 42*, 749–786.

Miller, J. D., Skinner, C. N., Safford, H. D., et al. (2012). Trends and causes of severity, size, and number of fires in northwest California, USA. *Ecological Applications, 22*, 184–203.

Mishra, V., & Lettenmaier, D. P. (2011). Climatic trends in major U.S. urban areas, 1950–2009. *Geophysical Research Letters, 38*, L16401. doi:10.1029/2011GL048255.

Mishra, V., Michael, J. P., Boyles, R., et al. (2012). Reconciling the spatial distribution of the surface temperature trends in the southeastern United States. *Journal of Climate, 25*, 3610–3618.

Morgan, P., Heyerdahl, E. K., & Gibson, C. E. (2008). Multi-season climate synchronized forest fires throughout the 20th century, Northern Rockies, USA. *Ecology, 89*, 717–728.

Moser, S. C., & Luers, A. L. (2008). Managing climate risks in California: The need to engage resource managers for successful adaptation to change. *Climatic Change, 87*, S309–S322.

Mote, P. W., Hamlet, A. F., Clark, M. P., & Lettenmaier, D. P. (2005). Declining mountain snowpack in western North America. *Bulletin of the American Meteorological Society, 86*, 39–49.

National Integrated Drought Information System (NIDIS). (n.d.). U.S. drought portal. http://www.drought.gov/drought

National Oceanic and Atmospheric Administration, National Climate Data Center (NCDC). (n.d.). http://www.ncdc.noaa.gov/sotc

Nowak, J. T. (2004). Southern pine beetle prevention and restoration. *Forest Landowners Conference Proceedings, 63*, 21–22.

Pederson, G. T., Gray, S. T., Woodhouse, C. A., et al. (2011). The unusual nature of recent snowpack declines in the North American Cordillera. *Science, 333*, 332–335.

Peterson, D. L., Millar, C. I., Joyce, L. A. et al. (2011). *Responding to climate change on national forests: A guidebook for developing adaptation options* (Gen. Tech. Rep. PNW-GTR-855, 109pp). Portland: U.S. Department of Agriculture, Forest Service, Pacific Northwest Research Station.

Shaw, J. D., Steed, B. E., & DeBlander, L. T. (2005). Forest inventory and analysis (FIA) annual inventory answers the question: What is happening to pinyon-juniper woodlands? *Journal of Forestry, 103*, 280–285.

Sullivan, B. K., & Doan, L. (2012, October 29). Sandy brings hurricane-force gusts after New Jersey landfall. *Washington Post*. http://washpost.bloomberg.com/Story?docId=1376-MCMWP11A1I4H01-2DUORIV7RUREVIT7O7L4UFGTVF

Tans, P., & Keeling, R. (n.d.). *Recent Mauna Loa CO_2*. National Oceanic and Atmospheric Administration/Earth System Research Laboratory and Scripps Institute of Oceanography. http://www.esrl.noaa.gov/gmd/ccgg/trends

U.S. Environmental Protection Agency (USEPA). (n.d.(a)). *Climate change indicators in the United States, U.S. and global temperature*. http://www.epa.gov/climatechange/science/indicators/weather-climate/temperature.html

U.S. Environmental Protection Agency (USEPA). (n.d.(b)). *Greenhouse gas emissions, carbon dioxide emissions*. http://www.epa.gov/climatechange/ghgemissions/gases/co2.html

Walsh, K. J. E., & Ryan, B. F. (2000). Tropical cyclone intensity increase near Australia as a result of climate change. *Journal of Climate, 13*, 3029–3036.

Westerling, A. L., Hidalgo, H. G., Cayan, D. R., & Swetnam, T. W. (2006). Warming and earlier spring increase western U.S. forest wildfire activity. *Science, 313*, 940–943.

Wikipedia. (n.d.). *List of costliest Atlantic hurricanes*. http://en.wikipedia.org/wiki/List_of_costliest_Atlantic_hurricanes. Accessed 18 Dec 2012.

Williams, J. W., & Jackson, S. T. (2007). Novel climates, no-analog communities, and ecological surprises. *Frontiers in Ecology and the Environment, 5*, 475–482.

Wiseman, M. G. (2012, October 26). How US drought damaged economy as well as crops. *Bloomberg Businessweek News*. http://www.businessweek.com/ap/2012-10-26/how-us-drought-damaged-economy-as-well-as-crops

Wolken, J. M., Hollingsworth, T. N., Rupp, T. S., et al. (2011). Evidence and implications of recent and projected climate change in Alaska's forest ecosystems. *Ecosphere, 2*(11):124. doi:10.1890/ES11-00288.1

Chapter 2
Projected Changes in Future Climate

Chelcy F. Miniat and David L. Peterson

2.1 Methods for Projecting Future Climate

Most of the climate projections used to describe future climatic conditions in this book are based on model "ensembles," which are syntheses of the output of various global climate models (GCMs). The book also includes output from four GCMs:

- **CCSM2 (Community Climate System Model, version 2)**—U.S. National Center for Atmospheric Research (http://www.CESM.NCAR.edu).
- **CSIRO Mk3**—Australian Commonwealth Scientific Industrial Research Organisation (Gordon et al. 2002).
- **Hadley (versions 1–3)**—United Kingdom Hadley Center (Burke et al. 2006).
- **PCM (Parallel Climate Model)**—U.S. National Center for Atmospheric Research (Washington et al. 2000).

The book also uses terminology that refers to standard greenhouse gas (GHG) emission scenarios as described by the Intergovernmental Panel on Climate Change (IPCC). Emission scenarios cited in the book are described below, in which A scenarios have higher GHG emissions and higher projected temperature increases than B scenarios.

- **A2**—A2 scenarios represent a more divided world, characterized by independently operating, self-reliant nations; continuously increasing population, and regionally oriented economic development.

C.F. Miniat
Southern Research Station, U.S. Forest Service, Otto, NC, USA
e-mail: crford@fs.fed.us

D.L. Peterson (✉)
Pacific Northwest Research Station, U.S. Forest Service, Seattle, WA, USA
e-mail: peterson@fs.fed.us

- **A1F1**—A1 scenarios represent a more integrated world, characterized by rapid economic growth, a global population that reaches nine billion in 2050 and then gradually declines, quick spread of new and efficient technologies, a world in which income and way of life converge between regions, and extensive social and cultural interactions worldwide. A1F1 emphasizes the use of fossil fuels.
- **A1B**—Same as A1F1, except it emphasizes a balance of energy sources.
- **B1**—B1 scenarios represent a more integrated, ecologically friendly world, characterized by rapid economic growth as in A1, but with rapid changes toward a service and information economy, population rising to nine billion in 2050 and then declining as in A1, reductions in material intensity and the introduction of clean and resource efficient technologies, and an emphasis on global solutions to economic, social, and environmental instability.
- **B2**—B2 scenarios represent a more divided but more ecologically friendly world, characterized by continuously increasing population but at a slower rate than in A2; emphasis on local rather than global solutions to economic, social, and environmental instability; intermediate levels of economic development; and less rapid and more fragmented technological change than in A1 and B1.

The forthcoming Fifth IPCC Assessment, scheduled for publication in 2014, will use representative concentration pathways (RCPs) rather than the emission scenarios that were used in the Fourth Assessment (Solomon et al. 2007). The RCPs are four GHG concentrations (not emissions), named after a possible range of radiative forcing (increased irradiance caused by GHGs) values at the earth's surface in the year 2100: RCP2.6, RCP4.5, RCP6, and RCP8.5, which represent 2.6, 4.5, 6.0 and 8.5 W m^{-2}, respectively (Moss et al. 2008). Current radiative forcing is approximately 1.6 W m^{-2}, which is equivalent to a global-scale warming effect of 800 terawatts (8×10^{14} W).

2.2 Projected Future Climate in the United States

2.2.1 Temperature

Trends in temperature and precipitation from weather stations show that the United States has warmed over the past 100 years, but the trends differ by region (Backlund et al. 2008). The southeastern United States has cooled slightly (<0.7 °C), and Alaska has warmed the most (~4.5 °C); other northern and western U.S. regions also show a warming trend (~1.5 °C). Here we discuss projected changes in future climate based on output from an ensemble of 15 global climate models (GCMs) (Kunkel et al. 2013). All model runs used future scenarios of economic growth, population growth, and greenhouse gas emissions scenarios that were intended to represent the high (A2) and low (B1) ends of future emissions (see Sect. 2.1).

Average annual air temperatures across the continental United States are likely to steadily increase over the next century under the two emission scenarios (Fig. 2.1).

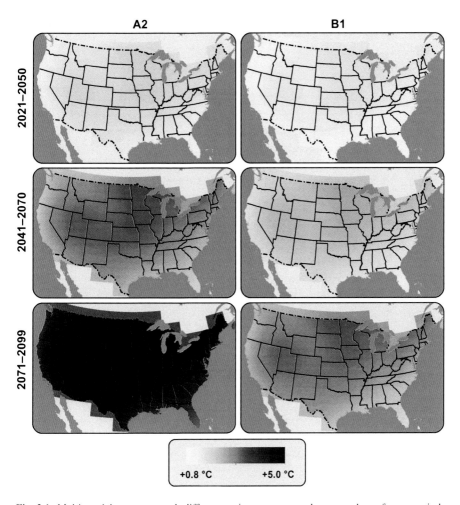

Fig. 2.1 Multi-model mean annual differences in temperature between three future periods compared to 1971–2000, from 15 GCMs using two emission scenarios (*A2* and *B1*). The A2 scenario is for higher emissions than for B1 (see text). For most interior states, models project a 1.4–1.9 °C temperature increase, rising to 2.5–3.6 °C for 2051–2071, and to greater than 4.2 °C for 2071–2099, depending on the emission scenario. Data from the World Climate Research Programme's (WCRP's) Coupled Model Intercomparison Project phase 3 (CMIP3) multi-model dataset

Compared to 1971 through 2000, average annual air temperature will likely increase from 0.8 to 1.9 °C by 2050, from 1.4 to 3.1 °C by 2070, and from 2.5 to 5.3 °C by 2099. The range of these estimated temperatures is bounded by the B1 and A2 emission scenarios. Within each scenario, the magnitude of increase depends on latitude and proximity to coastal areas. More warming is projected in northern and interior areas; the largest temperature increases are projected for the upper Midwest, and the smallest increases are projected for peninsular Florida. The magnitude of annual warming is modified by differences in seasonal temperature increases. For the A2 scenario, winter is projected to have the most pronounced warming across the United States, with increases up to 3.6 °C in the northern United States and smaller increases in the South. During the summer, greater warming is projected for more interior locations (up to 3.6 °C warming across the central United States from Kentucky to Nevada). The least amount of warming is expected for autumn (1.9–3.1 °C) and spring (1.4–2.5 °C).

In addition to overall warming during the twenty-first century, both the number of days when maximum temperatures exceed 35 °C and when heat waves occur (defined as the number of consecutive days with maximum temperatures exceeding 35 °C) will increase (Fig. 2.2). For the A2 scenario, the Southeast will likely experience an additional month of days with maximum temperatures exceeding 35 °C, and the Pacific Northwest and Northeast will experience 10 more of these days per year. In addition, the United States will likely experience longer heat waves. In the Southwest, the average length of the annual longest heat wave is projected to increase by 20 days or more. Little or no change is predicted for this metric of heat waves in the Northwest, Northeast, and northern parts of the Great Plains and Midwest, but increases in less extreme levels of heat waves are projected for these regions. Most other areas will likely see heat waves of 2–20 additional days.

2.2.2 Precipitation

Much of the eastern and southern United States now receives more precipitation than 100 years ago, whereas other areas, especially in the Southwest, now receive less (Backlund et al. 2008). Precipitation differs even more than temperature across the United States and through seasons and years. As a result, long-term trends in precipitation are less apparent. Observed data from the past century across the United States show that mean annual precipitation has significant interannual variability, with two particularly dry decades (1930s and 1950s) followed by a few relatively wet decades (1970–1999); the overall result is a century-long trend of increasing precipitation (Groisman et al. 2004).

Using the same multi-model approach as for temperature (see Sect. 2.2.1), projections for the twenty-first century across the entire United States indicate little or no change in precipitation, although variance among models is high. Some models predict a significantly drier future (at least in some regions), and others a significantly wetter future. Agreement among projections for precipitation is high for some models (Solomon et al. 2007). For example, general consensus exists that annual precipitation in the Southwest will decrease by 6–12 % (Fig. 2.3), whereas precipitation in the northern states will increase by 6–10 % (Easterling et al. 2000a, b; Groisman et al. 2004; Huntington 2006; Pachuri and Reisinger 2007; Solomon et al. 2007).

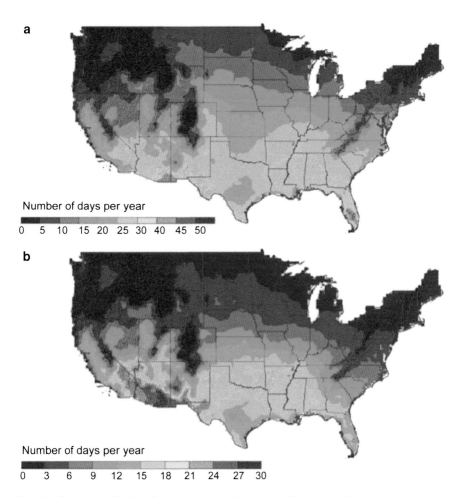

Fig. 2.2 Spatial distribution of the mean change in the annual number of days with a maximum temperature above 35 °C (**a**), and in the annual number of consecutive days with a maximum temperature above 35 °C (**b**) between 1971 and 2000 and 2041–2070. Models project that much of the southeastern and southwestern United States will experience more days with maximum temperature above 35 °C, and more consecutive days above that temperature. Results are for the high (A2) emission scenario only, from the North American Regional Climate Change Assessment Program multi-model means (n = 9 GCMs) (From 2012 draft version of Kunkel et al. (2013). On file with: Ken Kunkel, NOAA's National Climate Center, 151 Patton Avenue, Asheville, NC (USA) 28801)

Higher precipitation and increased frequency and magnitude of droughts and floods have occurred in some regions of the United States over the last 50 years (Easterling et al. 2000a, b; Groisman et al. 2004; Huntington 2006; Pachuri and Reisinger 2007; Solomon et al. 2007). In most GCMs, as the climate warms, the frequency of extreme precipitation events increases globally, producing an intensification of the hydrologic cycle (Huntington 2006). In fact, the upper 99th

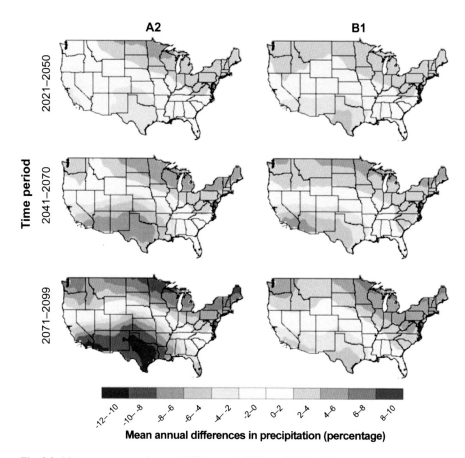

Fig. 2.3 Mean percentage of annual differences in U.S. precipitation between three future periods relative to a 1971–2000 reference period. The Northeast, northern Midwest and Pacific Northwest are projected to have slightly more precipitation, and the Southwest is projected to have 2–12 % less precipitation, depending on the emission scenario, location, and time period. Means are for 15 GCMs (From Kunkel et al. (2013))

percentile of the precipitation distribution is projected to increase by 25 % when atmospheric CO_2 concentration of the Earth reaches around 600 ppm (Allen and Ingram 2002). The timing and spatial distribution of extreme precipitation events are among the most uncertain aspects of future climate scenarios (Karl et al. 1995; Allen and Ingram 2002).

2.2.3 Drought

As the climate warms from increasing GHGs, both the proportion of land experiencing drought and the duration of drought events will likely increase

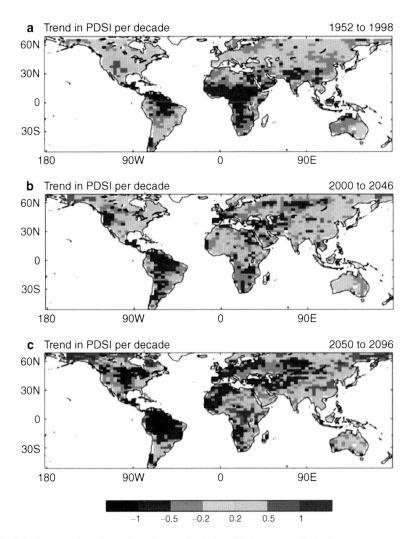

Fig. 2.4 Trend in the Palmer Drought Severity Index (*PDSI*) per decade for (**a**) observed data and the mean of the (**b**) first half and (**c**) second half of the twenty-first century. The PDSI is projected to decrease by 0.5–1 unit per decade for the period 2050–2096. For the PDSI, −1.9 to 1.9 is near normal, −2 to −2.9 is moderate drought, −3 to −3.9 is severe drought and less than −4 is extreme drought. Projections are made by HadCM3 with the A2 emission scenario (Figure from Burke et al. (2006), © British Crown Copyright 2006, Met Office, with permission)

(Burke et al. 2006). Projected spatial distribution of changes in drought over the twenty-first century for the A2 scenario indicates significant drying over the United States (Fig. 2.4). The Palmer Drought Severity Index is projected to decrease by 0.3 per decade (indicating more drought) globally for the first half of the twenty-first century. Relative to historical data, the amount of land surface subject to annual drought is projected to increase in 2010–2020 from 1 to 3 % for extreme droughts,

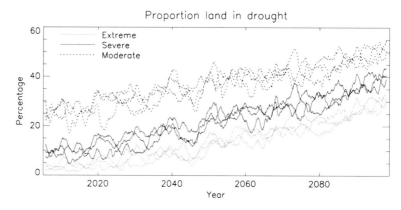

Fig. 2.5 The projected average annual proportion of the global land surface in drought each month shows drought increasing over the current century. Drought is defined as extreme, severe, or moderate, which represents 1, 5 and 20 %, respectively, of the land surface in drought under present-day conditions. Results from the three simulations are from HadCM3 with the A2 emission scenario (Figure from Burke et al. (2006), © British Crown Copyright 2006, Met Office, with permission)

from 5 to 10 % for severe droughts, and from 20 to 28 % for moderate droughts (Fig. 2.5). This drying trend continues throughout the twenty-first century, and by the 2090s, the amount of land area in drought is projected to increase for extreme, severe, and moderate droughts to 30, 40 and 50 %, respectively. The number of drought events is projected to double for extreme and severe droughts, but remain stable for moderate drought. The duration of all forms of drought events also increases.

2.3 Sea Level Rise

Global sea level rise results from changing the ocean's water volume because of changes in temperature, salinity, ice melting, and land surface runoff. Global sea level responds to climate cycles of alternating glacial and interglacial conditions over millions of years (Kawamura et al. 2007). Mean sea level rose by 120 m since the most recent ice age, at a rate of about 1 m per century. Sea level has remained relatively stable for the last 6,000 years, and observed data indicate a global mean increase of 0.17 m per century (Grinsted et al. 2010).

Satellite altimetry records show that mean sea level rise since the middle of the nineteenth century has not been uniform. The Pacific Coast of the United States showed little sea level rise, consistent with tide gage records (see discussion in Parris et al. 2011). In contrast, sea level rise in the Gulf of Mexico has averaged 3.2 mm year^{-1} since 1992. The trajectory of spatially explicit trends in the future is a topic of active research. For example, the spatial trend in the Pacific is thought to be a combination of wind stress patterns associated with the Pacific Decadal Oscillation

(PDO) and El Niño-Southern Oscillation (ENSO). Because of regular phase shifts in PDO (20–40 years) and ENSO (3–7 years), it is unlikely that the observed sea level rise trends will continue with the same magnitude and direction.

For various emission scenarios, as temperature increases, several factors (e.g., polar ice sheet melting) contribute to sea level rise (Parris et al. 2011). Different emission scenarios result in disparate projections of sea level rise by 2100, ranging from 0.2 m for the lowest scenario to 2.0 m for the highest scenario. For two intermediate scenarios, sea level is projected to increase from 0.5 m (B1) to 1.2 m (A2). This wide range of projections reflects high uncertainty in how sea level rise may affect coastal forests and terrestrial and aquatic ecosystems in coastal regions.

2.4 Using Climate Projections to Estimate Effects on Forests

No standard approach exists for linking climate projections with potential effects on forest species and ecosystems. Therefore, users of climate information must develop their own approach for accessing that information and applying it to assessments of vulnerability of natural resources to climate change. Different providers of climate information use different GCMs to develop projections of climate, typically at different time increments until 2100, and they often use different emission scenarios (B1, A2, etc.) (see Sect. 2.1). This can make it challenging for resource managers to identify appropriate data for specific applications.

Despite the diversity of GCMs and emission scenarios, most temperature projections are similar until around 2050, so one can have confidence in most model output during this time period. Beyond 2050, model output diverges considerably, especially as a function of emission scenario. All models project significant temperature increases with some confidence, but precipitation projections are more variable (some increase, some decrease) and are less reliable than for temperature projections. Some users select output from a model in which they have confidence. Others select output from multiple models—typically temperature projections that are high, moderate, and low—thus bracketing a range of potential future climates and effects on resources. This latter approach is a reasonable way to project a range of possible futures for natural resources. In addition, the current trajectory of GHG emissions is at the high end of the IPCC emission scenarios, so it may be more realistic to base climate projections on the "A" scenarios. It is always important to document which models and scenarios were used for any particular application and to understand how their basic assumptions affect climate projections.

In recent years, considerable effort has been invested in downscaling GCM output to smaller geographic areas, with the intention of providing more site-specific climate projections. Downscaled climate data may or may not be useful, depending on the spatial scale of interest for different natural resources, and downscaled data do not reduce the uncertainty of climate projections. For example, simply knowing that mean annual temperature will increase 2–4 °C by the year 2060 is probably sufficient for estimating effects on forest growth, wildfire, and insects and for developing appropriate adaptation options.

References

Allen, M. R., & Ingram, W. J. (2002). Constraints on future changes in climate and the hydrologic cycle. *Nature, 419*, 224–232.

Backlund, P., Janetos, A., Schimel, D., et al. (2008). *The effects of climate change on agriculture, land resources, water resources, and biodiversity in the United States* (Final report, synthesis, and assessment product 4.3, 362pp) Washington, DC: U.S. Department of Agriculture.

Burke, E. J., Brown, S. J., & Christidis, N. (2006). Modeling the recent evolution of global drought and projections for the twenty-first century with the Hadley centre climate model. *Journal of Hydrometeorology, 7*, 1113–1125.

Easterling, D. R., Evans, J. L., Groisman, P. Y., et al. (2000a). Observed variability and trends in extreme climate events: A brief review. *Bulletin of the American Meteorological Society, 81*, 417–425.

Easterling, D. R., Meehl, G. A., Parmesan, C., et al. (2000b). Climate extremes: Observations, modeling, and impacts. *Science, 289*, 2068–2074.

Gordon, H. B., Rotstayn, L. D., McGregor, J. L., et al. (2002). The CSIRO Mk3 climate system model (Tech. Paper 60, 130pp). Aspendale: Commonwealth Scientific Industrial Research Organisation Atmospheric Research.

Grinsted, A., Moore, J. C., & Jevrejeva, S. (2010). Reconstructing sea level from paleo and projected temperatures 200 to 2100 AD. *Climate Dynamics, 34*, 461–472.

Groisman, P. Y., Knight, R. W., Karl, T. R., et al. (2004). Contemporary changes of the hydrological cycle over the contiguous United States: Trends derived from in situ observations. *Journal of Hydrometeorology, 5*, 64–85.

Huntington, T. G. (2006). Evidence for intensification of the global water cycle: Review and synthesis. *Journal of Hydrology, 319*, 83–95.

Karl, T. R., Knight, R. W., & Plummer, N. (1995). Trends in high-frequency climate variability in the twentieth century. *Nature, 377*, 217–220.

Kawamura, K., Parrenin, F., Lisieck, L., et al. (2007). Northern Hemisphere forcing of climatic cycles in Antarctica over the past 360,000 years. *Nature, 448*, 912–916.

Kunkel, K. E., Stevens, L. E., Stevens, S. E., et al. (2013). Climate of the contiguous United States. In: *Regional climate trends and scenarios for the U.S. national climate assessment* (Tech. Rep. NESDIS 14209, 85pp, Chapter 9). Washington, DC: U.S. Department of Commerce; National Oceanic and Atmospheric Administration; National Environmental Satellite, Data, and Information Service.

Moss, R., Babiker, M., Brinkman, S., et al. (2008). *Towards new scenarios for analysis of emissions, climate change, impacts, and response strategies* (132pp). Geneva: Intergovernmental Panel on Climate Change.

Pachauri, R. K., & Reisinger, A. (Eds.). (2007). *Climate change 2007: Synthesis report: Contribution of working groups I, II and III to the fourth assessment report of the Intergovernmental Panel on Climate Change* (104pp). Geneva: Intergovernmental Panel on Climate Change.

Parris, A., et al. (2011, November 3). *Sea level change scenarios for the U.S. national climate assessment (NCA), Version 0*. Washington, DC: U.S. Global Change Research Program, National Climate Assessment.

Solomon, S., Qin, D., Manning, M., et al. (2007). *Climate change 2007: The physical science basis—contribution of Working Group I to the fourth assessment report of the Intergovernmental Panel on Climate Change* (996pp). Cambridge: Cambridge University Press.

Washington, W. M., Weatherly, J. W., Meehl, G. A., et al. (2000). Parallel climate model (PCM) control and transient simulations. *Climate Dynamics, 16*, 755–774.

Part II
Effects of Climatic Variability and Change

Chapter 3
Forest Processes

Michael G. Ryan, James M. Vose, Paul J. Hanson, Louis R. Iverson,
Chelcy F. Miniat, Charles H. Luce, Lawrence E. Band, Steven L. Klein,
Don McKenzie, and David N. Wear

3.1 Introduction

Some of the changes to U.S. forests will be directly caused by the effects of an altered climate, such as increases in atmospheric carbon dioxide (CO_2) temperature (T), and nitrogen (N) deposition on tree growth, mortality, and regeneration.

M.G. Ryan
Natural Resource Ecology Laboratory, Colorado Sate University, Fort Collins, CO, USA
e-mail: mike.ryan@colostate.edu

J.M. Vose (✉)
Southern Research Station, U.S. Forest Service, Raleigh, NC, USA
e-mail: jvose@fs.fed.us

P.J. Hanson
Environmental Sciences Division, Oak Ridge National Laboratory, Oak Ridge, TN, USA
e-mail: hansonpj@ornl.gov

L.R. Iverson
Northern Research Station, U.S. Forest Service, Delaware, OH, USA
e-mail: liverson@fs.fed.us

C.F. Miniat
Southern Research Station, U.S. Forest Service, Otto, NC, USA
e-mail: crford@fs.fed.us

C.H. Luce
Rocky Mountain Research Station, U.S. Forest Service, Boise, ID, USA
e-mail: cluce@fs.fed.us

L.E. Band
Department of Geography, University of North Carolina, Chapel Hill, NC, USA
e-mail: lband@email.unc.edu

S.L. Klein
National Health and Environmental Effects Research Laboratory, Western Ecology Division,
U.S. Environmental Protection Agency, Corvallis, OR, USA
e-mail: klein.steve@epa.gov

Other changes will be indirectly caused by climate-induced changes in disturbances, such as droughts, fire, insect outbreaks, pathogens, and storms (see Chap. 4). In this chapter, we document current knowledge of the potential direct of climate change on biogeochemical cycling (i.e., carbon [C], nutrients, and water) and forest tree distributions.

3.2 Carbon and Nutrient Cycling

The United States has about 303 million ha of forest land, about 8 % of the world's total (see Chap. 5). Forest C stocks and uptake or loss rates differ greatly across a wide range in environmental conditions, land use, land-use history, and current human influences (see Chap. 7). Forests of the conterminous United States cover about 281 million ha and contain 45,988 Tg C. Estimates of the amount of CO_2 emissions (1,500 Tg C in 2009) offset by forests and forest products in the United States vary from 10 to 20 % depending on assumptions and accounting methods (McKinley et al. 2011), with 13 % being the commonly used estimate as of 2011 (USEPA 2011). Ninety-four percent of forest C storage comes from growth on current forest lands, with the remaining 6 % from a net positive conversion of other land uses to forests. Regional differences in forest C pools and storage rates are reported in McKinley et al. (2011); Woodbury et al. (2007), and U.S. Environmental Protection Agency (USEPA) (2011). Updates of the inventories used to estimate these pools and storage rates may be important to capture C losses in recent large fires, bark beetle outbreaks, and drought mortality. Components such as dead wood and C in soil are either sparsely measured or are only estimated (Woodbury et al. 2007).

These forest C storage estimates are similar to those reported in a global study of forest sinks derived from the same sources (Pan et al. 2011). An analysis using eddy covariance flux measurements, satellite observations, and modeling estimated annual C storage in the conterminous United States as 630 Tg C year^{-1} (Xiao et al. 2011), largely from forests and savannas; most agricultural lands either store little additional C or lose C (USEPA 2011). The large discrepancy between the biometric USEPA estimates and those of Xiao et al. (2011) is probably caused by: (1) woodland encroachment (Van Auken 2000; Pacala et al. 2001; McKinley and Blair 2008) not measured by the U.S. Forest Service Forest Inventory and Analysis used for the USEPA reporting, and (2) poor performance of eddy

---

D. McKenzie
Pacific Northwest Research Station, U.S. Forest Service, Seattle, WA, USA
e-mail: donaldmckenzie@fs.fed.us

D.N. Wear
Southern Research Station, U.S. Forest Service, Raleigh, NC, USA
e-mail: dwear@fs.fed.us

covariance measurements in estimating ecosystem respiration (Barford et al. 2001; Bolstad et al. 2004; Kutsch et al. 2008; Wang et al. 2010). Other estimates for the conterminous United States are $1,200 \pm 400$ Tg C year^{-1} from inversion analysis (Butler et al. 2010) and 500 ± 400 Tg C year^{-1} from three-dimensional atmospheric CO_2 sampling (Crevoisier et al. 2010).

3.2.1 Response of Forest C Cycling to Changing Environmental Conditions

Carbon storage in forest ecosystems results from the balance between growth of wood, foliage, and roots and their death or shedding and subsequent decomposition. Temperature, atmospheric CO_2 concentration, ecosystem water balance, and N cycling all interact to alter photosynthesis and growth. For example, higher temperatures can benefit growth, but the most benefit would come with adequate nutrition and soil water. Disturbance rapidly changes the balance between production and decomposition, but chronic changes in temperature, precipitation, CO_2, and N deposition over large areas can also alter C balance over longer time periods.

Experiments and measurements provide insights into forest C balance. Atmospheric concentrations of CO_2, currently near 400 ppm, are expected to rise to 700–900 ppm by 2100, depending on future emission scenarios and any changes in atmospheric uptake by terrestrial and aquatic ecosystems. Experimental results confirm that the primary direct effect of elevated CO_2 on forest vegetation is increased photosynthesis (Norby et al. 2005), but individual studies show that photosynthetic enhancement, growth and C storage are moderated by water and nutrient availability (Finzi et al. 2006; Johnson 2006; Norby et al. 2010; Garten et al. 2011). Free-air CO_2 enrichment studies (Norby and Zak 2011) show that (1) elevated CO_2 does not increase leaf area in forests, (2) net primary production (NPP) is enhanced under elevated CO_2 only when water and nutrient supplies are abundant, (3) water use is reduced through stomatal closure (Leuzinger and Körner 2007; Warren et al. 2011), and (4) CO_2-promoted increases in photosynthesis and NPP do not always increase forest C storage.

Elevated atmospheric CO_2 will likely increase forest productivity, although the magnitude of increase will be affected by how elevated CO_2 will affect belowground processes (Lukac et al. 2009), mature trees, and wetlands. For example, recent study suggests that, a 19 % increase in CO_2 over the past 50 years may have increased quaking aspen (*Populus tremuloides* Michx.) growth more than 50 % (Cole et al. 2010). Elevated CO_2 commonly enhances soil CO_2 efflux, suggesting that some of the additional photosynthesis is rapidly cycled back to the atmosphere (Bernhardt et al. 2006). An increase in labile C in soil may increase decomposition and potentially reduce soil C storage (Hofmockel et al. 2011). In a mature forest, sustained increases in photosynthesis in response to elevated CO_2 (Bader et al. 2009) did not increase wood growth (Körner et al. 2005), soil respiration (Bader and Körner 2010), or root

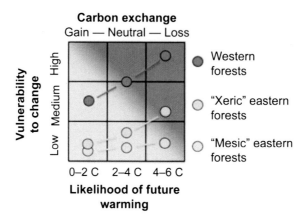

Fig. 3.1 Risk analysis diagram for the forest C cycle. Western forests are considered inherently limited by water demands that exceed precipitation supplies during substantial portions of the year. Xeric Eastern forests include those growing on shallow or coarse-textured soils or those present at the prairie-forest transition zone that experience water deficits in some years. Mesic Eastern forests experience severe water deficits only in occasional years and for relatively brief periods

or soil C storage (Asshoff et al. 2006; Bader et al. 2009). In wetlands, elevated CO_2 can increase CO_2 and methane efflux (Ellis et al. 2009), but these fluxes interact with hydrologic cycling and potential species changes (Fenner et al. 2007).

In temperate and boreal forests, modest increases in temperature tend to increase growth (Way et al. 2010) (Fig. 3.1). Warming will probably enhance upland forest growth for ecosystems with ample water, through changes in plant development and a longer growth season (Hänninen et al. 2007; Bronson et al. 2009; Gunderson et al. 2012). Growth in water-limited ecosystems will probably be reduced (Hu et al. 2010; Arend et al. 2011), and net C storage may be reduced (Cai et al. 2010). Warming will also enhance microbial decomposition and nutrient mineralization in soils (Melillo et al. 2002), increasing plant nutrient availability (Melillo et al. 2011), but the long-term tradeoff between soil C loss and nutrient-enhanced productivity is unknown. A longer growing season may increase the possibility of damage to trees from late frost events (Gu et al. 2008; Augspurger 2009).

Eastern forests, particularly on deep soils, are well buffered against substantial reductions in precipitation; forest growth, soil C storage, and nutrient availability show little effect of a chronic 12-year, 33 % reduction in precipitation (Hanson et al. 2007; Froberg et al. 2008; Johnson et al. 2008). Forests that rely on snowmelt for water will probably grow less in drier conditions (Boisvenue and Running 2010; Hu et al. 2010), and more frequent droughts in Western forests will reduce tree growth, vigor, and survival (McDowell et al. 2008; McDowell 2011). Precipitation amount may be more important for forest productivity than its frequency and intensity (Gerten et al. 2008).

In areas where N deposition increases, it may enhance ecosystem C storage by increasing forest productivity (Churkina et al. 2009; de Vries 2009) and decreasing decomposition of soil organic matter (Janssens et al. 2010), but those gains may be offset by the concurrent release of nitrous oxide, a potent greenhouse gas (Zaehle et al. 2011). The potential for enhancing C gain would be low in regions where N deposition is already high (e.g., the Northeast) and high in regions where N deposition is low (e.g., the Southwest). Tree species have a wide range of susceptibility to tropospheric ozone, which also varies regionally, and damage caused by ozone is not completely offset by elevated CO_2 (Karnosky et al. 2005).

Modeling has also been used to provide insights into forest C balance. Forests in different regions will probably respond differently to climate change because of variation in species composition, water and nutrient availability, soil depth and texture, and strength of other environmental factors such as ozone and N deposition. Understanding how these multiple factors will interact is difficult to test experimentally or measure, so modeling approaches are often used. In the eastern United States, model output suggests that productivity or forest C storage will increase with projected changes in climate, N, and CO_2, especially if precipitation increases, promoting higher photosynthesis under increased temperature. For example, upland oak forests in Tennessee are projected to increase their current C storage rate by 20 % for the climate and atmosphere projected for 2100 (CO_2 concentration 770 ppm, ozone concentration 20 ppb higher than today's level, 4 °C temperature increase, 20 % winter precipitation increase) (Hanson et al. 2005). Globally, temperate forest and grassland NPP is projected to increase 25–28 % for CO_2 concentration of 550 ppm (Pinsonneault et al. 2011), an estimate that includes expected changes in climate. Based on a four-model simulation of the effects of increased temperature and CO_2 and altered precipitation, Eastern forests showed increases in net C storage rates and net ecosystem production (Luo et al. 2008), and forest productivity increased specifically in the Northeast (Campbell et al. 2009).

For the western United States, models vary in their projections of productivity and C storage in forests. For example, changes in climate and CO_2 are projected to turn Rocky Mountain forests into a C source by 2090 (Biome-BGC model) (Boisvenue and Running 2010), and decrease forest C storage for boreal aspen (Grant et al. 2006), whereas other models project increased C storage for Western forests (CENTURY Model) (Smithwick et al. 2009; Melillo et al. 2011; Pinsonneault et al. 2011). Carbon in northern bogs, peat lands, and permafrost regions may be lost with a warming climate (increasing methane production), depending on hydrology and other factors (Heijmans et al. 2008; Ise et al. 2008; Koven et al. 2011). Global model simulations of climate change and ecosystem productivity (Friend 2010; Pinsonneault et al. 2011) project higher C storage for both eastern and western United States forests, with the larger increase in the East. It is important to note that none of these simulations consider the effects of altered disturbance regimes. For example, in the West, climate-driven increased fire and bark beetle outbreaks are likely to reduce forest C storage (Westerling et al. 2006, 2011; Metsaranta et al. 2010), jeopardizing the current U.S. forest sink (see Chap. 4). Recent large fires have already turned Arizona and Idaho forests from a C sink to a C source (USEPA 2011). Reduced tree vigor caused by drought and elevated temperature has promoted bark beetle outbreaks, resulting in short-term C loss for some forests in the West (Allen et al. 2010). Limited data suggest that mortality, perhaps related to climate, has increased slightly in some older forests in the West (van Mantgem et al. 2009). Little information on tree mortality trends exist for the eastern United States, but tree mortality in some forests in this region are sensitive to air pollution (Dietze and Moorcroft 2011). Tree regeneration after disturbance, which is critical for maintaining forest cover and associated C stocks (McKinley et al. 2011), is uncertain for Western montane forests in a warmer climate (Bonnet et al. 2005), especially if fire severity increases (Haire and McGarigal 2010).

3.2.2 Effects on Nutrient Cycling

Carbon cycling responses to elevated CO_2 and warming will be linked to nutrient availability, especially N. Biological processes that convert nutrients held in organic matter to available mineral forms are generally temperature dependent. Experimental soil-warming studies confirm that N mineralization will increase in response to higher temperatures (Melillo et al. 2011), with an average increase in net N mineralization of about 50 % (Rustad et al. 2001). These effects may be transient, however, because the supply of mineralizable substrates may not keep pace with opportunities for mineralization. Soil-warming studies are limited by methodological constraints that make it difficult to scale results to ecosystems or incorporate system interactions. However, modeling approaches that scale to the ecosystem and incorporate interactions have generally confirmed patterns observed in soil-warming experiments (Campbell et al. 2009). Recent studies have used observed climatic variability and corresponding measures of stream N in forested watersheds to infer changes in N cycling processes. For example, in the western United States, recent warming temperatures have melted glacial ice, subsequently flushing N from microbially active sediments (Baron et al. 2009). In the eastern United States, Brookshire et al. (2011) found that seasonal variation in stream nitrate was coupled with recent warming, and used modeling to project that higher temperature will increase future N export threefold more than will projected changes in N deposition.

Altered species composition can affect belowground nutrient cycling processes (Lovett et al. 2006; Knoepp et al. 2011). For example, forests with beech bark disease have increased litter decomposition, decreased soil C:N ratio, and increased extractable nitrate in the soil and soil solution (Lovett et al. 2010). In eastern hemlock (*Tsuga canadensis* [L.] Carrière) stands infested with hemlock woolly adelgid (*Adelges tsugae* Annand), litter N is increased, and N mineralization is accelerated even before tree mortality is observed (Stadler et al. 2006; Orwig et al. 2008). Defoliation by insects also alters N pools and fluxes in forests (Lovett et al. 2002).

3.3 Forest Hydrological Processes

Climate change will have both indirect and direct effects on forest water hydrologic processes. Indirect effects, which work primarily through effects on forest evapotranspiration (ET), are associated with changes in atmospheric CO_2, increased temperature, altered soil water availability, changes in species composition, and changes in disturbance regimes or management that alter forest structure and composition. Direct effects are associated with more rainfall and more intense storms in some regions (see Chap. 4). These in turn increase base flows in streams (particularly intermittent streams), increase flood risk, accelerate erosion, and increase the potential for both landslides and increased inter-storm periods and drought, along with climate-related changes in infiltration rate. Indirect and direct effects are interdependent.

3.3.1 Forest Evapotranspiration and Streamflow

Forest ET may be responding to changing climate (Labat et al. 2004; Walter et al. 2004; Gedney et al. 2006), but studies disagree about the direction of the change. Over relatively large areas and long temporal scales, streamflow is the balance between rainfall input and ET. Hence, the rainfall not used in ET is available for streamflow and groundwater recharge, and in many forest ecosystems, ET strongly influences streamflow and groundwater recharge. Walter et al. (2004) concluded that ET has been increasing across most of the United States at a rate of 10.4 mm per decade (inferred from U.S. Geological Survey records of precipitation and river discharge in six major basins). In contrast, river discharge throughout the East has been increasing at a rate of 4 % for each 1 °C increase in temperature (Labat et al. 2004), suggesting a reduction in ET. Different response patterns are not unexpected because ET is affected by several co-occurring and often counteracting climatic, physiological, and structural variables. For example, increased discharge (Labat et al. 2004) has been attributed to the physiological effect of CO_2 on water use efficiency (thereby decreasing ET), and not to the effect of changing land use (Gedney et al. 2006).

3.3.2 Elevated Atmospheric CO_2

Over long time scales, higher CO_2 concentrations decrease stomatal density and aperture, both of which reduce transpiration (Franks and Beerling 2009; Prentice and Harrison 2009). Observational and experimental studies confirm long-term and large-scale changes in leaf stomatal conductance in response to elevated CO_2 (Lammertsma et al. 2011; Warren et al. 2011). As leaf stomatal conductance decreases, ecosystem ET can also decline; however, any decline depends on stand age, species composition, and leaf area. Empirical studies linking reduced stomatal conductance to reduced stand-level ET have not yet been possible, and most researchers have used modeling to make this linkage.

Warren et al. (2011) applied the Forest BGC model to data from several elevated CO_2 studies and projected that ET was reduced by 11 % in older stands that did not experience an increase in leaf area. In younger stands, ET increased because of stimulation of leaf area, although field studies have not yet identified an increase in stand leaf area with elevated CO_2 (Norby and Zak 2011). In a modeling study of deciduous forests in the northeastern United States, the projected effect of elevated CO_2 on ET was modest, ranging from a 4 % decrease to an 11 % increase (Ollinger et al. 2008). In Mediterranean forest systems, changes in ET are also expected to be modest with increased temperature and CO_2, ranging from no change to a 10 % decrease (Tague et al. 2009). Although the effects of elevated CO_2 on ET remain uncertain, the direct effects will likely be modest (± 10 %) compared to changes expected for other variables that affect ET, such as precipitation variability (Leuzinger and Körner 2010).

Higher temperature and thus increased vapor pressure deficit (VPD) between the inside of the leaf and the surrounding air may offset the water use efficiency effects of elevated CO_2. As the air becomes drier, transpiration typically increases following an exponential saturation curve, with the rate of increase continually slowed by reduced stomatal opening. Most studies show that a physiological effect of reduced stomatal conductance in response to elevated CO_2 is observed only when the canopy air is very humid (low VPD). In a study of six deciduous tree species, elevated CO_2 reduced transpiration by 22 %, but only at low VPD (Cech et al. 2003). These results suggest that the physiological effects of elevated CO_2 on ecosystem water balance may depend on precipitation and atmospheric humidity.

Warming has changed the timing of foliage green-up and senescence, but the effects of these phenological changes on ET are complex and poorly understood. Warming-induced lengthening of the growth season could increase ET and offset the reduction in stomatal conductance from elevated CO_2, but these effects are difficult to generalize across species and regions (Hänninen and Tanino 2011). The frost-free season across the United States has lengthened by about 2 weeks, resulting in a longer, warmer growing season, however, growth cessation in autumn might come earlier with increasing temperatures for some boreal and temperate tree species (Kunkel et al. 2004). For other tree species, spring budburst might be delayed by warmer temperatures (Zhang et al. 2007), perhaps because of insufficient chilling hours (Schwartz and Hanes 2010). In higher latitudes where chilling requirements are still being met, green-up is occurring sooner. Thus, springtime ET in the lower latitudes could be delayed, whereas ET in the higher latitudes could be advanced.

The potential increase in ET owing to a lengthened growing season can be constrained by water availability and drought in the growing season (Zhao and Running 2010). Water directly limits ET (lower water availability reduces transpiration), and many regions of the United States have experienced more frequent precipitation extremes, including droughts, over the last 50 years (Easterling et al. 2000b; Groisman et al. 2004; Huntington 2006; Solomon et al. 2007).

3.3.3 Changing Species Composition

Evapotranspiration is affected by the plant and tree species that comprise the canopy cover of a forest ecosystem. In general, pine forests are more responsive to climatic variation than are deciduous forests (Stoy et al. 2006; Ford et al. 2011); however, even within the same forest, growing-season transpiration rates among canopy species (adjusted for differences in tree size) can vary by as much as fourfold, and co-occurring species can differ considerably in their responsiveness to climatic variation (Ford et al. 2011). Characteristics of the xylem and sapwood, which vary by species, are among the most important determinants of stand transpiration in both observational (Vose and Ford 2011; Wullschleger et al. 2001) and theoretical studies (Enquist et al. 1998; Meinzer et al. 2005). Therefore, shifts in hydroclimate may be accommodated by changes in canopy leaf area, phenology, or species-based hydraulic efficiency.

Increased drought severity and frequency may contribute to changes in forest species composition in two ways. First, drought plays an important role in tree mortality (Allen et al. 2010); as soil water availability declines, forest trees either reduce stomatal conductance to reduce water loss (drought avoidance), or they experience progressive hydraulic failure (Anderegg et al. 2011) and eventually die. Second, some native insect outbreaks, and the mortality they cause, are also triggered by drought. Third, as temperature increases, plant metabolism increases exponentially, and if high temperature coincides with drought stress in forests, C starvation and mortality can occur quickly (Adams et al. 2009). For example, Adams et al. (2009) projected a fivefold increase in pinyon pine (*Pinus edulis* Engelm.) mortality from an increase of 4.3 °C, based on historical drought frequency. If drought frequency increases as expected, the projected mortality could be even higher.

Insect outbreaks and fire will be the likely primary forces behind rapid changes in forest composition and structure, although direct studies of these effects on hydrology are limited (Tchakerian and Couslon 2011). Potential biogeophysical effects from tree-killing biotic disturbances include (1) increased surface albedo, which will reduce the absorption of solar radiation, (2) decreased transpiration until the new forest is reestablished, and (3) decreased surface roughness, which affects atmospheric drag (Bonan 2008). After disturbances that cause widespread tree mortality, streamflow increases, the annual hydrograph advances, and low flows increase; at the same time, snow accumulation increases and snowmelt is more rapid after needle drop (Boon 2012; Pugh and Small 2011). Collectively, these studies show strong, but mostly indirect evidence that large-scale forest mortality will alter water cycling processes; however, the magnitude and duration of responses will differ among species and across regions.

3.3.4 Snowmelt

As a result of a warming climate, snow cover in North America has decreased in duration, extent, and depth over the last few decades, with increased interannual variability (Mote et al. 2005; Pagano and Garen 2005; Regonda et al. 2005; Barnett et al. 2008; Luce and Holden 2009). Reduced snowpack depth, persistence, and duration affect water stress, disturbance, erosion, and biogeochemical cycling in forest ecosystems. In arid and semiarid forests, early and reduced snowmelt leads to increased water stress in the late growing season, increased fire frequency, and higher susceptibility to insect attack (Breshears et al. 2005; Adams et al. 2012; Holden et al. 2011; Westerling et al. 2011). The rapid flush of water to the soil in spring snowmelt can release solutes that have been slowly accumulating as a result of subnival biogeochemical cycling (Williams et al. 2009). These spring pulses can provide the major input of nutrients to aquatic ecosystems. Reductions in the spring flush, and increased rain in winter and early spring, can change the timing of N release from northern forests. Higher frequency and magnitude of rain-on-snow events may also increase soil erosion, sedimentation, and landslides.

3.3.5 Soil Infiltration

Forest ecosystems typically support high infiltration capacities because of large soil pores developed by root systems and soil fauna, so surface runoff is uncommon. However, high-intensity precipitation or snowmelt events can rapidly move water in the soil to the unsaturated zone or groundwater, or into a local stream, particularly in steep terrain (Troch et al. 2009; Brooks et al. 2011). Increased storm intensity projected for the future may increase peak streamflow and flooding through this process.

3.3.6 Carbon and Water Tradeoffs

Expanding C sequestration or wood-based bioenergy markets to offset fossil fuel emissions may affect water resources (Jackson et al. 2005), depending on the specific management activity and scale of implementation. Planting fast-growing species for bioenergy production (or C sequestration) may reduce water availability (Jackson et al. 2005), but these reductions may be minor if the planting area is small relative to the watershed size. In wetter regions, where interception represents a higher proportion of ET, evergreen species may have a bigger effect on site water balance. In drier regions, where transpiration represents the greatest proportion of ET, species that use large amounts of water, such as *Populus* or *Eucalyptus*, would have a larger effect on the hydrologic cycle (Farley et al. 2005). Shortening rotation length might increase streamflow because the proportion of time during which the stand is at canopy closure (when leaf area index is highest and streamflow is lowest) will be reduced. In a global analysis of forest plantations, Jackson et al. (2005)

found the biggest reductions in streamflow in plantations that were 15–20 years old. This would be exacerbated if short-rotation forests are irrigated, which might be considered a necessity in some areas of the western United States where drought frequency and intensity are expected to increase.

3.4 Tree Species Distribution

The ranges of plant and animal species have always shifted through time (e.g., Davis and Shaw 2001), but in recent decades, some species may be moving faster than in the past (Parmesan and Yohe 2003; Chen et al. 2010; Dobrowski et al. 2011). For example, in a meta-analysis of 764 species range changes (mostly insects and no tree species), the average rate of northward migration was 16.9 km per decade (Chen et al. 2011). In contrast, an earlier meta-analysis, using 99 species of birds, butterflies, and alpine herbs, reported a northward migration of 6.1 km per decade (Parmesan and Yohe 2003). There is also evidence of upward elevation migration of tree species (Beckage et al. 2008; Holzinger et al. 2008; Lenoir et al. 2008).

Woodall et al. (2009) used forest inventory data to investigate surrogates for migration among 40 tree species in the eastern United States, comparing mean latitude of biomass of larger trees (>2.5 cm diameter) relative to mean latitude of seedling density (<2.5 cm diameter) across each species range of latitude. For many species, this analysis indicated higher regeneration success at the northern edge of their ranges. Compared to mean latitude of tree biomass, mean latitude of seedlings was significantly farther north (>20 km) for the northern study species, southern species showed no shift, and general species showed southern expansion. Density of seedlings relative to tree biomass of northern tree species was nearly ten times higher in northern latitudes than in southern latitudes. These results suggest that Eastern tree species have moved northward, with rates approaching 100 km per century for some species. Pollen records suggest migration rates for tree species during the Holocene were 2–2.5 km per decade (Davis 1989), a time when species were not slowed by forest fragmentation (Iverson et al. 2004a, b).

Vegetation change can be projected for the future using two types of predictive models: (1) empirical, species distribution models that establish statistical relationships between species or life forms and (often numerous) predictor variables, and (2) process-based models, which simulate vegetation dynamics at the taxonomic resolution of species or life forms. There are well-recognized tradeoffs between using these different models to assess potential changes in species habitats resulting from projections of environmental change (Thuiller et al. 2008). When both approaches yield similar results for a particular area, confidence in model projections is improved. Demographic studies inform species distribution models (SDMs), and migration models are sometimes incorporated in process-based models.

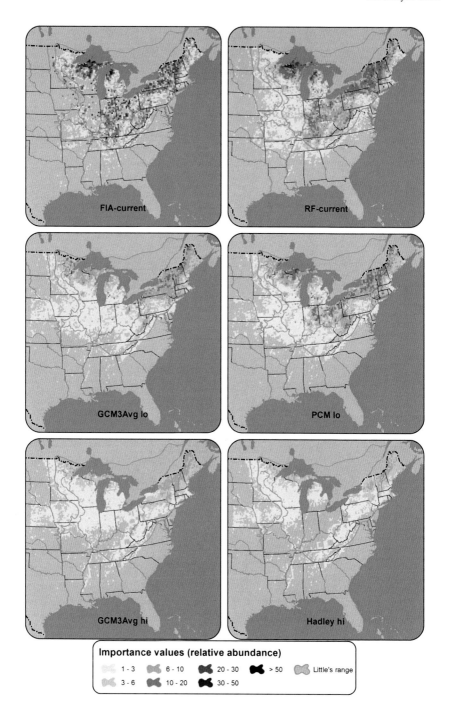

3.4.1 Modeling Species Distribution and Abundance

3.4.1.1 Species Distribution Models

Species distribution models, which extrapolate species distributions in space and time, are based on statistical models of habitat suitability (Franklin 2009) and built with observations of species occurrences along with environmental variables thought to influence habitat suitability and "equilibrium" species distribution. Predictive mapping of suitable habitat (but not whether a species will reach those habitats) in space and time are therefore possible. The SDMs have limitations, including assumptions that (1) selected variables reflect the niche requirements of a species, (2) species are in equilibrium with their suitable habitat, (3) species will be able to disperse to their suitable locations, (3) projections can be made for novel climates and land covers, (4) effects of adaptation and evolution are minimal, and (5) the effects of biotic interactions (including human interactions) are minimal (Ibáñez et al. 2006; Pearson et al. 2006). However, SDMs can provide glimpses of probable futures useful for incorporating future conditions into conservation and management practices.

Species distribution models project a northward movement of tree species habitat in North America from 400 to 800 km by 2100 depending on the assumptions used in projecting future climate (Iverson et al. 2008; McKenney et al. 2011). Species distribution model projections also differ based on future scenarios and with time. For example, under a scenario of high greenhouse gas emissions (Hadley A1F1), about 66 species would gain and 54 species would lose at least 10 % of their suitable habitat under climate change. A lower emission pathway would result in both fewer losers and gainers. Sugar maple (*Acer saccharum* Marsh.) would lose a large proportion of its habitat under the warmest scenario (Lovett and Mitchell 2004; Iverson et al. 2008) (Fig. 3.2), but would still maintain a presence of habitat in most areas. When multiple species are compiled to create "forest types," models project a loss of suitable habitat for spruce-fir (*Picea-Abies*), white-red-jack pine (*Pinus strobus* L., *P. resinosa* Aiton, *P. banksiana* Lamb.), and aspen-birch (*Populus-Betula*), but an expansion of suitable habitat for oak-hickory (*Quercus-Carya*) (Iverson and Prasad 2001; Iverson et al. 2008) (Fig. 3.3).

Fig. 3.2 Maps of current and potential future suitable habitat for sugar maple in the United States show potential northward movement of habitat by 2100. In addition to showing the range of sugar maple in Little (1971), the map includes the current inventory estimate of abundance from U.S. Forest Service Forest Inventory and Analysis (FIA-current) sampling and the modeled current distribution (RF-current). Model projections for future climate are: (1) low emission scenario (B1) using the average of three global climate models (GCM3 Avg lo), (2) low emission scenario (B1) using the National Center for Atmospheric Research Parallel Climate Model (PCM lo), (3) high emission scenario (A1F1) using the average of three global climate models (GCM3 Avg hi), (4) high emission scenario (A1F1) using the HadleyCM3 model (Hadley hi) (Data from Prasad and Iverson (1999-ongoing))

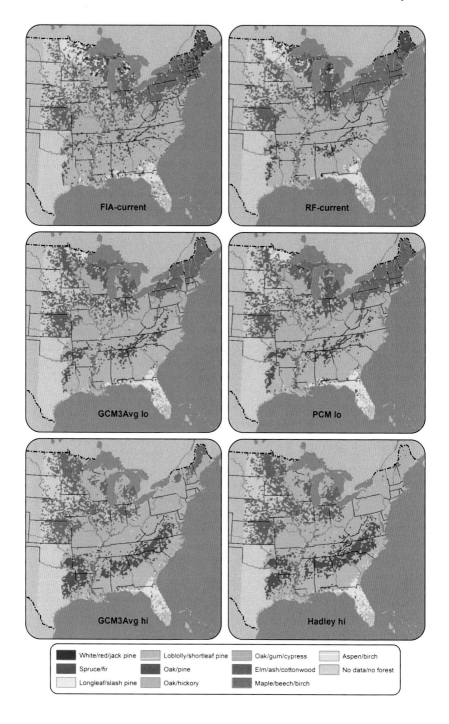

3.4.1.2 Process Models

To model species composition changes, a fully process-driven approach might be preferable to isolate mechanisms and to create "what-if" scenarios. However, such an approach is presently difficult because of the (1) necessity of detailed parameterization of species life histories and physiologies for a large number of species, (2) complexity of many interacting disturbance factors, and (3) necessary high-resolution modeling over very large areas (Lawler et al. 2006). Dynamic global vegetation models (DGVM) operate at scales from regional (hundreds of kilometers) to global; these models can aggregate species into life forms or plant functional types (PFTs), using structural or functional attributes such as needleleaf vs. broadleaf and evergreen vs. deciduous (Bachelet et al. 2003; Bonan et al. 2003; Neilson et al. 2005). Most of these models project shifts to more drought-tolerant and disturbance-tolerant species or PFTs for future climates. This general shift in vegetation may be offset by physiological changes induced by CO_2 fertilization, as suggested by a DGVM (MC1) that links water-use efficiency to CO_2-simulated expansion of forests into areas whose climate is currently too dry (Bachelet et al. 2003). This particular issue deserves further study to resolve the extent and duration of such mitigating effects of CO_2; these effects could change substantially depending on the outcome of climate change projections.

Ravenscroft et al. (2010) used the LANDIS model to simulate the potential effects of climate change to 2095 and found that mesic birch–aspen–spruce–fir and jack pine–black spruce (*Picea mariana* [Mill.] Britton, Sterns & Poggenb.) forest types would be substantially altered because of the loss of northerly species and the expansion of red maple (*Acer rubrum* L.) and sugar maple. Another promising modeling system that also includes climate variables is the Regional Hydro-Ecologic Simulation System (RHESSys) (Tague and Band 2004). Using this model in a Sierra Nevada mountain system, Christensen et al. (2008) found significant elevation differences in vegetation water use and sensitivity to climate, both of which will probably be critical to controlling responses and vulnerability of similar ecosystems under climate change. Transpiration at the lowest elevations was consistent across years because of topographically controlled high moistures, mid-elevation transpiration rates were controlled primarily by precipitation, and high-elevation transpiration rates were controlled primarily by temperature (Fig. 3.4).

Fig. 3.3 Maps of current and potential future suitable habitat for U.S. Forest Service forest types (named according to dominant species) in the eastern United States show potential northward movement of forest types by 2100. The map includes the current inventory estimate of abundance from U.S. Forest Service Forest Inventory and Analysis (FIA-current) sampling and modeled current distribution (RF-current). Model projections for future climate are: (1) low emission scenario (B1) using the average of three global climate models (GCM3 Avg lo), (2) low emission scenario (B1) using the National Center for Atmospheric Research Parallel Climate Model (PCM lo), (3) high emission scenario (A1F1) using the average of three global climate models (GCM3 Avg hi), (4) high emission scenario (A1F1) using the HadleyCM3 model (Hadley hi) (From Iverson et al. 2008)

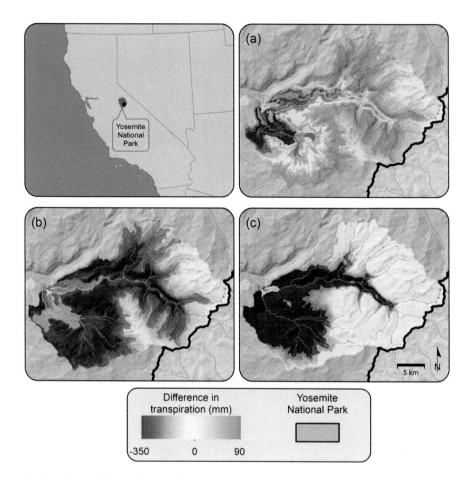

Fig. 3.4 Maps of Upper Merced River watershed, Yosemite Valley, California, showing areas with differences in transpiration in (**a**) warmest vs. coldest simulation years, (**b**) wet vs. average precipitation year, and (**c**) dry vs. average precipitation year. The largest decreases in transpiration between years are shown in *red*; increases between years are shown in *green* (From Christensen et al. 2008, with permission)

3.4.1.3 Demographic Studies

Demographic studies track individuals over time, rather than using periodic plot-level inventories, to fully understand the role of climate relative to other factors like competition, variation in physiology and function, and vulnerability to insects and pathogens. Demographic data sets are rare, but one study has tracked more than 27,000 individuals of 40 species over 6–11 years to address these interactions over a portion of the southeastern United States (Clark et al. 2011). This study found that the primary climatic controls are spring temperature (regulating species fecundity) and growing season moisture, particularly for species of *Pinus*, *Ulmus*, *Magnolia*, and *Fagus*. Pitch pine (*Pinus rigida* Mill.) tracked both spring temperature and summer drought, yellow poplar (*Liriodendron tulipifera* L.) tracked neither, and sweetgum (*Liquidambar styraciflua* L.) tracked summer drought but not spring temperature (Clark et al. 2011). Overall, the effect of competition on growth and mortality exceeded the effects of climatic variation for most species. Thus, demographic tracking can determine the vulnerability of individual trees to various factors, including climate change over time, variation in abiotic variables over space, and competition (Clark et al. 2011).

3.4.1.4 Dispersal and Migration Models

Each species affected by climate change will need to either migrate or be moved to a suitable habitat. Approaches used to model migration include reaction–diffusion models, phenomenological models, mechanistic models, and simulation models (Clark et al. 2003; Hardy 2005; Katul et al. 2005; Nathan et al. 2011). Recent advances in digital computation and more reliable data from seed dispersal studies have improved these models so that they can project the parameter values of seed dispersal curves as well as seed distributions. For example, Nathan et al. (2011) modeled 12 North American wind-dispersed tree species for current and projected future spread according to 10 key dispersal, demographic, and environmental factors affecting population spread. They found a low likelihood for any of the 12 species to spread 300–500 m per year, the rate of change that may be required under climate change (Loarie et al. 2009). The SHIFT model uses historical migration rates along with the strengths of the seed sources (abundance within the current range) and potential future sinks (abundance of potential suitable habitat). When model outputs of colonization potentials were combined with an SDM (DISTRIB) simulation of suitable habitat for five species—common persimmon (*Diospyros virginiana* L.), sweetgum, sourwood (*Oxydendrum arboreum* [L.] DC), loblolly pine, and southern red oak (*Quercus falcata* Michx.)—only 15 % of the newly suitable habitat had any likelihood of being colonized by those species within 100 years (Iverson et al. 2004a, b). These results suggest that a substantial lag will occur before species migrate into the new suitable habitat.

3.4.2 Assisted Migration

As noted above, models suggest that many tree species will be unable to migrate to suitable habitat within 100 years (Iverson et al. 2004a, b) and may face serious consequences if they cannot adapt to new climatic conditions. Assisted migration may help mitigate climate change by intentionally moving species to climatically suitable locations outside their natural range (McLachlan et al. 2007; Hoegh-Guldberg et al. 2008). Assisted migration has been controversial, with some advocating for it (Minteer and Collins 2010; Vitt et al. 2010) and some against

(Ricciardi and Simberloff 2009). Proponents state that these drastic measures are needed to save certain species that cannot adapt or disperse fast enough in response to rapid climate change. The main concern of opponents is that the placement of species outside their range may disturb native species and ecosystems when these "climate refugees" establish themselves in new environments. The uncertainty of climate in the future and the complexity and interactions associated with ecosystem response also argue against assisted migration.

One way to resolve the debate is to subdivide assisted migration into "rescue-assisted migration" and "forestry-assisted migration." Rescue-assisted migration moves species to minimize the risk of extinction and local extirpation in the face of climate change, and is the source of most of the controversy. Forestry-assisted migration is aimed more at maintaining high levels of productivity and diversity in widespread tree species that are commercially, socially, culturally, or ecologically valuable (Gray et al. 2011; Kreyling et al. 2011). Maintaining forest productivity and ecosystem services is generally the focus of forestry-assisted migration. Given the broad distribution of most tree species, and the relatively short distances proposed for tree seed migration, forestry-assisted migration typically involves transfers within or just beyond current range limits to locations where a population's bioclimatic envelope is expected to reside within the lifetime of the planted population (Gray et al. 2011). The introduction of genotypes to climatically appropriate locations may also contribute to overall forest health by establishing vigorous plantations across the landscape that are less susceptible to forest pests and pathogens (Wu et al. 2005). This approach may contribute to the continued flow of ecosystem services such as wildlife habitat, erosion prevention, and C uptake (Kreyling et al. 2011). If practiced in a manner in which genotypes are transferred within or just beyond current range limits, forestry-assisted migration may be a viable tool for adaptation to climate change, especially if limited to current intensively managed plantations.

3.5 Effects of Altered Forest Processes and Functions on Ecosystem Services

Ecosystem services link the effects of altered forest processes, conditions, and disturbance regimes to human well-being (World Resources Institute 2005). A broad range of utility and values derive from four broad types of ecosystem services: (1) provisioning or products from ecosystems, (2) regulation of ecosystem processes, (3) cultural or nonmaterial benefits, and (4) supporting services required for the production of all other ecosystem services (Joyce et al. 2008) (Fig. 3.5). Anticipated climate changes portend changes in all types of ecosystem services derived from forests. Because the assessment endpoint for ecosystem services is human well-being, we are ultimately concerned about the potential effects of climate change on the ecosystem services that forests provide.

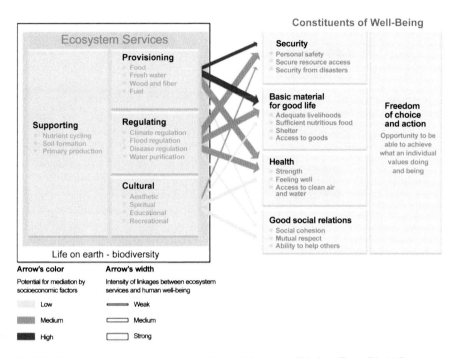

Fig. 3.5 Linkages between ecosystem services and human well-being (From World Resources Institute 2005)

Ecosystem services differ across temporal and spatial scales but are most often assessed and recognized at large spatial scales. Disturbances (natural and human) and stressors can control delivery of ecosystem services across variable timeframes. Ecosystem services occur in forests not as a single service but rather as a bundle of services. The bundle of services changes with time and in response to disturbance regimes and stressors. Vulnerability of ecosystem services to climate change will vary widely, depending not only on the service of concern (e.g., wood products or flood regulation) and location (defined by region), but also on the location in reference to human condition, such as rural versus urban settings. The value of the affected service multiplied by the likelihood of effect defines the risk to ecosystem services and provides a framework for understanding potential consequences and prioritizing actions.

Climate-related mechanisms of change in U.S. forests could alter ecosystem services in ways that are not yet fully understood, and estimating these effects introduces another layer of uncertainty. That is, climate regulates forest processes that control future forest conditions that determine future ecosystem services. Still, the potential effects of climate change on forest ecosystems could have profound and mostly disruptive consequences for ecosystem services with important implications for human well-being. Ecosystem services also depend on interactions with land use, human demographics, and economies, which may simultaneously adjust to climatic stimuli (see Chap. 5).

Forests in the United States consist of both managed (active) and unmanaged (passive) ecosystems (Ryan et al. 2008) held in public and private ownerships. Some public forests or wildlands are withdrawn from active management (e.g., national parks, state parks, wilderness areas, wild and scenic rivers), but most lands are managed for multiple-use goals (e.g., most national forests and Bureau of Land Management lands). Public land management in the United States is largely focused on non-market ecosystem services, including recreation, aesthetic values, and water quality. Most forest management for timber production occurs on private forest lands, using both capital-intensive (short rotation plantation silviculture) and land-extensive approaches (occasional harvesting followed by natural regeneration). Private lands also provide the full spectrum of ecosystem services, either by design through conservation easements, or as a byproduct of other management objectives (Butler et al. 2007). In many cases, private forest lands provide ecosystem services that accrue to broader social well-being without equitable financial compensation.

Ecosystem services provided under current climatic conditions differ across the assortment of public and private forests that are managed actively or passively. As a consequence of the regional distribution of anticipated future climate change, the provision of ecosystem services from these lands could also change and be modified by mitigation and adaptation strategies (see Chaps. 7 and 8). Social perception of risks to ecosystem services will be determined by the rate of change in these services (flows) (see Chap. 9), as well as by an understanding of mitigation and adaption strategies applied in response to climate change. Social systems will adapt to climate change and affect the condition of forests in the United States and throughout the world.

Several mechanisms of change in forest ecosystems have implications for ecosystem services. First, climate change could alter the amount and distribution of forest biomass in forests, either through shifts in productivity associated with atmospheric C concentrations or through altered forest disturbance regimes. Changes in forest biomass directly influence the supply of all wood products from lumber to fuel for electricity production (provisioning services), and they alter the amount of C stored in forest pools (a regulating service) (see Chap. 7). Future productivity and disturbance effects would probably be focused in the Rocky Mountain and intermountain West and Alaska, where only a small portion of U.S. timber production occurs. Declines in timber production would be small in the context of national markets, but they could represent substantial shares of local rural economic activity.

Changes in tree cover will affect microclimatic conditions (e.g., the cooling of urban heat islands), whereas shifts in C stocks through accumulation of biomass could affect changes in global climate trajectories. Projections of accelerated emissions related to elevated insect epidemics and fire activity in the Rocky Mountains and southwestern United States could represent a substantial effect on forest C storage, potentially shifting U.S. forests from net C sinks to net C emitters (Wear et al. 2012) (see Sect. 3.2 and Chap. 7).

The effects of climate change on forest productivity differ by region and contain sufficient uncertainty that their influence on timber markets and C stocks are difficult to project. However, if forest productivity were to increase in the eastern United States and decrease in the western United States (see Sect. 3.2), this could accelerate the shift in timber production from West to East, and especially to the Southeast.

Estimates of the economic consequences of insect and pathogen outbreaks focus on timber market effects (e.g., southern pine beetle [*Dendroctonus frontalis* Zimmerman] (Pye et al. 2011)) or the influence of tree mortality on property values (e.g., hemlock woolly adelgid (Holmes et al. 2010)). These measures of market effects for price-based services address one element of a complex of values affected by forest disturbances. In the case of forest insects, management decisions already account for a certain level of expected tree mortality, so the more relevant question is whether effects significantly exceed the "background" losses associated with endemic insects and pathogens. Property values define the effect of disturbance and related mortality on ecosystem services delivered to private property owners, but they cannot capture the "public good" aspects of changes to forest aesthetics for people who view forests. To illustrate, widespread tree mortality related to pine beetle epidemics on national forests can reduce the aesthetic values for millions of people. These "quality of life" effects represent real value losses, but they are difficult to quantify and may be transitory as regrowth occurs and society adjusts expectations regarding what constitutes a natural or aesthetically appealing condition.

Climate change could alter the complex of interactions between forest conditions and water flow and quality. Forest cover and condition constitute only one element of a complex system, so effects may be difficult to isolate, but forest condition appears to be strongly related to flood protection (a regulating service) and water quantity and quality (a provisioning service). More variable precipitation patterns (stronger drought and extreme rainfall events) increase the service value of forests in protecting against flooding and landslides, but they also change forest conditions in ways that reduce soil-protecting qualities. This negative feedback suggests potential for accelerated losses of flood protection services of forests. Reduced supplies of these services would coincide with strong growth in the demand for water services caused by population growth and associated water needs for personal and commercial uses.

The longer term and less certain effects of climate change on forest conditions discussed above suggest more forests in a state of disequilibrium with new species-climate associations. The notion of "novel" conditions suggests "unknowable" implications, especially regarding the supply and demand for ecosystem services and the reactions of private landowners and government to increasing scarcity of important services. However, economic factors will likely drive responses, and the risks of climate change to forests may open public dialogue regarding the costs and benefits of providing ecosystem services. Changes in forest policy may be needed to align producers and consumers of services on private and public forest lands (e.g., providing compensation for private landowners' provision of scarce ecosystem services).

Adaptation strategies in forests can build resistance to climate-related stressors, increase ecosystem resilience by minimizing the severity of climate change effects, or facilitate large-scale ecological transitions in response to changing environmental conditions (see Chap. 8). Adaptation and mitigation strategies for forests can alter the supply of ecosystem services and involve explicit tradeoffs between services. For example, thinning and fuel treatment to reduce the vulnerability of forests to disturbance regimes and stressors defines a specific tradeoff between short-term changes in C stocks and long-term stability of C emissions. Resistance, resilience, and transitions of forest ecosystems to new conditions are tiered to increasing levels of environmental change and time scales, and each adaptation strategy will result in a different bundle of ecosystem services.

References

Adams, H. D., Guardiola-Claramonte, M., Barron-Gafford, G. A., et al. (2009). Temperature sensitivity of drought-induced tree mortality portends increased regional die-off under global-change-type drought. *Proceedings of the National Academy of Sciences, USA, 106*, 7063–7066.

Adams, H. D., Luce, C. H., Breshears, D. D., et al. (2012). Ecohydrological consequences of drought- and infestation-triggered tree die-off: Insights and hypotheses. *Ecohydrology, 5*, 145–149.

Allen, C. D., Macalady, A. K., Chenchouni, H., et al. (2010). A global overview of drought and heat-induced tree mortality reveals emerging climate change risks for forests. *Forest Ecology and Management, 259*, 660–684.

Anderegg, R. L., Berry, J. A., Smith, D. D., et al. (2011). The roles of hydraulic and carbon stress in a widespread climate-induced forest die-off. *Proceedings of the National Academy of Sciences, USA, 109*, 233–237.

Arend, M., Kuster, T., Günthardt-Goerg, M. S., & Dobbertin, M. (2011). Provenance-specific growth responses to drought and air warming in three European oak species (*Quercus robur, Q. petraea* and *Q. pubescens*). *Tree Physiology, 31*, 287–297.

Asshoff, R., Zotz, G., & Körner, C. G. (2006). Growth and phenology of mature temperate forest trees in elevated CO_2. *Global Change Biology, 12*, 848–861.

Augspurger, C. K. (2009). Spring 2007 warmth and frost: Phenology, damage and refoliation in a temperate deciduous forest. *Functional Ecology, 23*, 1031–1039.

Bachelet, D., Neilson, R. P., Hickler, T., et al. (2003). Simulating past and future dynamics of natural ecosystems in the United States. *Global Biogeochemical Cycles, 17*, 14-1–14-21.

Bader, M. K.-F., & Körner, C. (2010). No overall stimulation of soil respiration under mature deciduous forest trees after 7 years of CO_2 enrichment. *Global Change Biology, 16*, 2830–2843.

Bader, M., Hiltbrunner, E., & Körner, C. (2009). Fine root responses of mature deciduous forest trees to free air carbon dioxide enrichment (FACE). *Functional Ecology, 23*, 913–921.

Barford, C. C., Wofsy, S. C., Goulden, M. L., et al. (2001). Factors controlling long- and short-term sequestration of atmospheric CO_2 in a mid-latitude forest. *Science, 294*, 1688–1691.

Barnett, T. P., Pierce, D. W., Hildago, H. G., et al. (2008). Human-induced changes in the hydrology of the Western United States. *Science, 319*, 1080–1083.

Baron, J. S., Schmidt, T. W., & Hartman, M. D. (2009). Climate-induced changes in high elevation stream nitrate dynamics. *Global Change Biology, 15*, 1777–1789.

Beckage, B., Osborne, B., Gavin, D. G., et al. (2008). A rapid upward shift of a forest ecotone during 40 years of warming in the Green Mountains of Vermont. *Proceedings of the National Academy of Sciences, USA, 105*, 4197–4202.

Bernhardt, E. S., Barber, J. J., Pippen, J. S., et al. (2006). Long-term effects of free air CO_2 enrichment (FACE) on soil respiration. *Biogeochemistry, 77*, 91–116.

Boisvenue, C., & Running, S. W. (2010). Simulations show decreasing carbon stocks and potential for carbon emissions in Rocky Mountain forests over the next century. *Ecological Applications, 20*, 1302–1319.

Bolstad, P. V., Davis, K. J., Martin, J., et al. (2004). Component and whole-system respiration fluxes in northern deciduous forests. *Tree Physiology, 24*, 493–504.

Bonan, G. B. (2008). Forests and climate change: Forcings, feedbacks, and the climate benefits of forests. *Science, 320*, 1444–1449.

Bonan, G. B., Levis, S., Sitch, S., et al. (2003). A dynamic global vegetation model for use with climate models: Concepts and description of simulated vegetation dynamics. *Global Change Biology, 9*, 1543–1566.

Bonnet, V. H., Schoettle, A. W., & Shepperd, W. D. (2005). Postfire environmental conditions influence the spatial pattern of regeneration for *Pinus ponderosa*. *Canadian Journal of Forest Research, 35*, 37–47.

Boon, S. (2012). Snow accumulation following forest disturbance. *Ecohydrology, 5*, 279–285.

Breshears, D. D., Cobb, N. S., Rich, P. M., et al. (2005). Regional vegetation die-off in response to global-change-type drought. *Proceedings of the National Academy of Sciences, USA, 102*, 15144–15148.

Bronson, D. R., Gower, S. T., Tanner, M., & Van Herk, I. (2009). Effect of ecosystem warming on boreal black spruce bud burst and shoot growth. *Global Change Biology, 15*, 1534–1543.

Brooks, P. D., Troch, P. A., Durcik, M., et al. (2011). Quantifying regional scale ecosystem response to changes in precipitation: Not all rain is created equal. *Water Resources Research, 47*, W00J08.

Brookshire, E. N. J., Gerber, S., Webster, J. R., et al. (2011). Direct effects of temperature on forest nitrogen cycling revealed through analysis of long-term watershed records. *Global Change Biology, 17*, 297–208.

Butler, B. J., Tyrrell, M., Feinberg, G., et al. (2007). Understanding and reaching family forest owners: Lessons from social marketing research. *Journal of Forestry, 105*, 348–357.

Butler, M. P., Davis, K. J., Denning, A. S., & Kawa, S. R. (2010). Using continental observations in global atmospheric inversions of CO_2: North American carbon sources and sinks. *Tellus Series B-Chemical and Physical Meteorology, 62*, 550–572.

Cai, T. E. B., Flanagan, L. B., & Syed, K. H. (2010). Warmer and drier conditions stimulate respiration more than photosynthesis in a boreal peatland ecosystem: Analysis of automatic chambers and eddy covariance measurements. *Plant, Cell and Environment, 33*, 394–407.

Campbell, J. L., Rustad, L. E., Boyer, E. W., et al. (2009). Consequences of climate change for biogeochemical cycling in forests of northeastern North America. *Canadian Journal of Forest Research, 39*, 264–284.

Cech, P. G., Pepin, S., & Körner, C. (2003). Elevated CO_2 reduces sap flux in mature deciduous forest trees. *Oecologia, 137*, 258–268.

Chen, P. Y., Welsh, C., & Hamann, A. (2010). Geographic variation in growth response of Douglas-fir to interannual climate variability and projected climate change. *Global Change Biology, 16*, 3374–3385.

Chen, I.-C., Hill, J. K., Ohlemüller, R., et al. (2011). Rapid range shifts of species associated with high levels of climate warming. *Science, 333*, 1024–1026.

Christensen, L., Tague, C. L., & Baron, J. S. (2008). Spatial patterns of simulated transpiration response to climate variability in a snow dominated mountain ecosystem. *Hydrological Processes, 22*, 3576–3588.

Churkina, G., Brovkin, V., von Bloh, K., et al. (2009). Synergy of rising nitrogen depositions and atmospheric CO_2 on land carbon uptake moderately offsets global warming. *Global Biogeochemical Cycles, 23*, GB4027.

Clark, J. S., Lewis, M., McLachlan, J. S., & HilleRisLambers, J. (2003). Estimating population spread: What can we forecast and how well? *Ecology, 84*, 1979–1988.

Clark, J. S., Bell, D. M., Hersh, M. H., & Nichols, L. (2011). Climate change vulnerability of forest biodiversity: Climate and resource tracking of demographic rates. *Global Change Biology, 17*, 1834–1849.

Cole, C. T., Anderson, J. E., Lindroth, R. L., & Waller, D. M. (2010). Rising concentrations of atmospheric CO_2 have increased growth in natural stands of quaking aspen (*Populus tremuloides*). *Global Change Biology, 16*, 2186–2197.

Crevoisier, C., Sweeney, C., Gloor, M., et al. (2010). Regional U.S. carbon sinks from three-dimensional atmospheric CO_2 sampling. *Proceedings of the National Academy of Sciences, USA, 107*, 18348–18353.

Davis, M. B. (1989). Lags in vegetation response to greenhouse warming. *Climatic Change, 15*, 75–82.

Davis, M. B., & Shaw, R. G. (2001). Range shifts and adaptive responses to quaternary climate change. *Science, 292*, 673–679.

De Vries, W. (2009). Assessment of the relative importance of nitrogen deposition and climate change on the sequestration of carbon by forests in Europe: An overview. *Forest Ecology and Management, 258*, vii–x.

Dietze, M. C., & Moorcroft, P. R. (2011). Tree mortality in the eastern and central United States: Patterns and drivers. *Global Change Biology, 17*, 3312–3326.

Dobrowski, S. Z., Thorne, J. H., Greenberg, J. A., et al. (2011). Modeling plant ranges over 75 years of climate change in California, USA: Temporal transferability and species traits. *Ecological Monographs, 81*, 241–257.

Easterling, D. R., Meehl, G. A., Parmesan, C., et al. (2000b). Climate extremes: Observations, modeling, and impacts. *Science, 289*, 2068–2074.

Ellis, T., Hill, P. W., Fenner, N., et al. (2009). The interactive effects of elevated carbon dioxide and water table draw-down on carbon cycling in a Welsh ombrotrophic bog. *Ecological Engineering, 35*, 978–986.

Enquist, B. J., Brown, J. H., & West, G. B. (1998). Allometric scaling of plant energetics and population density. *Nature, 395*, 163–165.

Farley, K. A., Jobbágy, E. G., & Jackson, R. B. (2005). Effects of afforestation on water yield: A global synthesis with implications for policy. *Global Change Biology, 11*, 1565–1576.

Fenner, N., Ostle, N. J., McNamara, N., et al. (2007). Elevated CO_2 effects on peatland plant community carbon dynamics and DOC production. *Ecosystems, 10*, 635–647.

Finzi, A. C., Moore, D. J. P., DeLucia, E. H., et al. (2006). Progressive nitrogen limitation of ecosystem processes under elevated CO_2 in a warm-temperate forest. *Ecology, 87*, 15–25.

Ford, C. R., Hubbard, R. M., & Vose, J. M. (2011). Quantifying structural and physiological controls on variation in canopy transpiration among planted pine and hardwood species in the southern Appalachians. *Ecohydrology, 4*, 183–195.

Franklin, J. (2009). *Mapping species distributions: spatial inference and prediction* (338pp). Cambridge: Cambridge University Press.

Franks, P. J., & Beerling, D. J. (2009). Maximum leaf conductance driven by CO_2 effects on stomatal size and density over geologic time. *Proceedings of the National Academy of Sciences, USA, 106*, 10343–10347.

Friend, A. D. (2010). Terrestrial plant production and climate change. *Journal of Experimental Botany, 61*, 1293–1309.

Froberg, M., Hanson, P. J., Todd, D. E., & Johnson, D. W. (2008). Evaluation of effects of sustained decadal precipitation manipulations on soil carbon stocks. *Biogeochemistry, 89*, 151–161.

Garten, C. T., Iversen, C. M., & Norby, R. J. (2011). Litterfall ^{15}N abundance indicates declining soil nitrogen availability in a free-air CO_2 enrichment experiment. *Ecology, 92*, 133–139.

Gedney, N., Cox, P. M., Betts, R. A., et al. (2006). Detection of a direct carbon dioxide effect in continental river runoff records. *Nature, 439*, 835–838.

Gerten, D., Luo, Y., Le Marie, G., et al. (2008). Modelled effects of precipitation on ecosystem carbon and water dynamics in different climatic zones. *Global Change Biology, 14*, 1–15.

Grant, R. F., Black, T. A., Gaumont-Guay, D., et al. (2006). Net ecosystem productivity of boreal aspen forests under drought and climate change: Mathematical ordsbi with Ecosys. *Agricultural and Forest Meteorology, 140*, 152–170.

Gray, L. K., Gylander, T., Mbogga, M. S., et al. (2011). Assisted migration to address climate change: Recommendations for aspen reforestation in western Canada. *Ecological Applications, 21*, 1591–1603.

Groisman, P. Y., Knight, R. W., Karl, T. R., et al. (2004). Contemporary changes of the hydrological cycle over the contiguous United States: Trends derived from in situ observations. *Journal of Hydrometeorology, 5*, 64–85.

Gu, L., Hanson, P. J., Post, W. M., et al. (2008). The 2007 eastern U.S. spring freeze: Increased cold damage in a warming world? *BioScience, 58*, 253–262.

Gunderson, C. A., Edwards, N. T., Walker, A. V., et al. (2012). Forest phenology and a warmer climate—Growing season extension in relation to climatic provenance. *Global Change Biology, 18*, 2008–2025.

Haire, S. L., & McGarigal, K. (2010). Effects of landscape patterns of fire severity on regenerating ponderosa pine forests (*Pinus ponderosa*) in New Mexico and Arizona, USA. *Landscape Ecology, 25*, 1055–1069.

Hänninen, H., & Tanino, K. (2011). Tree seasonality in a warming climate. *Trends in Plant Science, 16*, 412–416.

Hänninen, H., Slaney, M., & Linder, S. (2007). Dormancy release of Norway spruce under climatic warming: Testing ecophysiological models of bud burst with a whole-tree chamber experiment. *Tree Physiology, 27*, 291–300.

Hanson, P. J., Wullschleger, S. D., & Norby, R. J. (2005). Importance of changing CO_2, temperature, precipitation, and ozone on carbon and water cycles of an upland-oak forest: Incorporating experimental results into model simulations. *Global Change Biology, 11*, 1402–1423.

Hanson, P. J., Tschaplinski, T. J., Wullschleger, S. D., et al. (2007). The resilience of upland-oak forest canopy trees to chronic and acute precipitation manipulations. In D. S. Buckley & W. K. Clatterbuck (Eds.), *Proceedings 15th central hardwood forest conference* (E-General Technical Report SRS-101, pp. 3–12). Asheville: U.S. Department of Agriculture, Forest Service, Southern Research Station.

Hardy, C. C. (2005). Wildland fire hazard and risk: Problems, definitions, and context. *Forest Ecology and Management, 211*, 73–82.

Heijmans, M. M. P. D., Mauquoy, D., van Geel, B., & Berendse, F. (2008). Long-term effects of climate change on vegetation and carbon dynamics in peat bogs. *Journal of Vegetation Science, 19*, 307–320.

Hoegh-Guldberg, O., Hughes, L., McIntyre, S., et al. (2008). Assisted colonization and rapid climate change. *Science, 321*, 345–346.

Hofmockel, K. S., Zak, D. R., Moran, K. K., & Jastrow, J. D. (2011). Changes in forest soil organic matter pools after a decade of elevated CO_2 and O_3. *Soil Biology and Biochemistry, 43*, 1518–1527.

Holden, Z. A., Luce, C. H., Crimmins, M. A., & Morgan, P. (2011). Wildfire extent and severity correlated with annual streamflow distribution and timing in the Pacific Northwest, USA (1984–2005). *Ecohydrology*. doi:10.1002/eco.257.

Holmes, T. P., Liebhold, A. M., Kovacs, K. F., & Von Holle, B. (2010). A spatial-dynamic value transfer model of economic losses from a biological invasion. *Ecological Economics, 70*, 86–95.

Holzinger, B., Hülber, K., Camenisch, M., & Grabherr, G. (2008). Changes in plant species richness over the last century in the eastern Swiss Alps: Elevational gradient, bedrock effects and migration rates. *Plant Ecology, 195*, 179–196.

Hu, J., Moore, D. J. P., Burns, S. P., & Monson, R. K. (2010). Longer growing seasons lead to less carbon sequestration by a subalpine forest. *Global Change Biology, 16*, 771–783.

Huntington, T. G. (2006). Evidence for intensification of the global water cycle: Review and synthesis. *Journal of Hydrology, 319*, 83–95.

Ibáñez, I., Clark, J. S., Dietze, M. C., et al. (2006). Predicting biodiversity change: Outside the envelope, beyond the species-area curve. *Ecology, 87,* 1896–1906.

World Resources Institute. (2005). *Ecosystems and human well-being: opportunities and challenges for business and industry* (31p). Washington, DC: World Resources Institute.

Ise, T., Dunn, A. L., Wofsy, S. C., & Moorcroft, P. R. (2008). High sensitivity of peat decomposition to climate change through water-table feedback. *Nature Geoscience, 1,* 763–766.

Iverson, L. R., & Prasad, A. M. (2001). Potential changes in tree species richness and forest community types following climate change. *Ecosystems, 4,* 186–199.

Iverson, L. R., Schwartz, M. W., & Prasad, A. M. (2004a). How fast and far might tree species migrate under climate change in the eastern United States? *Global Ecology and Biogeography, 13,* 209–219.

Iverson, L. R., Schwartz, M. W., & Prasad, A. M. (2004b). Potential colonization of new available tree species habitat under climate change: An analysis for five eastern U.S. species. *Landscape Ecology, 19,* 787–799.

Iverson, L. R., Prasad, A. M., Matthews, S. N., & Peters, M. (2008). Estimating potential habitat for 134 eastern US tree species under six climate scenarios. *Forest Ecology and Management, 254,* 390–406.

Jackson, R. B., Jobbágy, E. G., Avissar, R., et al. (2005). Trading water for carbon with biological carbon sequestration. *Science, 310,* 1944–1947.

Janssens, I. A., Dieleman, W., Luyssaert, S., et al. (2010). Reduction of forest soil respiration in response to nitrogen deposition. *Nature Geoscience, 3,* 315–322.

Johnson, D. W. (2006). Progressive N limitation in forests: Review and implications for long-term responses to elevated CO_2. *Ecology, 87,* 64–75.

Johnson, D. W., Todd, D. E., & Hanson, P. J. (2008). Effects of throughfall manipulation on soil nutrient status: Results of 12 years of sustained wet and dry treatments. *Global Change Biology, 14,* 1661–1675.

Joyce, L. A., Blate, G. M., Littell, J. S., et al. (2008). National forests. In S. J. Julius & J. M. West (Eds.), *Preliminary review of adaptation options for climate-sensitive ecosystems and resources* (pp. 3-1–3-127). Washington, DC: U.S. Environmental Protection Agency.

Karnosky, D. F., Pregitzer, K. S., Zak, D. R., et al. (2005). Scaling ozone responses of forest trees to the ecosystem level in a changing climate. *Plant, Cell and Environment, 28,* 965–981.

Katul, G. G., Porporato, A., Nathan, R., et al. (2005). Mechanistic analytical models for long-distance seed dispersal by wind. *The American Naturalist, 166,* 368–381.

Knoepp, J. D., Vose, J. M., Clinton, B. D., & Hunter, M. D. (2011). Hemlock infestation and mortality: Impacts on nutrient pools and cycling in Appalachian forests. *Soil Science Society of America Journal, 75,* 1935–1945.

Körner, C., Asshoff, R., Bignucolo, O., et al. (2005). Carbon flux and growth in mature deciduous forest trees exposed to elevated CO_2. *Science, 309,* 1360–1362.

Koven, C. D., Ringeval, B., Friedlingstein, P., et al. (2011). Permafrost carbon-climate feedbacks accelerate global warming. *Proceedings of the National Academy of Sciences, USA, 108,* 14769–14774.

Kreyling, J., Bittner, T., Jaeschke, A., et al. (2011). Assisted colonization: A question of focal units and recipient localities. *Restoration Ecology, 19,* 433–440.

Kunkel, K. E., Easterling, D. R., Hubbard, K., & Redmond, K. (2004). Temporal variations in frost-free season in the United States: 1895–2000. *Geophysical Research Letters, 31,* L03201.

Kutsch, W. L., Kolle, O., Rebmann, C., et al. (2008). Advection and resulting CO_2 exchange uncertainty in a tall forest in central Germany. *Ecological Applications, 18,* 1391–1405.

Labat, D., Goddéris, Y., Probst, J. L., & Guyot, J. L. (2004). Evidence for global runoff increase related to climate warming. *Advances in Water Resources, 27,* 631–642.

Lammertsma, E. I., de Boer, H. J., Dekker, S. C., et al. (2011). Global CO_2 rise leads to reduced maximum stomatal conductance in Florida vegetation. *Proceedings of the National Academy of Sciences, USA, 108,* 4035–4040.

Lawler, J. J., White, D., Neilson, R. P., & Blaustein, A. R. (2006). Predicting climate-induced range shifts: Model differences and model reliability. *Global Change Biology, 12,* 1568–1584.

Lenoir, J., Gégout, J. C., Marquet, P. A., et al. (2008). A significant upward shift in plant species optimum elevation during the 20th century. *Science, 320,* 1768–1771.
Leuzinger, S., & Körner, C. (2007). Water savings in mature deciduous forest trees under elevated CO_2. *Global Change Biology, 13,* 2498–2508.
Leuzinger, S., & Körner, C. (2010). Rainfall distribution is the main driver of runoff under future CO_2-concentration in a temperate deciduous forest. *Global Change Biology, 16,* 246–254.
Little, E. L. (1971). *Atlas of United States trees, Vol. 1, conifers and important hardwoods* (Misc. Pub. 1146). Washington, DC: U.S. Department of Agriculture.
Loarie, S. R., Duffy, P. B., Hamilton, H., et al. (2009). The velocity of climate change. *Nature, 462,* 1052–1055.
Lovett, G. M., & Mitchell, M. J. (2004). Sugar maple and nitrogen cycling in the forests of eastern North America. *Frontiers in Ecology and the Environment, 2,* 81–88.
Lovett, G. M., Christenson, L. M., Groffman, P. M., et al. (2002). Insect defoliation and nitrogen cycling in forests. *BioScience, 52,* 335–341.
Lovett, G. M., Canham, C. D., Arthur, M. A., et al. (2006). Forest ecosystem responses to exotic pests and pathogens in eastern North America. *BioScience, 56,* 395–405.
Lovett, G. M., Arthur, M. A., Weathers, K. C., & Griffin, J. M. (2010). Long-term changes in forest carbon and nitrogen cycling caused by an introduced pest/pathogen complex. *Ecosystems, 13,* 1188–1200.
Luce, C. H., & Holden, Z. A. (2009). Declining annual streamflow distributions in the Pacific Northwest United States, 1948–2006. *Geophysical Research Letters, 36,* L16401.
Lukac, M., Lagomarsino, W., Moscatelli, M. C., et al. (2009). Forest soil carbon cycle under elevated CO_2—A case of increased throughput? *Forestry, 82,* 75–86.
Luo, Y. Q., Gerten, D., Le Maire, G., et al. (2008). Modeled interactive effects of precipitation, temperature, and CO_2 on ecosystem carbon and water dynamics in different climatic zones. *Global Change Biology, 14,* 1986–1999.
McDowell, N. G. (2011). Mechanisms linking drought, hydraulics, carbon metabolism, and vegetation mortality. *Plant Physiology, 155,* 1051–1059.
McDowell, N., Pockman, W. T., Allen, C. D., et al. (2008). Mechanisms of plant survival and mortality during drought: Why do some plants survive while others succumb to drought? *New Phytologist, 178,* 719–739.
McKenney, D. W., Pedlar, J. H., Rood, R. B., & Price, D. (2011). Revisiting projected shifts in the climate envelopes of North American trees using updated general circulation models. *Global Change Biology, 17,* 2720–2730.
McKinley, D. C., & Blair, J. M. (2008). Woody plant encroachment by *Juniperus virginiana* in a mesic native grassland promotes rapid carbon and nitrogen accrual. *Ecosystems, 11,* 454–468.
McKinley, D. C., Ryan, M. G., Birdsey, R. A., et al. (2011). A synthesis of current knowledge on forests and carbon storage in the United States. *Ecological Applications, 21,* 1902–1924.
McLachlan, J. S., Hellmann, J. J., & Schwartz, M. W. (2007). A framework for debate of assisted migration in an era of climate change. *Conservation Biology, 21,* 297–302.
Meinzer, F. C., Bond, B. J., Warren, J. M., & Woodruff, D. R. (2005). Does water transport scale universally with tree size? *Functional Ecology, 19,* 558–565.
Melillo, J. M., Steudler, P. A., Aber, J. D., et al. (2002). Soil warming and carbon-cycle feedbacks to the climate system. *Science, 298,* 2173–2176.
Melillo, J. M., Butler, S., Johnson, J., et al. (2011). Soil warming, carbon-nitrogen interactions, and forest carbon budgets. *Proceedings of the National Academy of Sciences, USA, 108,* 9508–9512. 620pp.
Metsaranta, J. M., Kurz, W. A., Neilson, E. T., & Stinson, G. (2010). Implications of future disturbance regimes on the carbon balance of Canada's managed forest (2010–2100). *Tellus Series B-Chemical and Physical Meteorology, 62,* 719–728.
Minteer, B. A., & Collins, J. P. (2010). Move it or lose it? The ecological ethics of relocating species under climate change. *Ecological Applications, 20,* 1801–1804.

Mote, P. W., Hamlet, A. F., Clark, M. P., & Lettenmaier, D. P. (2005). Declining mountain snowpack in western North America. *Bulletin of the American Meteorological Society, 86*, 39–49.

Nathan, R., Horvitz, N., He, Y., et al. (2011). Spread of North American wind-dispersed trees in future environments. *Ecology Letters, 14*, 211–219.

Neilson, R. P., Pitelka, L. F., Solomon, A. M., et al. (2005). Forecasting regional to global plant migration in response to climate change. *Bioscience, 55*, 749–759.

Norby, R. J., & Zak, D. R. (2011). Ecological lessons learned from free-air CO_2 enrichment (FACE) experiments. *Annual Review of Ecology, Evolution, and Systematics, 42*, 181–203.

Norby, R. J., DeLucia, E. H., Gielen, B., et al. (2005). Forest response to elevated CO_2 is conserved across a broad range of productivity. *Proceedings of the National Academy of Sciences, USA, 102*, 18052–18056.

Norby, R. J., Warren, J. M., Iversen, C. M., et al. (2010). CO_2 enhancement of forest productivity constrained by limited nitrogen availability. *Proceedings of the National Academy of Sciences, USA, 107*, 19368–19373.

Ollinger, S., Goodale, C., Hayhoe, K., & Jenkins, J. (2008). Potential effects of climate change and rising CO_2 on ecosystem processes in northeastern U.S. forests. *Mitigation and Adaptation Strategies for Global Change, 14*, 101–106.

Orwig, D. A., Cobb, R. C., D'Amato, A. W., et al. (2008). Multi-year ecosystem response to hemlock woolly adelgid infestation in southern New England forests. *Canadian Journal of Forest Research, 38*, 834–843.

Pacala, S. W., Hurtt, G. C., Baker, D., et al. (2001). Consistent land- and atmosphere-based U.S. carbon sink estimates. *Science, 292*, 2316–2320.

Pagano, T., & Garen, D. (2005). A recent increase in Western U.S. streamflow variability and persistence. *Journal of Hydrometeorology, 6*, 173–179.

Pan, Y., Birdsey, R. A., Fang, J., et al. (2011). A large and persistent carbon sink in the world's forests. *Science, 333*, 988–993.

Parmesean, C., & Yohe, G. (2003). A globally coherent fingerprint of climate change impacts across natural systems. *Nature, 521*, 37–42.

Pearson, R. G., Thuiller, W., Araújo, M. B., et al. (2006). Model-based uncertainty in species range prediction. *Journal of Biogeography, 33*, 1704–1711.

Pinsonneault, A. J., Matthews, H. D., & Kothavala, Z. (2011). Benchmarking climate-carbon model simulations against forest FACE data. *Atmosphere and Ocean, 49*, 41–50.

Prasad, A. M., & Iverson, L. R. (1999). *A climate change atlas for 80 forest tree species of the eastern United States Spatial database*. Delaware: U.S. Department of Agriculture Forest Service, Northeastern Research Station. http://www.fs.fed.us/ne/delaware/atlas/index.html.

Prentice, I. C., & Harrison, S. P. (2009). Ecosystem effects of CO_2 concentration: Evidence from past climates. *Climate of the Past, 5*, 297–307.

Pugh, E., & Small, E. (2011). The impact of pine beetle infestation on snow accumulation and melt in the headwaters of the Colorado River. *Ecohydrology*. doi:10.1002/eco.239.

Pye, J. M., Holmes, T. P., Prestemon, J. P., & Wear, D. N. (2011). Economic impacts of the southern pine beetle. In R. N. Coulson & K. D. Klepzig (Eds.), *Southern pine beetle II* (Gen. Tech. Rep. SRS-140, pp. 213–222). Asheville: U.S. Department of Agriculture, Forest Service, Southern Research Station.

Ravenscroft, C., Scheller, R. M., Mladenoff, D. J., & White, M. A. (2010). Forest restoration in a mixed-ownership landscape under climate change. *Ecological Applications, 20*, 327–346.

Regonda, S. K., Rajagopalan, B., Clark, M., & Pitlick, J. (2005). Seasonal cycle shifts in hydroclimatology over the Western United States. *Journal of Climate, 18*, 372–384.

Ricciardi, A., & Simberloff, D. (2009). Assisted colonization is not a viable conservation strategy. *Trends in Ecology and Evolution, 24*, 248–253.

Rustad, L. E., Campbell, J. L., Marion, G. M., et al. (2001). A meta-analysis of the response of soil respiration, net nitrogen mineralization, and aboveground plant growth to experimental warming. *Oecologia, 126*, 543–562.

Ryan, M. G., Archer, S. R., Birdsey, R. A., et al. (2008). Land resources: Forest and arid lands. In P. Backlund, A. Janetos, D. Schimel, et al., conv. lead authors (Eds.), *The effects of climate change on agriculture, land resources, water resources, and biodiversity in the United States. Synthesis and assessment product 4.3* (pp. 75–120). Washington, DC: U.S. Climate Change Science Program.

Schwartz, M. D., & Hanes, J. M. (2010). Continental-scale phenology: Warming and chilling. *International Journal of Climatology, 30*, 1595–1598.

Smithwick, E. A. H., Ryan, M. G., Kashian, D. M., et al. (2009). Modeling the effects of fire and climate change on carbon and nitrogen storage in lodgepole pine (*Pinus contorta*) stands. *Global Change Biology, 15*, 535–548.

Solomon, S., Qin, D., Manning, M., et al. (2007). *Climate change 2007: The physical science basis—Contribution of Working Group I to the fourth assessment report of the Intergovernmental Panel on Climate Change* (996pp). Cambridge: Cambridge University Press.

Stadler, B., Müller, T., & Orwig, D. (2006). The ecology of energy and nutrient fluxes in hemlock forests invaded by hemlock woolly adelgid. *Ecology, 87*, 1792–1804.

Stoy, P. C., Katul, G. G., Siqueira, M. B. S., et al. (2006). Separating the effects of climate and vegetation on evapotranspiration along a successional chronosequence in the southeastern US. *Global Change Biology, 12*, 2115–2135.

Tague, C. L., & Band, L. E. (2004). RHESSys: Regional hydro-ecologic simulation system: An object-oriented approach to spatially distributed modeling of carbon, water, and nutrient cycling. *Earth Interactions, 8*, 1–42.

Tague, C., Seaby, L., & Hope, A. (2009). Modeling the eco-hydrologic response of a Mediterranean type ecosystem to the combined impacts of projected climate change and altered fire frequencies. *Climatic Change, 93*, 137–155.

Tchakerian, M. D., & Couslon, R. N. (2011). Ecological impacts of southern pine beetle. In R. N. Coulson & K. D. Klepzig (Eds.), *Southern pine beetle II* (Gen. Tech. Rep. SRS-140, pp. 223–234). Asheville: U.S. Department of Agriculture, Forest Service, Southern Research Station.

Thuiller, W., Albert, C., Araujo, M. B., et al. (2008). Predicting global change impacts on plant species' distributions: Future challenges. *Perspectives in Plant Ecology, Evolution and Systematics, 9*, 137–152.

Troch, P. A., Martinez, G. F., Pauwels, V. R. N., et al. (2009). Climate and vegetation water use efficiency at catchment scales. *Hydrological Processes, 23*, 2409–2414.

U.S. Environmental Protection Agency (USEPA). (2011). *Draft inventory of U.S. greenhouse gas emissions and sinks: 1990–2009* (EPA 430-R-11-005). Washington, DC: United States Environmental Protection Agency, Office of Atmospheric Programs. http://epa.gov/climatechange/emissions/usinventoryreport.html. Accessed 12 Mar 2012.

Van Auken, O. W. (2000). Shrub invasions of North American semiarid grasslands. *Annual Review of Ecology and Systematics, 31*, 197–215.

van Mantgem, P. J., Stephenson, N. L., Byrne, J. C., et al. (2009). Widespread increase of tree mortality rates in the Western United States. *Science, 323*, 521–524.

Vitt, P., Havens, K., Kramer, A. T., et al. (2010). Assisted migration of plants: Changes in latitudes, changes in attitudes. *Biological Conservation, 143*, 18–27.

Vose, J. M., & Ford, C. R. (2011). Early successional forest habitats and water resources. In C. H. Greenberg, B. Collins, F. R. Thompson, III (Eds.), *Sustaining young forest communities: Ecology and management of early successional habitats in the central hardwood region, USA. New Springer Series: Managing forest ecosystems* (Vol. 21, pp. 253–269, Chapter 14). New York: Springer.

Walter, M. T., Wilks, D. S., Parlange, J.-Y., & Schneider, R. L. (2004). Increasing evapotranspiration from the conterminous United States. *Journal of Hydrometeorology, 5*, 405–408.

Wang, M., Guan, D.-X., Han, S.-J., & Wu, J.-L. (2010). Comparison of eddy covariance and chamber-based methods for measuring CO_2 flux in a temperate mixed forest. *Tree Physiology, 30*, 149–163.

Warren, J. M., Pötzelsberger, E., Wullschleger, S. D., et al. (2011). Ecohydrologic impact of reduced stomatal conductance in forests exposed to elevated CO_2. *Ecohydrology, 4*, 196–210.

Way, D. A., Ladeau, S. L., McCarthy, H. R., et al. (2010). Greater seed production in elevated CO_2 is not accompanied by reduced seed quality in *Pinus taeda* L. *Global Change Biology, 16*, 1046–1056.

Wear, D. N., Huggett, R., Li, R., et al. (2012). *Forecasts of forest conditions in regions of the United States under future scenarios: A technical document supporting the Forest Service 2010 RPA assessment*. Asheville: U.S. Department of Agriculture, Forest Service, Southern Research Station.

Westerling, A. L., Hidalgo, H. G., Cayan, D. R., & Swetnam, T. W. (2006). Warming and earlier spring increase western U.S. forest wildfire activity. *Science, 313*, 940–943.

Westerling, A. L., Turner, M. G., Smithwick, E. A. H., et al. (2011). Continued warming could transform Greater Yellowstone fire regimes by mid-21st century. *Proceedings of the National Academy of Sciences, USA, 108*, 13165–13170.

Williams, M. W., Seibold, C., & Chowanski, K. (2009). Storage and release of solutes from a subalpine snowpack: Soil and streamwater response, Niwot Ridge, Colorado. *Biogeochemistry, 95*, 77–94.

Woodall, C. W., Oswalt, C. M., et al. (2009). An indicator of tree migration in forests of the eastern United States. *Forest Ecology and Management, 257*, 1434–1444.

Woodbury, P. B., Smith, J. E., & Heath, L. S. (2007). Carbon sequestration in the U.S. forest sector from 1990 to 2010. *Forest Ecology and Management, 241*, 14–27.

Wu, H. X., Ying, C. C., & Ju, H.-B. (2005). Predicting site productivity and pest hazard in lodgepole pine using biogeoclimatic system and geographic variables in British Columbia. *Annals of Forest Science, 62*, 31–42.

Wullschleger, S. D., Hanson, P. J., & Todd, D. E. (2001). Transpiration from a multi-species deciduous forest as estimated by xylem sap flow techniques. *Forest Ecology and Management, 143*, 205–213.

Xiao, J., Zhuang, Q., Law, B. E., et al. (2011). Assessing net ecosystem carbon exchange of U.S. terrestrial ecosystems by integrating eddy covariance flux measurements and satellite observations. *Agricultural and Forest Meteorology, 151*, 60–69.

Zaehle, S., Ciais, P., Friend, A. D., & Prieur, V. (2011). Carbon benefits of anthropogenic reactive nitrogen offset by nitrous oxide emissions. *Nature Geoscience, 4*, 601–605.

Zhang, X., Tarpley, D., & Sullivan, J. T. (2007). Diverse responses of vegetation phenology to a warming climate. *Geophysical Research Letters, 34*, L19405.

Zhao, M., & Running, S. W. (2010). Drought-induced reduction in global terrestrial net primary production from 2000 through 2009. *Science, 329*, 940–943.

Chapter 4
Disturbance Regimes and Stressors

Matthew P. Ayres, Jeffrey A. Hicke, Becky K. Kerns, Don McKenzie, Jeremy S. Littell, Lawrence E. Band, Charles H. Luce, Aaron S. Weed, and Crystal L. Raymond

4.1 Introduction

Disturbances such as wildfire, insect outbreaks, pathogens, invasive species, drought, and storms are part of the ecological history of most forest ecosystems, influencing vegetation age and structure, plant species composition, productivity,

---

M.P. Ayres (✉) • A.S. Weed
Department of Biological Sciences, Dartmouth College, Hanover, NH, USA
e-mail: matt.ayres@dartmouth.edu; aaron.s.weed@dartmouth.edu

J.A. Hicke
Department of Geography, University of Idaho, Moscow, ID, USA
e-mail: jhicke@uidaho.edu

B.K. Kerns
Pacific Northwest Research Station, U.S. Forest Service, Corvallis, OR, USA
e-mail: bkerns@fs.fed.us

D. McKenzie
Pacific Northwest Research Station, U.S. Forest Service, Seattle, WA, USA
e-mail: donaldmckenzie@fs.fed.us

J.S. Littell
Alaska Climate Science Center, U.S. Geological Survey, Anchorage, AK, USA
e-mail: jlittell@usgs.gov

L.E. Band
Department of Geography, University of North Carolina, Chapel Hill, NC, USA
e-mail: lband@email.unc.edu

C.H. Luce
Rocky Mountain Research Station, U.S. Forest Service, Boise, ID, USA
e-mail: cluce@fs.fed.us

C.L. Raymond
Seattle City Light, Seattle, WA, USA
e-mail: crystal.raymond@seattle.gov

carbon (C) storage, water yield, nutrient retention, and wildlife habitat. Climate influences the timing, frequency, and magnitude of disturbances (Dale et al. 2001). As the climate continues to change, we expect increased disturbance through more frequent extreme weather events, including severe drought, wind storms, and ice storms. Indirect effects may amplify these changes, with conditions that favor wildfire, insects, pathogens, and invasive species.

If frequency and severity of disturbances increase in the future, they will almost certainly have a bigger impact on forest ecosystems than gradual changes in other forest processes in response to higher temperature (see Chap. 3). This will lead to rapid changes in forest structure and function. It will also create landscapes in which regeneration of vegetation will occur in a warmer environment, possibly with new competitive relationships among species. In this way, the indirect effects of climate change in forest ecosystems may be more important than direct effects.

4.2 Wildfire

Climate and fuels are the two most important factors controlling patterns of wildfire within forest ecosystems. Climate controls the frequency of weather conditions that promote fire, whereas the amount and arrangement of fuels influence fire intensity and spread. Climate influences fuels on longer time scales by shaping species composition and productivity (Marlon et al. 2008; Power et al. 2008), and large-scale climatic patterns such as the El Niño Southern Oscillation, Pacific Decadal Oscillation, Atlantic Multidecadal Oscillation, and Arctic Oscillation are important drivers of forest productivity and susceptibility to disturbance (Duffy et al. 2005; Collins et al. 2006; Fauria and Johnson 2006; Kitzberger et al. 2007).

Current and past land use, including timber harvest, forest clearing, fire suppression, and fire exclusion through grazing (Swetnam and Betancourt 1998; Allen et al. 2002) have affected the amount and structure of fuels in the United States. For example, in montane forests in the Southwest (Allen et al. 2002) and other dry forests in the interior West, removal of fine fuels by grazing and fire suppression have increased the number of trees and amount of fuels; these forest conditions have increased fire size and intensified fire behavior. In colder and wetter forests in the western United States, such as subalpine forests in Yellowstone National Park and forests in the maritime Northwest, grazing and fire suppression have not altered fire regimes as extensively. Forests in the northeastern United States (Foster et al. 2002) and the upper Midwest developed after widespread timber harvest, land clearing, and forest re-growth after land abandonment. Compared to other regions of the United States, forests in the Northeast and upper Midwest burn less often and with smaller fires. Forests in the southeastern United States are often managed for timber, and prescribed fire is generally more prevalent than uncontrolled ignitions (National Interagency Coordination Center 2011). Prescribed fire is applied every

2–4 years in some fire-dependent ecosystems in the Southeast (Mitchell et al. 2006). Fire suppression and deer herbivory in the central hardwoods section of the eastern United States have pushed the composition toward more mesic and fire intolerant species (e.g., from oak dominated to maple dominated) (Nowacki and Abrams 2008).

Weather remains the best predictor of how much area will burn, despite changes in land use and the resulting effects on fuels. Correlations between weather and either area burned by fire or number of large fires are similar for both pre-settlement fires and fires of the last few decades. These syntheses of fire-weather relationships for both pre-settlement and modern records exist in several subregions of the West (Northwest: Heyerdahl et al. 2002, 2008a; Hessl et al. 2004. Southwest: Swetnam and Betancourt 1998; Grissino-Mayer and Swetnam 2000. Northern Rocky Mountains: Heyerdahl et al. 2008b. Westwide: Westerling et al. 2003, 2006; Littell et al. 2009) and the East (Hutchinson et al. 2008). Pre-settlement fire-weather relationships are derived from trees scarred by fires or age classes of trees established after fire and independently reconstructed climate; modern fire-weather comparisons are derived from observed fire events and observed weather in seasons leading up to the fire. Drought and increased temperature promote large fires, but effects differ by forest and region (Westerling et al. 2003; Littell et al. 2009). Weather can also influence fire through higher precipitation, increasing understory vegetation growth, which later becomes fuel (Swetnam and Betancourt 1998; Littell et al. 2009). Increased temperature and altered precipitation also affect fuel moisture and the length of time during which wildfires can burn during a given year.

The potential effects of climate change on forest fire area have been assessed using statistical models that project area burned from climatic variables, and by using global climate models to project future climatic variables (Westwide: McKenzie et al. 2004; Spracklen et al. 2009; Littell et al. 2010. Pacific Northwest: Littell et al. 2010; Yellowstone region: Westerling et al. 2011). Estimated future changes in annual area burned in the West ranges from declines of 80 % to increases greater than 500 %, depending on the region, timeframe, methods, and climate model/emission scenario (Bachelet et al. 2001). Future fire potential is expected to increase in summer and autumn from low to moderate in eastern regions of the South, and from moderate to high levels in western regions of the South (Liu et al. 2010).

The risk posed by future fire activity in a changing climate can be assessed by its likely effects on human and ecological systems. At the wildland-urban interface (WUI), higher population and forest density have created forest conditions that are likely to experience more area burned and possibly greater fire severity than in the historical record. Fire risk is likely to increase in a warmer climate because of the longer duration of the fire season, and the greater availability of fuels if temperature increases and precipitation does not sufficiently increase to offset summer water balance deficit. Where fuels management is common, forest fuel reduction and restoration to pre-settlement tree density and surface fire regimes help mitigate fire hazard under current and future climatic conditions. Finally, future fire risk may depend on whether extreme fire weather conditions will change in step with monthly to seasonal climate changes. Even if fire weather and ignitions do not change, it is likely that risk driven only by seasonal climate changes will increase; particularly in the WUI and managed forests, where fire has been historically rare or fully suppressed. The current increase in annual area burned may be partially related to increased fuels in frequent-fire forest types, in addition to more frequent weather conditions conducive to fire.

The effects of climate change intersecting with increased fuel loads in frequent-fire forests will be an exceptional challenge for resource managers on both public and private lands. As noted above, active management is highly effective in reducing fuel quantity and continuity, thus reducing fire intensity and mortality in the forest overstory (see Chap. 9). Prescribed fire is applied routinely and extensively in pine forests in the Southeast, but funding for fuel treatment in Western forests is sufficient to treat only a small portion of the landscape that currently has elevated fuels. Fire suppression is currently a large proportion of federal agency budgets (approximately 50 % for the U.S. Forest Service). If area burned does in fact increase by 100 % or more in future decades, this will pose a major budgetary and policy issue, and create challenges in managing landscapes increasingly occupied by younger forests.

4.3 Insects and Pathogens

4.3.1 General Concepts

Biotic disturbances are natural features of forests that play key roles in ecosystem processes (Adams et al. 2010; Boon 2012; Hicke et al. 2012a). Epidemics by forest insects and pathogens affect more area and result in greater economic costs than other forest disturbances in the United States (Dale et al. 2001). By causing local to widespread tree mortality or reductions in forest productivity, insect and pathogen outbreaks have broad ecological and socioeconomic effects (Tkacz et al. 2010; Pfeifer et al. 2011). The first National Climate Assessment (Melillo et al. 2001) projected increased disturbance in forests, especially from insects, and especially from bark beetles, because of their high physiological sensitivity to climate, short generation times, high mobility, and explosive reproductive potential. These projections have been upheld, and current observations suggest that disturbances are occurring more rapidly than imagined a decade ago. Understanding how these disturbances are influenced by climate change is therefore critical for quantifying and projecting effects (Fig. 4.1).

The powerful general effect of temperature on insects and pathogens is well known (Gillooly et al. 2002). Clear examples exist of climatic effects on insects, yet the most important insects and pathogens in American forests remain poorly

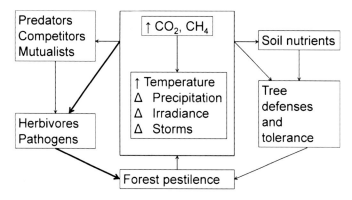

Fig. 4.1 General pathways by which atmospheric changes associated with increasing greenhouse gases can influence forest disturbance from insects and pathogens. CO_2 carbon dioxide, CH_4 methane

studied with respect to how interactions with climate will affect forests. Climatic warming can influence biotic disturbances of forests through effects on (1) the physiology of insects and pathogens that cause changes in their abundance and distribution, (2) tree defenses and tolerance, and (3) interactions of insects and pathogens with enemies, competitors, and mutualists (Fig. 4.1). Higher temperature has reduced winter mortality of insects, increased their range northward (Trân et al. 2007; Paradis et al. 2008; Safranyik et al. 2010), and increased their development rate during the growing season (Gillooly et al. 2002; Bentz et al. 2010). Temperature increases can also alter phenology, such as bringing leaf maturation into synchrony with insect feeding (Jepsen et al. 2011) or changing the life cycle synchrony of bark beetles, which depend on mass attack to overwhelm tree defenses (Powell et al. 2000; Friedenberg et al. 2007; Bentz et al. 2010).

A broader set of atmospheric drivers affects tree defenses against, and tolerance to, herbivores and pathogens (Lindroth 2010; Sturrock et al. 2011). Deficiencies of water or mineral nutrients can both increase and decrease tree defenses, depending on the severity of the deficiency, biochemical pathways, and the type of defense (Lombardero et al. 2000; Breshears et al. 2005; Worrall et al. 2008). In addition, tree mortality from severe drought may facilitate an increase in bark beetles, which then become sufficiently abundant to successfully attack healthy trees (Greenwood and Weisberg 2008; Raffa et al. 2008). Information on the effects of climate on tree-pathogen interactions is sparse, despite a theoretical expectation for temperature and moisture to have significant effects (Grulke 2011; Rohrs-Richey et al. 2011; Sturrock et al. 2011).

Outbreak dynamics of forest insects respond to interactions between herbivores and their enemies (Dwyer et al. 2004), and these interactions should be sensitive to temperature (Berggren et al. 2009; Klapwijk et al. 2012), but empirical studies are rare (Siegert et al. 2009). Similarly, for the many forest insects that involve mutualisms with fungi, it is logical that outbreak dynamics will be sensitive to climatic effects on the mutualism (Lombardero et al. 2000; Hofstetter et al. 2007; Six and Bentz 2007; Evans et al. 2011).

4.3.2 Climate and Biotic Disturbances

4.3.2.1 Bark Beetles

Multiple species of indigenous bark beetles affect millions of hectares of coniferous forests in North America. Major species include mountain pine beetle (*Dendroctonus ponderosae* Hopkins), the most important disturbance agent of pines in the western United States (Box 4.1; see Chap. 1); southern pine beetle (*D. frontalis* Zimmermann) in pine forests of the southeastern United States (Box 4.2), and spruce beetle (*D. rufipennis* Kirby). In the early 2000s, severe drought, coupled with several species of bark beetles, killed trees of several conifer species in the Southwest (Ganey and Vojta 2011), most notably pinyon pine (*Pinus edulis* Engelm.) attacked by pinyon ips (*Ips confusus* LeConte) across 1.2 million ha (Breshears et al. 2005).

> **Box 4.1: Mountain Pine Beetle and Five-Needle Pines**
>
> Five-needle pines, including whitebark (*Pinus albicaulis* Engelm.), limber (*P. flexilis* James), and bristlecone (*P. aristata* Engelm.) pines, play key roles in forest ecosystems of the western United States. They provide food resources for wildlife, affect snow distribution and melt, stabilize the soil, and provide cover for other vegetation, and are valued by the public for these services. However, these conifers are currently subjected to a climatically induced increase in biotic disturbance that is expected to continue in the coming decades. Mountain pine beetles (*Dendrotonus ponderosae* Hopkins) are attacking five-needle pines across the West; aerial surveys indicate that one million ha were affected by five-needle pine mortality during 1997 through 2010. Higher temperatures and drier conditions affect winter survival and development rate and population synchronization of beetles, as well as susceptibility of host trees.
>
> Similar epidemics occurred in the 1930s, also associated with a period of warmer years, but several differences exist between the mortality then and today. Most importantly, a cooler period followed the 1930s that was less suitable for the beetle. In contrast, the current warming trend has persisted for several decades, with resultant increases in climatic suitability for mountain pine beetle, and is expected to continue for decades to come. The recent beetle epidemics in five-needle pine stands are already more extensive than in the 1930s and are killing very old trees that survived previous outbreaks. Finally, white pine blister rust is predisposing whitebark pines to lethal attacks by
>
> (continued)

Box 4.1 (continued)

mountain pine beetle (*Cronartium ribicola* J.C. Fisch). Given the trajectory of future warming, strong ties between temperature and beetle epidemics, and extensive mortality that has already occurred in some areas, significant consequences are expected for these forests and the ecosystem services they provide.

Box 4.2: The Southern Pine Beetle Reaches the New Jersey Pinelands

The southern pine beetle (*Dendroctonus frontalis* Zimmermann) is the most destructive herbivore in the most productive forests of the United States. Like the closely related mountain pine beetle (*Dendroctounus ponderosae* Hopkins), it uses aggregation pheromones to coordinate mass attacks that overwhelm the resin defenses of otherwise healthy trees; virtually every attacked tree dies within weeks. It has multiple generations per year (at least four to five in the warm Gulf Coast region), so the aggregations that typically form in spring can expand throughout the year as growing "spots" of tree mortality within forest landscapes. Effective suppression of these epidemics involves locating the spots and cutting the infested trees. Effective prevention involves silvicultural thinning to reduce the occurrence of stands with high basal area (overstocked) that are especially suitable for beetle population growth. Monitoring, suppression, and prevention of southern pine beetle are integral to the management of pine ecosystems in the southeastern United States.

The northern distribution of southern pine beetle is constrained by the occurrence of lethal winter temperatures. As part of the first National Climate Assessment, it was estimated that an increase of 3 °C in minimum annual temperature would permit a northern expansion of about 180 km for this beetle. In fact, there was a regional increase of just over 3 °C from 1960 through 2005, and beetle populations are now epidemic in the New Jersey Pinelands, about 200 km north of forests with a long history of such epidemics. Warming winters did not cause the current epidemic but may have permitted it. Given the natural population dynamics of southern pine beetle and the projected absence of lethal winter temperatures, the New Jersey Pinelands has entered a new phase in which southern pine beetle will be influencing many aspects of forest ecology and management, as they have throughout the southeastern United States.

(*Photo shows an infestation of southern pine beetle in the New Jersey Pinelands in 2011. Aerial photo by Bob Williams, Land Dimensions. Close-up of beetle by Erich Vallery, U.S. Forest Service*)

(continued)

(continued)

The population dynamics of these native beetles are sensitive to climatic variation, and the extent of recent outbreaks have been facilitated by increasing temperatures during the last decade (Breshears et al. 2005; Raffa et al. 2008; Sherriff et al. 2011). Greater effects on forest ecosystems are anticipated from recent range expansions by beetles into areas with hosts that are new and may have low resistance (Cudmore et al. 2010). Mexican pine beetle (*D. mexicanus* Hopkins), previously known only in Mexico, has been recorded in the southwestern United States (Moser et al. 2005) and represents one of several species of Mexican bark beetles that may expand into U.S. forests in a warmer climate (Bentz et al. 2010; Salinas-Moreno et al. 2010). Climate change will continue to reshape the patterns of bark beetle outbreaks in U.S. forests, with outbreak tendencies increasing for some species in some regions and decreasing in others (Bentz et al. 2010; Littell et al. 2010; Evangelista et al. 2011).

4.3.2.2 Defoliating Insects

Defoliating insects are a continentally important biotic disturbance in American forests. For example, western spruce budworm (*Choristoneura occidentalis*

Freeman) is currently important in the West (USDA FS 2010), and Eastern boreal forests have been affected by many cycles of spruce budworm (*Archips fumiferana* Clemens) outbreaks (Candau and Fleming 2005). Other important defoliators include tussock moths, tent caterpillars, gypsy moths, and jack pine budworm (*A. pinus* Freeman). Most defoliating insects are indigenous to American forests, and many have cyclical outbreak dynamics involving predators, parasitoids, and pathogens (Dwyer et al. 2004).

Climatic effects on these predator–prey interactions remain largely unstudied (Klapwijk et al. 2012). In general, it is less clear (compared to what is known about bark beetles) how climatic patterns influence the frequency, extent, and geographic distribution of defoliators in American forests. There is limited evidence in some forest systems of climatic effects on winter populations (Thomson et al. 1984; Kemp et al. 1985; Williams and Liebhold 1995a; but see Reynolds et al. 2007; Thomson and Benton 2007), drought stress of host trees (Williams and Liebhold 1995b; Campbell et al. 2006), and phenological synchronization of larval emergence and bud break (Thomson et al. 1984). Considerable uncertainty remains about future responses of defoliators to climate change (Dukes et al. 2009; Rodenhouse et al. 2009). Hemlock woolly adelgid (*Adelges tsugae* Annand), a non-native, stem-feeding insect, has been spreading in the eastern United States (Box 4.3).

Box 4.3: Hemlock Woolly Adelgid

Invasive insects can cause extensive tree mortality owing to lack of genetic resistance in host trees and the absence of natural enemies. Thus, non-native insects and pathogens are likely to cause the loss of native tree species and produce other substantial effects on forests, wildlife, and biodiversity. Climate change can exacerbate the effects of established invasives by permitting their expansion into previously unsuitable climatic regions, as with the expansion of the hemlock woolly adelgid (*Adelges tsugae* Annand) into the northeastern United States.

Hemlock woolly adelgid, an aphid-like insect, was accidentally introduced from Japan some time before 1951 and has been a major biotic disturbance in American forests, killing eastern hemlock (*Tsuga canadensis* [L.] Carrière) and Carolina hemlock (*T. caroliniana* Engelm.) in advancing waves from its point of establishment in Virginia. Since establishment, this insect has largely eliminated hemlocks from a large swath of Eastern forests, including national icons such as the Shenandoah and Great Smoky Mountains National Parks. Consequences include lost value to property owners and persistent alterations to hydrological regimes, soil biogeochemistry, C stores, biodiversity, and forest composition, including permitting the establishment of undesirable invasive plants.

(continued)

(continued)

Hemlocks north of the infested regions have thus far been protected by winter temperatures that are lethal to hemlock woolly adelgid. However, these conditions are changing with the amelioration of extreme winter temperatures in the eastern United States, and projections under even conservative climatic scenarios predict the loss of hemlock through most of its current range.

4.3.2.3 Plant Pathogens

We identified 21 plant pathogens that are notable agents of disturbance in U.S. forests and may respond to climate change. Climatic effects on these pathogens are generally not well studied, but we expect that some of these pathogens will be affected directly by climatic influences on sporulation and infection, indirectly by predisposing trees to infection, or both (Sturrock et al. 2011). For pathogens that involve associations with insects, climatic effects on the animal associates may also be important.

A few cases of climate-pathogen interactions have been documented. For example, Swiss needle cast (*Phaeocryptopus gaeumannii* T. Rohde), a native foliar pathogen in the Northwest, is influenced by winter warming and spring precipitation. Climatic projections suggest an increase in Swiss needle cast distribution and severity (Stone et al. 2008). The susceptibility of alder to a cankering pathogen is related to the phenology of the plant, the pathogen, and water availability (Grulke 2011; Rohrs-Richey et al. 2011). Outbreaks of some virulent invasive pathogens can also be enhanced by climate (e.g., sudden oak death; Sturrock et al. 2011), whereas others are not very sensitive to climate (Garnas et al. 2011b).

The potential effects of climate change on root pathogens are difficult to project (Ayres and Lombardero 2000), but it will be important to understand this relationship because endemic root diseases are widespread and often have a major influence on forest dynamics and management. One would expect root diseases to be affected by both the distribution of host species and the effects of a changing climate on susceptibility of host species and prevalence of fungal pathogens. If a warmer climate increases physiological stress in a particular tree species, then it may be less resistant to some root diseases, potentially causing lower tree vigor, higher mortality in mature trees and seedlings, and lower C storage. Although some initial modeling of future changes in root pathogens has been attempted (*Armillaria* spp.; Klopfenstein et al. 2009), geographic specificity for host-pathogen relationships is highly uncertain based on current knowledge.

4.3.2.4 Non-native and Emerging Insects and Pathogens

Invasive, non-native insects and pathogens are becoming an increasingly important component of forest disturbance (Lovett et al. 2006; Seppälä et al. 2009), and warming and precipitation shifts associated with climate change can affect forest vulnerability (Paradis et al. 2008; Sturrock et al. 2011). For example, the geographic range and incidence of dothistroma needle blight (*Dothistroma septosporum* [Dorog.] M. Morelet and *D. pini* Hulbary), which reduces growth of many conifers by causing premature needle defoliation, may shift with changing precipitation patterns (Woods et al. 2010).

At present, the primary cause of biological invasions is global commerce. However, increasing temperatures are generally expanding the geographic zones where potential invasive species could survive and reproduce if they arrive, for example, at ports of entry on the Eastern seaboard and in the Great Lakes waterway. The potential for global, climate-driven increases in invasion risks has prompted international organizations to discuss changes in trade restrictions to manage associated phytosanitation risks (Standards and Trade Development Facility 2009).

Outbreaks of lesser known forest insects have recently occurred in U.S. forests. Aspen leaf miner (*Phyllocnistis populiella* Chambers), which reduces longevity of aspen leaves, has damaged 2.5 million ha of quaking aspen (*Populus tremuloides* Michx.) in Alaska since 1996 (Wagner et al. 2008). Large areas of willows were damaged during two eruptive outbreaks of the willow leafblotch miner (*Micurapteryx salicifoliella* Chambers) in the 1990s in two major river drainages in Alaska (Furniss et al. 2001); outbreaks of this leaf miner had not been previously reported. Substantial defoliation by Janet's looper (*Nepytia janetae* Rindge) of stressed trees in Southwestern spruce-fir forests was preceded by uncharacteristically warm winters; defoliation by Janet's looper encouraged attack by opportunistic bark beetles. These examples demonstrate that previously rare native insects that displayed new eruptive behavior and caused notable forest disturbances.

4.3.3 Effects and Interactions with Other Disturbances

Through their effects on tree growth and mortality, insects and pathogens have broad effects on ecosystem processes (see Chap. 3). Insects and pathogens, by virtue of their host preferences, can alter tree species composition within stands, remove most host trees from some landscapes (Lovett et al. 2006), and modify forest types (e.g., from conifers to hardwoods) (Veblen et al. 1991; Orwig et al. 2002; Collins 2011). Forests typically shift toward younger, smaller trees after biotic disturbances (Ylioja et al. 2005; Garnas et al. 2011a; Tchakerian and Couslon 2011), which can affect wildlife habitat and biodiversity by quickly modifying multiple trophic levels (Chan-McLeod 2006; Drever et al. 2009). Both positive and negative effects occur depending on species, time since disturbance, surviving vegetation, ecosystem type, and spatial extent of outbreak.

Fire and biotic disturbances interact in several ways. Fires lead to younger stands that may be less susceptible to attack, and dead trees provide a food resource for some insects and pathogens (Parker et al. 2006). Insect-killed trees influence fuels and therefore fire behavior, although the effect depends on a number of factors, including the number of attacked trees within a stand and time since outbreak (e.g., Ayres and Lombardero 2000; Hicke et al. 2012b; Jenkins et al. 2008; Simard et al. 2011), and fire-induced increases in tree defenses can mitigate bark beetle risks (Lombardero and Ayres 2011).

Extreme soil water deficits (drought) that reduce tree growth might also reduce tree defenses to insects and pathogens (Bentz et al. 2009; Sturrock et al. 2011), although previous studies suggest there may be either no effect (Gaylord et al. 2007; McNulty et al. 1997) or the opposite effect (Lombardero et al. 2000). Drought also facilitates population increases of western bark beetles. Some aggressive species such as mountain pine beetle are able to maintain epidemics after return to normal conditions, whereas others such as pinyon ips decline with alleviation of drought stress (Raffa et al. 2008).

Insects and pathogens clearly affect the economic value of forests that are intended for harvest for wood products, and direct economic effects occur for tree removal and replacement, such as the $10 billion spent after emerald ash borer (*Agrilus planipennis* Fairmaire) infestations (Kovacs et al. 2010, 2011). A more complete valuation of socioeconomic effects is challenging because it is difficult to quantify all ecosystem services, especially those with non-market values (Holmes et al. 2010). Regions with dead and dying trees have reduced aesthetic value (Sheppard and Picard 2006) and may have reduced housing prices (Holmes et al. 2010; Price et al. 2010).

4.4 Invasive Plants

4.4.1 Introduction

Invasive plants are recent introductions of non-native, exotic, or nonindigenous species that are (or have the potential to become) successfully established or naturalized, and that spread into new localized natural habitats or ecoregions with the potential to cause economic or environmental harm (Lodge et al. 2006). This definition of "invasive" (1) does not consider native species that have recently expanded their range, such as juniper (*Juniperus* spp.) in the western United States (Miller and Wigand 1994; Miller et al. 2005), (2) involves defined temporal and spatial scales, and (3) considers social values related to economic and environmental effects.

An estimated 5,000 nonnative plant species exist in U.S. natural ecosystems (Pimentel et al. 2005) (Table 4.1). The effects of invasive plants include reduced native biodiversity, altered species composition, loss of habitat for dependent

Table 4.1 Common invasive plant species and environmental impacts for forests and woodlands in the United States

Species	Common name	Growth form	Environmental impacts
Acer platanoides L.	Norway maple	Tree	Reduces abundance and diversity of native species; alters community structure (shading of understory)
Ailanthus altissima Desf.	Tree of heaven	Tree	Alters ecosystem processes (increases soil N, alters successional trajectories); displaces native vegetation; allelopathic; roots can damage buildings and sewer lines; risk to human health (pollen allergies, sap-caused dermatitis)
Alliaria petiolata (M. Bieb.) Cavara and Grande	Garlic mustard	Biennial forb	Reduces abundance and diversity of native species; potentially allelopathic
Berberis thunbergii DC.	Japanese barberry	Shrub	Displaces native shrubs; changes soil properties (soil microbial composition, nitrate concentration); alters successional patterns; potentially increases fire risk (increased biomass)
Bromus tectorum L.	Cheatgrass	Annual grass	Decreases community diversity; increases fire frequency and severity; alters successional patterns and nutrient cycling
Celastrus orbiculatus Thunb.	Oriental bittersweet	Vine	Alters soil chemistry (e.g., increased pH, increased calcium), plant succession, and stand structure (e.g., shades out understory, increases continuity of overstory vegetation); decreases native plant diversity; reduces productivity
Centaurea solstitialis L.	Yellow star-thistle	Annual forb	Displaces native plants, reduces native wildlife and forage; decreases native diversity; depletes soil moisture, altering water cycle; reduces productivity in agricultural systems (lowers yield and forage quality)
Centaurea stoebe L.	Spotted knapweed	Biennial/perennial	Reduces plant richness, diversity, cryptogam cover, soil fertility; reduces forage production; increases bare ground and surface water runoff, and can lead to stream sedimentation; allelopathic
Cirsium arvense (L.) Scop.	Canada thistle	Perennial forb	Possible allelopathy; displaces native vegetation; alters community structure and composition; reduces diversity; reduces forage and livestock production
Cytisus scoparius (L.) Link	Scotch broom	Shrub	Interferes with conifer establishment; reduces growth and biomass of trees; alters community composition and structure (increases stand density, often creating monospecific stands); alters soil chemistry (increases N); toxic to livestock

(continued)

Table 4.1 (continued)

Species	Common name	Growth form	Environmental impacts
Hedera helix L.	English ivy	Vine	Alters community structure; displaces native ground flora; weakens or kills host trees; potentially reduces water quality and increases soil erosion
Imperata cylindrica (L.) P. Beauv.	Cogongrass	Grass	Alters ecosystem structure (e.g., decreases growth and increases mortality of young trees) and function and decreases diversity; shortens fire return intervals and increases fire intensity; interferes with pine and oak regeneration
Ligustrum sinense Lour.	Chinese privet	Shrub	Interferes with native hardwood regeneration; alters species composition and community structure (forms dense monospecific stands)
Lonicera japonica Thunb.	Japanese honeysuckle	Vine	Alters forest structure and species composition; inhibits pine regeneration and weakens or kills host trees; suppresses native vegetation; provides food for wildlife; early- and late-season host for agricultural pests
Lygodium japonicum (Thunb.) Sw.	Japanese climbing fern	Climbing fern	Reduces native understory vegetation; potentially weakens or kills host trees; interferes with overstory tree regeneration
Microstegium vimineum (Trin.) A. Camus	Japanese stiltgrass, Nepalese browntop	Annual grass	Reduces ecosystem function (alters soil characteristics and microfaunal composition, decreases diversity, alters stand structure); reduces timber production; possibly allelopathic
Pueraria montana (Lour.) Merr.	Kudzu	Vine	Potentially eliminates forest cover; overtops, weakens, and kills host trees; reduces timber production; increases winter fire hazard
Triadica sebifera (Willd.) Maesen and S.M. Alemeida ex Sanjappa and Predeep	Chinese tallow, tallowtree	Tree	Displaces native species and reduces diversity; increases soil nutrient availability; reduces fire frequency and intensity

species (e.g., wildlife), changes in biogeochemical cycling, changes in ecosystem water use, and altered disturbance regimes. Billions of dollars are spent every year to mitigate invasive plants or control their effects (Pimentel et al. 2005). Negative environmental effects are scale-dependent (Powell et al. 2011), with some subtle beneficial properties (Sage et al. 2009) on ecosystem function (Myers et al. 2000; Zavaleta et al. 2001). For example, some consider species in the genus *Tamarix* to be among the most aggressively invasive and detrimental plants in the United States (Stein and Flack 1996), but others point out benefits, including sediment stabilization and the creation of vertebrate habitat in riparian areas that can no longer support native vegetation (Cohn 2005).

The spatial extent of many invasive plants at any point in time has been difficult to determine, limiting assessment of overall consequences. One assessment (Duncan et al. 2004) for the western United States indicates that 16 invasive plants account for most current invasive plant problems. *Centaurea* species are particularly widespread and persistent in the West. Cogongrass (*Imperata cylindrica* [L.] Raeusch.), which has invaded extensive forested areas of the Southeast, is considered to be one of the most problematic invasive plants in the world (Box 4.4). Mountain ecosystems tend to have fewer invasive plant species than other regions because of a short growing season, limited settlement history, relatively low frequency of seed sources, and prevalence of closed-canopy conifer forests that limit light in the understory and acidify the soil (Parks et al. 2005).

Box 4.4: Invasive Grasses, Fire, and Forests

Cogongrass (*Imperata cylindrica* [L.] P. Beauv.) in the Southeast and cheatgrass (*Bromus tectorum* [L.]) in the West are invaders that alter fire regimes and are some of the most important ecosystem-altering species on the planet. Cogongrass threatens native ecosystems and forest plantations, generally invading areas after a disturbance (e.g., mining, timber harvest, highway construction, natural fire). It is a major problem for forest industry, invading and persisting in newly established pine plantations. In sandhill plant communities, cogongrass provides horizontal and vertical fuel continuity, shifting surface fire regimes to crown fire regimes and increasing fire-caused mortality in longleaf pine (*Pinus palustris* Mill.), potentially shifting a species-diverse pine savanna to a grassland dominated by cogongrass. Cogongrass does not tolerate low temperatures, but increased warming could increase the threat of cogongrass invasion into new areas. In a warmer climate, cogongrass

(continued)

(continued)

is expected to greatly increase in the Gulf Coast region. (*Photo shows an infestation of cogongrass in a longleaf pine upland in central Florida.* [*Photo by James R. Meeker, U.S. Forest Service, available from Forestry Images,* http://www.forestryimages.org/browse/detail.cfm?imgnum=3970058])

Cheatgrass is widely distributed in western North America and dominates many steppe communities. After disturbance, this species can invade low-elevation forests, creating surface fuel continuity from arid lowlands into forested uplands. After establishment, cheatgrass contributes to fine, highly combustible fuel components that dry out early in the year, thus increasing the length of the fire season. Future changes in the climatic habitat of cheatgrass will depend on precipitation as well as temperature. If precipitation decreases, especially in summer, cheatgrass will likely expand, whereas increased precipitation may reduce suitable habitat. Elevated CO_2 increases cheatgrass productivity, a phenomenon that may already be contributing to the vigor and spread of this species. Increased productivity causes higher fuel loads, potentially resulting in more frequent, higher intensity fires and altered fire regimes.

4.4.2 *Interactions Between Climate Change and Plant Invasion*

Plant invasions can be influenced by warmer temperatures, earlier springs and earlier snowmelt, reduced snowpack, changes in fire regimes, elevated nitrogen (N) deposition, and elevated CO_2 concentrations. The responses of invasive plants to climate change should be considered separately from those of native species, because invasive plants (1) have characteristics that may differ from native species, (2) can be highly adaptive (Sexton et al. 2002), (3) have life history characteristics that facilitate rapid population expansion, and (4) often require different management approaches than for native species (Hellmann et al. 2008). Successful invasion of areas dominated by native plants depends on environment, disturbance, resource availability, biotic resistance, and propagule pressure (Davis et al. 2000; D'Antonio et al. 2001; Levine et al. 2004; Eschtruth and Battles 2009; Pauchard et al. 2009). Climate change may influence all of these drivers of invasion, with high variability across space and time.

4.4.2.1 Temperature, Precipitation, and CO_2

Climate change will alter the abiotic conditions under which plant species can establish, survive, reproduce, and spread. These effects are expected to increase plant stress and decrease survival in the drier, warmer, and lower elevation portions of species ranges (Allen and Breshears 1998). Abiotic factors probably constrain the range of many invasive plants and limit their successful establishment (Alpert et al. 2000; Pauchard et al. 2009). With climate change, however, new habitat, once too cold or wet, may become available, enabling plants to survive outside their historical ranges and expand beyond their current ranges.

Many native plants are projected to move northward or upward in elevation with climate change. Examples of invasive plants projected to follow this pattern are rare, but information on species tolerances provides insight on potential responses. For example, the northern limit of Japanese barberry (*Berberis thunbergii* DC.), an invasive shrub in the eastern United States, is determined by low temperature tolerance, the southern limit by cold stratification requirements for germination, and the western limit by drought tolerance (Silander and Klepeis 1999). The widespread invasive tree of heaven (*Ailanthus altissima* [P. Mill.] Swingle) is limited by cold and prolonged snow cover to lower mountain slopes, but it may be able to colonize during several successive years of mild climate, conditions that may become more common under climate change (Miller 1990). Soil water availability and regional changes in climatic water balance may be important for plant invasions, particularly at lower elevations (Chambers et al. 2007; Crimmins et al. 2011). Species growth, productivity, and reproduction may also change as climatic conditions change. For example, invasive plants may be better able to adjust to rapid changes in abiotic conditions by tracking seasonal temperature trends and shifting their phenologies (e.g., earlier spring warming) (Willis et al. 2010).

Increased productivity in response to elevated CO_2 has been documented under controlled conditions for several invasive plant species, including cheatgrass (*Bromus tectorum* L.) (Box 4.4), Canada thistle (*Cirsium arvense* [L.] Scop.), spotted knapweed (*Centaurea melitensis* L.), yellow star-thistle (*C. solstitialis* L.), and kudzu (*Pueraria montana* [Lour.] Merr.) (Dukes et al. 2011; Ziska and Dukes 2011;

Ziska and George 2004). Response to CO_2 enrichment is less predictable when plants are grown in the field (Dukes and Mooney 1999; Ziska and Dukes 2011), where response may be limited by nutrients and water availability. Carbon dioxide enrichment can also increase water use efficiency, which can partially ameliorate conditions associated with decreased water availability, particularly for C_3 plants (Eamus 1991). This phenomenon may be partially responsible for global patterns of encroachment of C_3 plants in grasslands dominated by C_4 plants or mixed species (Bond and Midgley 2000).

4.4.2.2 Disturbance and Resource Availability

Disturbances such as fire, landslides, volcanic activity, logging, and road building open forest canopies, reduce competition, and expose mineral soil, increasing light and nutrient availability. Invasive plants are generally well adapted to use increased resources. Fluctuating resource availability, coinciding with available propagules, facilitates regeneration and establishment of invasive species associated with forest development after disturbance (Halpern 1989; Davis et al. 2000; Parks et al. 2005). Opportunities for invasions may also be created by forest thinning, fuel treatments, and biofuel harvesting (Bailey et al. 1998; Silveri et al. 2001; Nelson et al. 2008). However, the spatial extent of invasions may be limited (Nelson et al. 2008), especially for shade intolerant species in closed-canopy Western forests.

The reintroduction of fire is a high priority for restoration and management of fire-adapted forests such as ponderosa pine (*Pinus ponderosa* Douglas ex P. Lawson & C. Lawson), longleaf pine (*P. palustris* Mill.), and loblolly pine (*P. taeda* L.). Invasive plants, especially annual grasses (Box 4.4), can spread rapidly after fire, particularly in high-severity burns (D'Antonio 2000; Kerns et al. 2006; Keeley and McGinnis 2007). Forest sites treated with prescribed fire, which are often near the wildland-urban interface and roads, are also well positioned for invasive plant introduction and spread (Keeley et al. 2003).

The success of plant invasions is regulated by competition from resident plants (Levine 2000; Seabloom et al. 2003), and land managers can alter post-disturbance (logging, fire) invasive establishment by seeding to increase native plant competition. Although native plant competition can be overwhelmed by invasive plant seed abundance (D'Antonio et al. 2001; Lonsdale 1999), resistance related to soil properties is more likely to withstand seed abundance. Native plant competition with invasive plants can also be affected by the effects of predation, herbivory, and pathogens associated with native species. Native plant competition may change as temperature and ambient CO_2 increase; numerous studies have documented that weedy plants are more productive in an elevated CO_2 environment (Ziska and George 2004).

Propagule pressure, which includes seed size, numbers, and temporal and spatial patterns, is perhaps the most important driver of successful invasions in forest ecosystems (Tilman 1997; Colautti et al. 2006; Eschtruth and Battles 2009; Simberloff 2009). For invasive plants, propagule pressure is largely controlled by factors other than climate. For example, the most critical factors projecting plant invasion in eastern hemlock (*Tsuga canadensis* [L.] Carrière) forests in the eastern United States are overstory canopy disturbance and propagule pressure (Eschtruth and Battles 2009). However, little is known about how biotic and abiotic resistance factors interact with propagule supply to influence exotic plant invasion (D'Antonio et al. 2001; Lonsdale 1999).

Atmospheric CO_2 may influence seed production through enhanced flowering under elevated CO_2, increasing the probability that a smaller seed can establish a viable population (Simberloff 2009). Of greater concern is how climate change may alter human activities that transfer seeds. For example, climate change could alter tourism and commerce, enhance survival of seeds during transport (Hellmann et al. 2008), and shift recreation to higher elevations. Changes in atmospheric circulation patterns could also alter wind-dispersed species, allowing new species to arrive in areas that previously had few seeds.

Climate change will affect invasive plants in forests because of the potential for increased ecological disturbance, effects of warming on species distributions, enhanced competitiveness of invasive plants owing to elevated CO_2, and increased stress to native species and ecosystems (Dukes and Mooney 1999; Breshears et al. 2005; Pauchard et al. 2009; Ziska and Dukes 2011). Warming will increase the risk of invasion in temperate mountainous regions because cold temperature has tended to limit the establishment of invasive plants.

Empirical models suggest that a warmer climate could result in both range expansion and contraction for common invasive plants (Sasek and Strain 1990; Pattison and Mack 2008; Bradley et al. 2009; Kerns et al. 2009), although these types of species distribution models do not account for species ecophysiology and biotic interactions. Process-based models may ultimately prove more robust for prediction, although model parameters are quantified from experimental data or the research literature, which themselves have uncertainties.

For management responses to plant invasions to be cost effective and successful, assertive action is needed in the early phase of invasion. A potentially useful approach is a climate change-based modification of the Early Detection and Rapid Response System (National Invasive Species Council 2001). For example, risk assessment could be done over broader geographic areas than has been performed in the past (Hellman et al. 2008). Unfortunately, some biocontrol methods may no longer be effective in a warmer climate (Hellmann et al. 2008), and some herbicides are less effective on plants grown in elevated CO_2 (Ziska and Teasdale 2000). The successful control of invasive plants over large forest landscapes will depend on knowledge about resistance of native species to invasion and our ability to limit propagule pressure.

4.5 Erosion, Landslides, and Precipitation Variability

Based on analysis of recent climate records and the projections of climate change simulations, hydroclimatic extremes will become more prominent with a warming climate (O'Gorman and Schneider 2009; Trenberth et al. 2009), with potential increases in flood frequency, droughts and low flow conditions, saturation events, landslide occurrence, and erosion. Ecosystems are expected to differ in their response to changes in precipitation intensity and inter-storm length because of differences in geomorphic conditions, climate, species assemblages, and susceptibility to drought. For erosion, these differences may be predictable with a general mass balance framework, but other processes are poorly understood, such as the effect of drought on tree mortality, vegetation resistance to insects and pathogens, and subsequent feedbacks to erosion processes. The indirect effects of disturbances (e.g., fire, insect infestations, pathogens) to shifts in water balance will complicate the response of erosion. Changing species composition will also potentially affect forest ecosystem water balance (see Chap. 3).

4.5.1 Erosion and Landslides

Changes in precipitation intensity, and in the magnitude and frequency of precipitation events that saturate soil and cause runoff, will interact with mass wasting and erosion. Potential annual increases and decreases in precipitation will directly contribute to the amount of water available to drive mass wasting at seasonal and event scales. Increases in extremes of precipitation intensity (Easterling et al. 2000; Karl and Knight 1998), rain-on-snow during mid-winter melt (Hamlet and Lettenmaier 2007; Wenger et al. 2011), and transport of moisture in atmospheric rivers (Ralph et al. 2006; Dettinger 2011) can increase pore water pressure on hillslopes, thus increasing the risk of landslides, erosion, and gully formation for individual storms. Seasonal to annual changes in precipitation will contribute to soil moisture and groundwater levels, which can amplify or mitigate individual events.

Direct effects of some climatic changes on sediment yield and mass wasting may be overshadowed by longer term, indirect effects through vegetation response (Istanbulluoglu and Bras 2006; Collins and Bras 2008; Goode et al. 2011). Although decreasing precipitation in some places might suggest reduced risks of erosion or landslides, this change may have indirect effects on mortality and thinning of vegetation and fire risk, which could in turn increase erosion and landslides through lower root reinforcement of soil and higher exposure of soil to precipitation. For example, paleoclimatic and paleoecological evidence links periods of drought and severe fire to severe erosion events (Briffa 2000; Meyer and Pierce 2003; Whitlock et al. 2003; Pierce et al. 2004; Marlon et al. 2006). At shorter time scales, years of widespread fire are linked to severely dry and warm years (e.g., McKenzie et al. 2004; Morgan et al. 2008; Littell et al. 2009). As we shift toward a drier and warmer climate in the western United States, more areas are likely to burn annually (e.g., Littell et al. 2009; Spracklen et al. 2009), with resulting postfire debris flows (Meyer and Pierce 2003; Luce 2005; Shakesby and Doerr 2006; Moody and Martin 2009; Cannon et al. 2010). Breshears et al. (2005) documented drought-induced canopy mortality of ponderosa pine, followed by erosional loss of topsoil and nutrients, with subsequent species replacement by pinyon pine and juniper. These types of state transitions may indicate the type of complex feedbacks that will lead to permanent shifts in dominant vegetation, rather than to recovery following disturbance.

Adjustment of canopy density and root distributions to longer inter-storm periods may increase the efficiency of use of rain or snowmelt (Hwang et al. 2009; Brooks et al. 2011). The response of both annual runoff and runoff from extreme events may be amplified or mitigated by forest canopy adjustment to temperature, moisture, N, and atmospheric CO_2. Increased precipitation intensity and amount, combined with lower root biomass from a drier climate, can yield more unstable slopes. Shifts in species dominance can also cause changes in root depth and cohesion (Hales et al. 2009). The spatial pattern of unstable slope conditions that can lead to landslides is influenced by interactions among the lateral redistribution of soil water in large events, the resulting pattern of high pore pressures with topographic slope, and root cohesive strength (Band et al. 2011).

4.5.2 Drought and Water Supply

Projections of drought extent over the next 75 years show that the proportion of global land mass experiencing drought will double from 15 to 30 % (Burke et al. 2006), and on most land masses, dry season precipitation is expected to decline by 15 % (Solomon et al. 2009). Projections for the largest declines in the United States are in the Southwest, strongly affecting water supply (Barnett and Pierce 2008; Rajagopalan et al. 2009). As noted above, lower precipitation will probably increase both forest mortality (Allen et al. 2010; Holden et al. 2011a) and fire risk (Westerling et al. 2011); however, forest mortality may not substantially mitigate runoff reductions associated with decreased precipitation (Adams et al. 2012). Historical observations of interannual variability in precipitation in the western United States have shown substantial increases in variability in the last 50 years (Luce and Holden 2009; Pagano and Garen 2005), even in areas not projected to show precipitation declines. Short-term severe droughts have consequences for vegetation (Holden et al. 2011b; van Mantgem et al. 2009) and water supply.

Although there has been interest in using forest harvest to augment water supplies, most increases in water yield after harvest occur in wet years (Brown et al. 2005; Ford et al. 2011), which may be less frequent in the future. In addition, water yield increases in snow environments occur earlier in the year, exacerbating flow timing issues caused by climate change (Troendle et al. 2010). Finally, in warmer and moister locations, increases in water yields can be replaced by decreases as young vegetation reestablishes within a few years (Brown et al. 2005; Ford et al. 2011).

4.6 Disturbance Interactions

4.6.1 Disturbances and Thresholds

Understanding interactions among disturbance regimes is a significant challenge for projecting the effects of climate change on forest ecosystems (Bigler et al. 2005; Busby et al. 2008) (Box 4.5). For example, how will massive outbreaks of bark beetles, which kill trees by feeding on cambial tissues, increase the potential for large severe wildfires in a warming climate (Box 4.1)? Interactions between processes can amplify or mute the overall effects of changes in complex forest ecosystems. The predominance of negative and positive feedbacks within and between processes will determine the stability or instability of the system.

> **Box 4.5: Response of Western Mountain Ecosystems to Climatic Variability and Change: The Western Mountain Initiative**
>
> The Western Mountain Initiative (WMI) uses paleoecological studies, contemporary studies, and modeling to understand responses to climatic variability and change in mountainous landscapes in the 11 large conterminous Western states (http://westernmountains.org). Initiated in 1991, the WMI consists of ten scientific laboratories in two federal agencies and four universities.
>
> Research has documented how climatic variability and change affect long-term patterns of snow, glaciers, and water geochemistry; forest productivity, vigor, and demography; and changing patterns of treeline dynamics and forest disturbances. Empirical and simulation modeling indicates that major changes in hydrologic function and ecological disturbance will occur in a warming climate. WMI data show that extreme disturbances have rapidly altered the structure and function of forest ecosystems over the past decade (Peterson et al. 2012).
>
> WMI research on disturbance interactions and their effects on ecosystem processes indicates that *synergistic interactions between disturbances produce larger effects than would occur from an individual disturbance*, especially when combined with chronic stressors such as air pollution, periodic drought, and reduced snowpack. For example, bark beetle outbreaks have been linked to increased likelihood of stand-replacing fire and changes in fire behavior, with the nature of the effect depending on the time since outbreak. Combined with increasing climatic stress on tree populations and growth, disturbance interactions can alter forest structure and function faster than could be expected from species redistribution or disturbance alone. Simultaneous climatically driven shifts in the locations of species optima, ecosystem productivity, disturbance regimes, and interactions between them
>
> (continued)

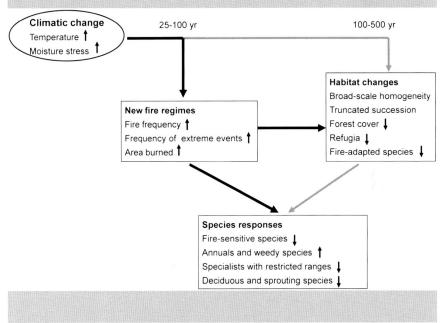

(continued)

can reset forest succession over large areas and short timeframes. (*Figure is a conceptual model of relative time scales for disturbance versus climate change alone to alter ecosystems. The focus is on fire, but the same logic applies to insect outbreaks. Adapted from McKenzie et al.* (2004))

Disturbance interactions may rapidly bring ecosystems to thresholds (Groffman et al. 2006). For example, Allen and Breshears (1998) and Breshears et al. (2005) documented rapid dieback of pinyon pine across the arid Southwest. Mature trees were pushed over a threshold by a combination of "global-change type drought" (Breshears et al. 2005) and an opportunistic bark beetle invasion. Regeneration of pinyon pine will determine whether this mortality represents a threshold for the ecosystem. Characteristic patterns of patchiness or continuity may indicate thresholds that have been approached or crossed (Scheffer et al. 2009). For example, the invasion of sagebrush (*Artemisia tridentata* Nutt.) steppe by cheatgrass (Fischer et al. 1996) and of the Sonoran Desert by buffelgrass (*Cenchrus ciliaris* L.) (Esque et al. 2007) provide fuel continuity and the potential for much more extensive wildfires than non-invaded areas with patchy fuels.

A notable threshold response to multiple stressors is the reproductive cycle of mountain pine beetle (Logan and Powell 2001) (see Sect. 4.3), whose outbreaks have killed mature trees across millions of hectares of pine in western North America. Within particular ranges of winter temperatures and growing-season degree days, the reproductive cycle is synchronized to the seasonal cycle, permitting maximum survival and epidemic population size. This "adaptive seasonality," combined with drought-caused and age-related vulnerability of the host species, may promote an abrupt increase in mortality of lodgepole pine (Hicke et al. 2006).

Conceptually, thresholds are fairly well understood. Modeling of thresholds has by necessity taken place in simplified (often virtual) ecosystems, and a major challenge remains to apply such sophistication to real-world systems outside the specific examples chosen by modelers to test their hypotheses. A larger challenge will always be the unpredictability of the occurrence of contingent, interacting events that push systems across thresholds.

4.6.2 Stress Complexes: From Conceptual to Quantitative Models

In the context of the effects of climate change on ecosystems, sensitivity to disturbance interactions is extended to environmental drivers not usually identified as disturbances. For example, extreme temperatures, drought, and air pollution put forest ecosystems under stress, which may increase their vulnerability to "true" disturbances such as fire, insect outbreaks, and pathogens. Following McKenzie et al. (2009), we refer to interacting stresses as stress complexes and present three examples from the Sierra Nevada, Alaska, and the Southeast.

A striking feature of mixed conifer forests in the southern Sierra Nevada and southern California is ambient air pollution, particularly elevated ozone, which affects plant vigor by reducing net photosynthesis and therefore growth (Peterson et al. 1991) and is often concentrated at middle and upper elevations (Brace and Peterson 1998). Air pollution exacerbates drought stress from warmer temperatures, which amplifies biotic stresses such as insects and pathogens (Ferrell 1996). The stress complex for California forests is represented in Fig. 4.2; interacting disturbances form the core of drivers of ecosystem change, modified by climate, management, and air pollution.

Alaska has experienced massive fires in the last decade, including the five largest fires in the United States. Over 2.5 million ha burned in the interior in 2004. Concurrently (1990s), massive outbreaks of the spruce bark beetle occurred on and near the Kenai Peninsula in south-central Alaska (Berg et al. 2006) (Fig. 4.3). Although periodic outbreaks have occurred throughout the historical record, both in south-central Alaska and the southwestern Yukon, these most recent outbreaks may be unprecedented in both extent and percentage mortality (over 90 % in many places) (Berg et al. 2006). Both wildfire and bark beetle outbreaks are associated with warmer temperatures in recent decades (Duffy et al. 2005; Werner et al. 2006). At the same time, major hydrological changes are underway from the cumulative effects of warming. Permafrost degradation is widespread in central Alaska, shifting ecosystems from birch forests to wetland types such as bogs and fens (Jorgenson et al. 2001). If broad-scale water balances become increasingly negative, peatlands may begin to support upland forest species (Klein et al. 2005).

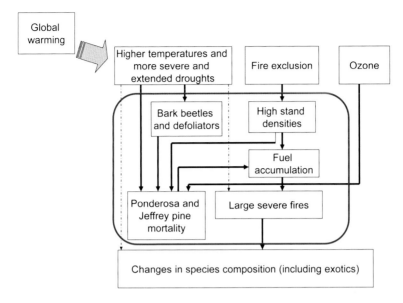

Fig. 4.2 Conceptual model of stress complexes in mixed conifer forests of the southern Sierra Nevada and southern California. The effects of insects and fire disturbance regimes (*red box*) and of fire exclusion are exacerbated by higher temperature. Stand-replacing fires and drought-induced mortality both contribute to species changes and invasive species (Modified from McKenzie et al. 2009)

Fig. 4.3 Mortality of white spruce from bark beetle attack on the Kenai Peninsula, Alaska (Photo by W.M. Ciesla, Forest Health Management International, Bugwood.org)

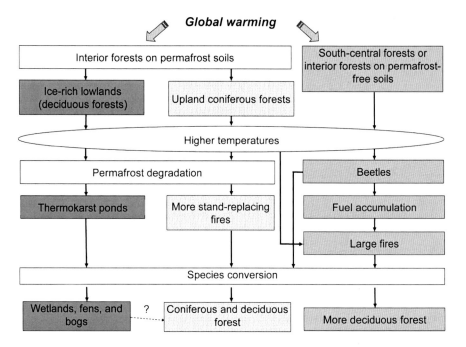

Fig. 4.4 Conceptual model of stress complexes in the interior and coastal forests of Alaska. Rapid increases in the severity of disturbance regimes (insects and fire) are triggered by a warmer climate. Stand-replacing fires, massive mortality from insects, and permafrost degradation contribute to species changes and conversion to deciduous life forms (Modified from McKenzie et al. 2009)

The stress complex for Alaska is represented conceptually in Fig. 4.4; upland and lowland ecosystems may follow parallel but contrasting paths toward new structure and species composition.

Much of the forested landscape in the southeastern United States is adapted to frequent fire, and prescribed fire is a mainstay of ecosystem management. Fire-adapted inland forests overlap geographically with coastal areas affected by hurricanes and potentially by sea-level rise (Ross et al. 2009), such that interactions between wildfires and hurricanes are synergistic (Fig. 4.5). For example, dry-season (prescribed) fires may have actually been more severe than wet-season (lightning) fires in some areas, causing structural damage via cambium kill and subsequent increased vulnerability to hurricane damage (Platt et al. 2002). The stress complex for the Southeast is represented conceptually in Fig. 4.6.

4.6.3 Uncertainties

Current knowledge about multiple stressors is mainly qualitative, despite case studies in various ecosystems that have measured the effects of interactions and

4 Disturbance Regimes and Stressors

Fig. 4.5 Interactions between wildfire and hurricanes are synergistic in the southern United States. Figure depicts a longleaf pine/saw palmetto flatwoods stand on the Atlantic coastal plain, 2.5 years after a hurricane and with a previous history of prescribed fire (Courtesy of the Fire and Environmental Research Applications team, U.S. Forest Service, Digital Photo Series)

even followed them over time (Hicke et al. 2012b). In the three examples above, the directional effects of warming-induced stressors may be clear (e.g., in California, species composition shifts to those associated with frequent fire). However, the magnitudes of these effects are not, nor are the potentially irreversible crossings of ecological thresholds. Given the complexity and diversity of potential interacting stressors in U.S. forests, a fruitful way to advance quantitative knowledge may be with explicit simulations with models of "intermediate complexity" to ascertain the sensitivity of ecosystems to uncertainties associated with key parameters (e.g., the thickness of the arrows in Figs. 4.2, 4.4 and 4.6). As the climate continues to warm, new empirical data will incrementally help to quantify disturbance and stressor interactions, providing greater certainty about the nature of stress complexes in forest ecosystems.

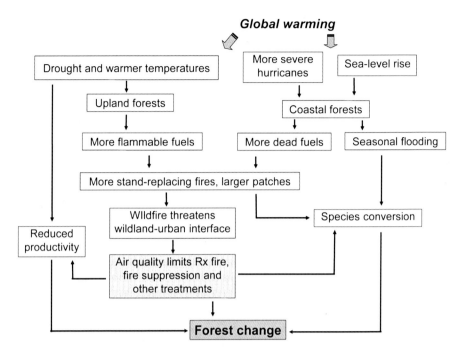

Fig. 4.6 Conceptual model of stress complexes in the interior and coastal forests of the Southeast. Increases in the severity of hurricanes are triggered by global warming as sea level rises. Warmer and drier climate in uplands leads to longer periods with flammable fuels. Changes in fire and hydrologic regimes, and responses to them, lead to species change and altered C dynamics

References

Adams, H. D., Macalady, A. K., Breshears, D. D., et al. (2010). Climate-induced tree mortality: Earth system consequences. *EOS, Transactions of the American Geophysical Union, 91*, 153.

Adams, H. D., Luce, C. H., Breshears, D. D., et al. (2012). Ecohydrological consequences of drought- and infestation-triggered tree die-off: Insights and hypotheses. *Ecohydrology, 5*, 145–149.

Allen, C. D., & Breshears, D. D. (1998). Drought-induced shift of a forest-woodland ecotone: Rapid landscape response to climate variation. *Proceedings of the National Academy of Sciences, USA, 95*, 14839–14842.

Allen, C. D., Savage, M., Falk, D. A., et al. (2002). Ecological restoration of Southwestern ponderosa pine ecosystems: A broad perspective. *Ecological Applications, 12*, 1418–1433.

Allen, C. D., Macalady, A. K., Chenchouni, H., et al. (2010). A global overview of drought and heat-induced tree mortality reveals emerging climate change risks for forests. *Forest Ecology and Management, 259*, 660–684.

Alpert, P., Bone, E., & Holzapfel, C. (2000). Invasiveness, invasibility and the role of environmental stress in the spread of non-native plants. *Perspectives in Plant Ecology, Evolution and Systematics, 3*, 52–66.

Ayres, M. P., & Lombardero, M. J. (2000). Assessing the consequences of global change for forest disturbance from herbivores and pathogens. *The Science of the Total Environment, 262*, 263–286.

Bachelet, D., Neilson, R. P., Lenihan, J. M., & Drapek, R. J. (2001). Climate change effects on vegetation distribution and carbon budget in the United States. *Ecosystems, 4*, 164–185.

Bailey, J. D., Mayrohn, C., Doescher, P. S., et al. (1998). Understory vegetation in old and young Douglas-fir forests of western Oregon. *Forest Ecology and Management, 112*, 289–302.

Band, L. E., Hwang, T., Hales, T. C., et al. (2011). Ecosystem processes at the watershed scale: Mapping and modeling ecohydrological controls of landslides. *Geomorphology, 137*, 159–167.

Barnett, T. P., & Pierce, D. W. (2008). When will Lake Mead go dry? *Water Resources Research, 44*, W03201.

Bentz, B., Logan, J., MacMahon, J., et al. (2009). *Bark beetle outbreaks in western North America: Causes and consequences* (44pp). Salt Lake City: University of Utah Press.

Bentz, B. J., Régnière, J., Fettig, C. J., et al. (2010). Climate change and bark beetles of the Western United States and Canada: Direct and indirect effects. *BioScience, 60*, 602–613.

Berg, E. E., Henry, J. D., Fastie, C. L., et al. (2006). Spruce beetle outbreaks on the Kenai Peninsula, Alaska, and Kluane National Park and Reserve. Yukon Territory: Relationship to summer temperatures and regional differences in disturbance regimes. *Forest Ecology and Management, 227*, 219–232.

Berggren, A., Böjrkman, C., Bylund, H., & Ayres, M. P. (2009). The distribution and abundance of animal populations in a climate of uncertainty. *Oikos, 118*, 1121–1126.

Bigler, C., Kulakowski, D., & Veblen, T. T. (2005). Multiple disturbance interactions and drought influence fire severity in Rocky Mountain subalpine forests. *Ecology, 86*, 3018–3029.

Bond, W. J., & Midgley, G. F. (2000). A proposed CO_2-controlled mechanism of woody plant invasion in grasslands and savannas. *Global Change Biology, 6*, 865–869.

Boon, S. (2012). Snow accumulation following forest disturbance. *Ecohydrology, 5*, 279–285.

Brace, S., & Peterson, D. L. (1998). Spatial patterns of tropospheric ozone in the Mount Rainier region of the Cascade Mountains, U.S.A. *Atmospheric Environment, 32*, 3629–3637.

Bradley, B. A., Oppenheimer, M., & Wilcove, D. S. (2009). Climate change and plant invasions: Restoration opportunities ahead? *Global Change Biology, 15*, 1511–1521.

Breshears, D. D., Cobb, N. S., Rich, P. M., et al. (2005). Regional vegetation die-off in response to global-change-type drought. *Proceedings of the National Academy of Sciences, USA, 102*, 15144–15148.

Briffa, K. R. (2000). Annual climate variability in the Holocene: Interpreting the message of ancient trees. *Quaternary Science Reviews, 19*, 87–105.

Brooks, P. D., Troch, P. A., Durcik, M., et al. (2011). Quantifying regional scale ecosystem response to changes in precipitation: Not all rain is created equal. *Water Resources Research, 47*, W00J08.

Brown, A. E., Zhang, L., McMahon, T. A., et al. (2005). A review of paired catchment studies for determining changes in water yield resulting from alterations in vegetation. *Journal of Hydrology, 310*, 28–61.

Burke, E. J., Brown, S. J., & Christidis, N. (2006). Modeling the recent evolution of global drought and projections for the twenty-first century with the Hadley Centre climate model©. *Journal of Hydrometeorology, 7*, 1113–1125.

Busby, P. E., Motzkin, G., & Foster, D. R. (2008). Multiple and interacting disturbances lead to *Fagus grandifolia* dominance in coastal New England. *Journal of the Torrey Botanical Society, 135*, 346–359.

Campbell, R., Smith, D. J., & Arsenault, A. (2006). Multicentury history of western spruce budworm outbreaks in interior Douglas-fir forests near Kamloops, British Columbia. *Canadian Journal of Forest Research, 36*, 1758–1769.

Candau, J.-N., & Fleming, R. A. (2005). Landscape-scale spatial distribution of spruce budworm defoliation in relation to bioclimatic conditions. *Canadian Journal of Forest Research, 35*, 2218–2232.

Cannon, S. H., Gartner, J. E., Rupert, M. G., et al. (2010). Predicting the probability and volume of postwildfire debris flows in the intermountain western United States. *Geological Society of America Bulletin, 122*, 127–144.

Chambers, J. C., Roundy, B. A., Blank, R. R., et al. (2007). What makes Great Basin sagebrush ecosystems invasible by *Bromus tectorum*? *Ecological Monographs, 77*, 117–145.

Chan-McLeod, A. C. A. (2006). A review and synthesis of the effects of unsalvaged mountain-pine-beetle-attacked stands on wildlife and implications for forest management. *BC Journal of Ecosystems and Management, 7*, 119–132.

Cohn, J. P. (2005). Tiff over tamarisk: Can a nuisance be nice, too? *Bioscience, 55*, 648–654.

Colautti, R. I., Grigorovich, I. A., & MacIsaac, H. J. (2006). Propagule pressure: A null model for biological invasions. *Biological Invasions, 8*, 1023–1037.

Collins, D. B. G., & Bras, R. L. (2008). Climate control of sediment yield in dry lands following climate and land cover change. *Water Resources Research, 44*, W10405.

Collins, B. M., Omi, P. N., & Chapman, P. L. (2006). Regional relationships between climate and wildfire-burned area in the Interior West, USA. *Canadian Journal of Forest Research, 36*, 699–709.

Collins, B. J., Rhoades, C. C., Hubbard, R. M., & Battaglia, M. A. (2011). Tree regeneration and future stand development after bark beetle infestation and harvesting in Colorado lodgepole pine stands. *Forest Ecology and Management, 261*, 2168–2175.

Crimmins, S. M., Dobrowski, S. Z., Greenberg, J. A., et al. (2011). Changes in climate water balance drive downhill shifts in plant species optimum elevations. *Science, 331*, 324–327.

Cudmore, T. J., Bjorkland, N., Carroll, A. L., & Lindgren, B. S. (2010). Climate change and range expansion of an aggressive bark beetle: Evidence of higher beetle reproduction in naïve host tree populations. *Journal of Applied Ecology, 47*, 1036–1043.

D'Antonio, C. M. (2000). Chapter 4: Fire, plant invasions, and global changes. In H. A. Mooney & R. J. Hobbs (Eds.), *Invasive species in a changing world* (pp. 65–94). Washington, DC: Island Press.

D'Antonio, C., Levine, J., & Thomsen, M. (2001). Ecosystem resistance to invasion and the role of propagule supply: A California perspective. *Journal of Mediterranean Ecology, 2*, 233–245.

Dale, V. H., Joyce, L. A., McNulty, S., et al. (2001). Climate change and forest disturbances. *BioScience, 51*, 723–734.

Davis, M. A., Grime, J. P., & Thompson, K. (2000). Fluctuating resources in plant communities: A general theory of invasibility. *Journal of Ecology, 88*, 528–534.

Dettinger, M. (2011). Climate change, atmospheric rivers, and floods in California—A multimodel analysis of storm frequency and magnitude changes. *Journal of the American Water Resources Association, 47*, 514–523.

Drever, M. C., Goheen, J. R., & Martin, K. (2009). Species-energy theory, pulsed resources, and regulation of avian richness during a mountain pine beetle outbreak. *Ecology, 90*, 1095–1105.

Duffy, P. A., Walsh, J. E., Graham, J. M., et al. (2005). Impacts of large-scale atmospheric-ocean variability on Alaskan fire season severity. *Ecological Applications, 15*, 1317–1330.

Dukes, J. S., & Mooney, H. A. (1999). Does global change increase the success of biological invaders? *Trends in Ecology & Evolution, 14*, 135–139.

Dukes, J. S., Pontius, J., Orwig, D., et al. (2009). Responses of insect pests, pathogens, and invasive plant species to climate change in the forests of northeastern North America: What can we predict? *Canadian Journal of Forest Research, 39*, 231–248.

Dukes, J. S., Chiariello, N. R., Loarie, S. R., & Field, C. B. (2011). Strong response of an invasive plant species (*Centaurea solstitialis* L.) to global environmental changes. *Ecological Applications, 21*, 1887–1894.

Duncan, C. A., Jachetta, J. J., Brown, M. L., et al. (2004). Assessing the economic, environmental, and societal losses from invasive plants on rangeland and wildlands. *Weed Technology, 18*, 1411–1416.

Dwyer, G., Dushoff, J., & Yee, S. H. (2004). The combined effects of pathogens and predators on insect outbreaks. *Nature, 430*, 341–345.

Eamus, D. (1991). The interaction of rising CO_2 and temperatures with water use efficiency. *Plant, Cell & Environment, 14*, 843–852.

Easterling, D. R., Evans, J. L., Groisman, P. Y., et al. (2000). Observed variability and trends in extreme climate events: A brief review. *Bulletin of the American Meteorological Society, 81*, 417–425.

Eschtruth, A. K., & Battles, J. J. (2009). Assessing the relative importance of disturbance, herbivory, diversity, and propagule pressure in exotic plant invasion. *Ecological Monographs, 79*, 265–280.

Esque, T. C., Schwalbe, C. R., Lissow, J. A., et al. (2007). Buffelgrass fuel loads in Saguaro National Park, Arizona, increase fire danger and threaten native species. *Park Science, 24*, 33–37.

Evangelista, P. H., Kumer, S., Stohlgren, T. J., & Young, N. E. (2011). Assessing forest vulnerability and the potential distribution of pine beetles under current and future climate scenarios in the Interior West of the US. *Forest Ecology and Management, 262*, 307–316.

Evans, L. M., Hofstetter, R. W., Ayres, M. P., & Klepzig, K. D. (2011). Temperature alters the relative abundance and population growth rates of species within the *Dendroctonus frontalis* (Coleoptera: Curculionidae) community. *Environmental Entomology, 40*, 824–834.

Fauria, M. M., & Johnson, E. A. (2006). Large-scale climatic patterns control large lightning fire occurrence in Canada and Alaska forest regions. *Journal of Geophysical Research, 111*, G04008.

Ferrell, G. T. (1996). Chapter 45: The influence of insect pests and pathogens on Sierra forests. In *Sierra Nevada Ecosystem Project: Final report to Congress, Vol. II, Assessments and scientific basis for management options* (pp. 1177–1192). Davis: University of California, Centers for Water and Wildland Resources.

Fischer, R. A., Reese, K. P., & Connelly, J. W. (1996). An investigation on fire effects within xeric sage grouse brood habitat. *Journal of Range Management, 49*, 194–198.

Ford, C. R., Laseter, S. H., Swank, W. T., & Vose, J. M. (2011). Can forest management be used to sustain water-based ecosystem services in the face of climate change? *Ecological Applications, 21*, 2049–2067.

Foster, D. R., Clayden, S., Orwig, D. A., et al. (2002). Oak, chestnut and fire: Climatic and cultural controls of long-term forest dynamics in New England, USA. *Journal of Biogeography, 29*, 1359–1379.

Friedenberg, N. A., Powell, J. A., & Ayres, M. P. (2007). Synchrony's double edge: Transient dynamics and the Allee effect in stage structured populations. *Ecology Letters, 10*, 564–573.

Furniss, M. M., Holsten, E. H., Foote, M. J., & Bertram, M. (2001). Biology of a willow leafblotch miner, *Micrurapteryx salicifoliella*, (Lepidoptera: Gracillariidae) in Alaska. *Environmental Entomology, 30*, 736–741.

Ganey, J. L., & Vojta, S. C. (2011). Tree mortality in drought-stressed mixed-conifer and ponderosa pine forests, Arizona, USA. *Forest Ecology and Management, 261*, 162–168.

Garnas, J. R., Ayres, M. P., Liebhold, A. M., & Evans, C. (2011a). Subcontinental impacts of an invasive tree disease on forest structure and dynamics. *Journal of Ecology, 99*, 532–541.

Garnas, J. R., Houston, D. R., Ayres, M. P., & Evans, C. (2011b). Disease ontogeny overshadows effects of climate and species interactions on population dynamics in a nonnative forest disease complex. *Ecography, 35*, 412–421.

Gaylord, M. L., Kolb, T. E., Wallin, K. F., & Wagner, M. R. (2007). Seasonal dynamics of tree growth, physiology, and resin defenses in a northern Arizona ponderosa pine forest. *Canadian Journal of Forest Research, 37*, 1173–1183.

Gillooly, J. F., Charnov, E. L., West, G. B., et al. (2002). Effects of size and temperature on developmental time. *Nature, 417*, 70–73.

Goode, J. R., Luce, C. H., & Buffington, J. M. (2011). Enhanced sediment delivery in a changing climate in semi-arid mountain basins: Implications for water resource management and aquatic habitat in the northern Rocky Mountains. *Geomorphology, 139/140*, 1–15.

Greenwood, D. L., & Weisberg, P. J. (2008). Density-dependent tree mortality in pinyon-juniper woodlands. *Forest Ecology and Management, 255*, 2129–2137.

Grissino-Mayer, H. D., & Swetnam, T. W. (2000). Century-scale climate forcing of fire regimes in the American Southwest. *The Holocene, 10*, 213–220.

Groffman, P. M., Baron, J. S., Blett, T., et al. (2006). Ecological thresholds: The key to successful environmental management or an important concept with no practical application? *Ecosystems, 9*, 1–13.

Grulke, N. E. (2011). The nexus of host and pathogen phenology: Understanding the disease triangle with climate change. *New Phytologist, 189*, 8–11.

Hales, T. C., Ford, C. R., Hwang, T., et al. (2009). Topographic and ecologic controls on root reinforcement. *Journal of Geophysical Research, 114*, F03013.

Halpern, C. B. (1989). Early successional patterns of forest species: Interactions of life history traits and disturbance. *Ecology, 70*, 704–720.

Hamlet, A. F., & Lettenmaier, D. P. (2007). Effects of 20th century warming and climate variability on flood risk in the western U.S. *Water Resources Research, 43*, W06427.

Hellmann, J. J., Byers, J. E., Bierwagen, B. G., & Dukes, J. S. (2008). Five potential consequences of climate change for invasive species. Special section. *Conservation Biology, 22*, 534–543.

Hessl, A. E., McKenzie, D., & Schellhaas, R. (2004). Drought and Pacific Decadal Oscillation linked to fire occurrence in the inland Pacific Northwest. *Ecological Applications, 14*, 425–442.

Heyerdahl, E. K., Brubaker, L. B., & Agee, J. K. (2002). Annual and decadal climate forcing of historical regimes in the interior Pacific Northwest, USA. *The Holocene, 12*, 597–604.

Heyerdahl, E. K., McKenzie, D., Daniels, L. D., et al. (2008a). Climate drivers of regionally synchronous fires in the inland Northwest (1651–1900). *International Journal of Wildland Fire, 17*, 40–49.

Heyerdahl, E. K., Morgan, P., & Riser, J. P. (2008b). Multi-season climate synchronized historical fires in dry forests (1650–1900), Northern Rockies, U.S.A. *Ecology, 89*, 705–716.

Hicke, J. A., Logan, J. A., Powell, J., & Ojima, D. S. (2006). Changes in temperature influence suitability for modeled mountain pine beetle (*Dendroctonus ponderosae*) outbreaks in the Western United States. *Journal of Geophysical Research, 11*, G02019.

Hicke, J. A., Allen, C. D., Desai, A. R., et al. (2012a). Effects of biotic disturbances on forest carbon cycling in the United States and Canada. *Global Change Biology, 18*, 7–34.

Hicke, J. A., Johnson, M. C., Hayes, J. L., & Preisler, H. K. (2012b). Effects of bark beetle-caused tree mortality on wildfire. *Forest Ecology and Management, 271*, 81–90.

Hofstetter, R. W., Dempsey, T. D., Klepzig, K. D., & Ayres, M. P. (2007). Temperature-dependent effects on mutualistic, antagonistic, and commensalistic interactions among insects, fungi and mites. *Community Ecology, 8*, 47–56.

Holden, Z. A., Abatzoglou, J. T., Luce, C. H., & Baggett, L. S. (2011a). Empirical downscaling of daily minimum air temperature at very fine resolutions in complex terrain. *Agricultural and Forest Meteorology, 151*, 1066–1073.

Holden, Z. A., Luce, C. H., Crimmins, M. A., & Morgan, P. (2011b). Wildfire extent and severity correlated with annual streamflow distribution and timing in the Pacific Northwest, USA (1984–2005). *Ecohydrology*. doi:10.1002/eco.257.

Holmes, T. P., Liebhold, A. M., Kovacs, K. F., & Von Holle, B. (2010). A spatial-dynamic value transfer model of economic losses from a biological invasion. *Ecological Economics, 70*, 86–95.

Hutchinson, T. F., Long, R. P., Ford, R. D., & Sutherland, E. K. (2008). Fire history and the establishment of oaks and maples in second-growth forests. *Canadian Journal of Forest Research, 38*, 1184–1198.

Hwang, T., Band, L., & Hales, T. C. (2009). Ecosystem processes at the watershed scale: Extending optimality theory from plot to catchment. *Water Resources Research, 45*, W11425.

Istanbulluoglu, E., & Bras, R. L. (2006). On the dynamics of soil moisture, vegetation, and erosion: Implications of climate variability and change. *Water Resources Research, 42*, W06418.

Jenkins, M. J., Hebertson, E., Page, W., & Jorgensen, C. A. (2008). Bark beetles, fuels, fires and implications for forest management in the Intermountain West. *Forest Ecology and Management, 254*, 16–34.

Jepsen, J. U., Kapari, L., & Hagen, S. B. (2011). Rapid northwards expansion of a forest insect pest attributed to spring phenology matching with sub-Arctic birch. *Global Change Biology, 17*, 2071–2083.

Jorgenson, M. T., Racine, C. H., Walters, J. C., & Osterkamp, T. E. (2001). Permafrost degradation and ecological changes associated with a warming climate in central Alaska. *Climatic Change, 48*, 551–571.

Karl, T. R., & Knight, R. W. (1998). Secular trends of precipitation amount, frequency, and intensity in the United States. *Bulletin of the American Meteorological Society, 79*, 231–241.

Keeley, J. E., & McGinnis, T. W. (2007). Impact of prescribed fire and other factors on cheatgrass persistence in a Sierra Nevada ponderosa pine forest. *International Journal of Wildland Fire, 16*, 96–106.

Keeley, J. E., Lubin, D., & Fotheringham, C. J. (2003). Fire and grazing impacts on plant diversity and alien plant invasions in the southern Sierra Nevada. *Ecological Applications, 13*, 1355–1374.

Kemp, W. P., Everson, D. O., & Wellington, W. G. (1985). *Regional climatic patterns and western spruce budworm outbreaks* (Tech. Bull. 1693, 31pp). Washington, DC: U.S. Department of Agriculture, Forest Service, Canada/United States Spruce Budworms Program.

Kerns, B. K., Thies, W. G., & Niwa, C. G. (2006). Season and severity of prescribed burn in ponderosa pine forests: Implications for understory native and exotic plants. *Ecoscience, 13*, 44–55.

Kerns, B. K., Naylor, B. J., Buonopane, M., et al. (2009). Modeling tamarisk (*Tamarix* spp.) habitat and climate change effects in the Northwestern United States. *Invasive Plant Science and Management, 2*, 200–215.

Kitzberger, T., Brown, P. M., Heyerdahl, E. K., et al. (2007). Contingent Pacific-Atlantic Ocean influence on multicentury wildfire synchrony over western North America. *Proceedings of the National Academy of Sciences, USA, 104*, 543–548.

Klapwijk, M. J., Ayres, M. P., Battisti, A., & Larsson, S. (2012). Assessing the impact of climate change on outbreak potential. In P. Barbosa, D. L. Letourneau, & A. A. Agrawal (Eds.), *Insect outbreaks revisited* (pp. 429–450). New York: Wiley-Blackwell.

Klein, E., Berg, E. E., & Dial, R. (2005). Wetland drying and succession across the Kenai Peninsula Lowlands, south-central Alaska. *Canadian Journal of Forest Research, 35*, 1931–1941.

Klopfenstein, N. B., Kim, M.-S., Hanna, J. W., et al. (2009). *Approaches to predicting potential impacts of climate change on forest disease: An example with Amillaria root disease* (Res. Pap. RMRS-RP-76, 10pp). Fort Collins: U.S. Department of Agriculture, Forest Service, Rocky Mountain Research Station.

Kovacs, K. F., Haight, R. G., McCollough, D. G., et al. (2010). Cost of potential emerald ash borer damage in U.S. communities, 2009–2019. *Ecological Economics, 69*, 569–578.

Kovacs, K. F., Mercader, R. J., Haight, R. G., et al. (2011). The influence of satellite populations of emerald ash borer on projected economic costs in U.S. communities, 2010–2020. *Journal of Environmental Management, 92*, 2170–2181.

Levine, J. M. (2000). Species diversity and biological invasions: Relating local process to community pattern. *Science, 288*, 852–854.

Levine, J. M., Adler, P. B., & Yelenik, S. G. (2004). A meta-analysis of biotic resistance to exotic plant invasions. Review. *Ecology Letters, 7*, 975–989.

Lindroth, R. L. (2010). Impacts of elevated atmospheric CO_2 and O_3 on forests: Phytochemistry, trophic interactions, and ecosystem dynamics. *Journal of Chemical Ecology, 36*, 2–21.

Littell, J. S., McKenzie, D., Peterson, D. L., & Westerling, A. L. (2009). Climate and wildfire area burned in western U.S. ecoprovinces, 1916–2003. *Ecological Applications, 19*, 1003–1021.

Littell, J. S., Oneil, E. E., McKenzie, D., et al. (2010). Forest ecosystems, disturbance, and climatic change in Washington State, USA. *Climatic Change, 102*, 129–158.

Liu, Y., Stanturf, J., & Goodrick, S. (2010). Trends in global wildfire potential in a changing climate. *Forest Ecology and Management, 259*, 685–697.

Lodge, D. M., Williams, L., MacIsaac, H. J., et al. (2006). Biological invasions: Recommendations for U.S. policy and management. *Ecological Applications, 16*, 2035–2054.

Logan, J. A., & Powell, J. A. (2001). Ghost forests, global warming, and the mountain pine beetle (Coleoptera: Scolytidae). *American Entomologist, 47*, 160–173.

Lombardero, M. J., & Ayres, M. P. (2011). Factors influencing bark beetle outbreaks after forest fires on the Iberian Peninsula. *Environmental Entomology, 40*, 1007–1018.

Lombardero, M. J., Ayres, M. P., Ayres, B. D., & Reeve, J. D. (2000). Cold tolerance of four species of bark beetle (Coleoptera: Scolytidae) in North America. *Environmental Entomology, 29*, 421–432.

Lonsdale, W. M. (1999). Global patterns of plant invasions and the concept of invasibility. *Ecology, 80*, 1522–1536.

Lovett, G. M., Canham, C. D., Arthur, M. A., et al. (2006). Forest ecosystem responses to exotic pests and pathogens in eastern North America. *BioScience, 56*, 395–405.

Luce, C. H. (2005). Land use and land cover effects on runoff processes: Fire. In M. G. Anderson (Ed.), *Encyclopedia of hydrological sciences* (pp. 1831–1838). Hoboken: Wiley.

Luce, C. H., & Holden, Z. A. (2009). Declining annual streamflow distributions in the Pacific Northwest United States, 1948–2006. *Geophysical Research Letters, 36*, L16401.

Marlon, J., Bartlein, P. J., & Whitlock, C. (2006). Fire-fuel-climate linkages in the northwestern USA during the Holocene. *The Holocene, 16*, 1059–1071.

Marlon, J. R., Bartlein, P. J., Carcaillet, C., et al. (2008). Climate and human influences on global biomass burning over the past two millennia. *Nature Geoscience, 1*, 697–702.

McKenzie, D., Gedalof, Z., Peterson, D. L., & Mote, P. (2004). Climatic change, wildfire and conservation. *Conservation Biology, 18*, 890–902.

McKenzie, D., Peterson, D. L., & Littell, J. J. (2009). Global warming and stress complexes in forests of western North America. In A. Bytnerowicz, M. J. Arbaugh, A. R. Riebau, & C. Andersen (Eds.), *Wildland fires and air pollution* (pp. 319–337). Amsterdam/London: Elsevier. Developments in Environmental Science 8. Chapter 15.

McNulty, S. G., Lorio, P. L., Ayres, M. P., & Reeve, J. D. (1997). Predictions of southern pine beetle populations under historic and projected climate using a forest ecosystem model. In R. A. Mickler & S. Fox (Eds.), *The productivity and sustainability of southern forest ecosystems in a changing environment* (pp. 617–634). New York: Springer.

Melillo, J. M., Janetos, A. C., & Karl, T. R. (2001). *Climate change impacts on the United States: The potential consequences of climate variability and change.* Cambridge: Cambridge University Press.

Meyer, G. A., & Pierce, J. L. (2003). Climatic controls on fire-induced sediment pulses in Yellowstone National Park and central Idaho: a long-term perspective. *Forest Ecology and Management, 178*, 89–104.

Miller, J. H. (1990). Ailanthus altissima (Mill.) Swingle ailanthus. In R. M. Burns, & B. H. Honkala (Tech. Coords.), *Silvics of North America: Vol. 2. Hardwoods. Agriculture handbook 654* (pp. 101–104). Washington, DC: U.S. Department of Agriculture, Forest Service.

Miller, R. F., & Wigand, P. E. (1994). Holocene changes in semiarid pinyon-juniper woodlands. *Bioscience, 44*, 465–474.

Miller, R. F., Bates, J. D., Svejcar, T. J., et al. (2005). *Biology, ecology, and management of western juniper (Juniperus occidentalis)* (Tech. Bull. 152, 82pp). Corvallis: Oregon State University, Agricultural Experiment Station.

Mitchell, R. J., Hiers, J. K., O'Brien, J. J., et al. (2006). Silviculture that sustains: The nexus between silviculture, frequent prescribed fire, and conservation of biodiversity in longleaf pine forests of the southeastern United States. *Canadian Journal of Forest Research, 36*, 2724–2736.

Moody, J. A., & Martin, D. A. (2009). Synthesis of sediment yields after wildland fire in different rainfall regimes in the Western United States. *International Journal of Wildland Fire, 18*, 96–115.

Morgan, P., Heyerdahl, E. K., & Gibson, C. E. (2008). Multi-season climate synchronized widespread forest fires throughout the 20th century, Northern Rockies, USA. *Ecology, 89*, 717–728.

Moser, J. C., Fitzgibbon, B. A., & Klepzig, K. D. (2005). The Mexican pine beetle, *Dendroctonus mexicanus*: First record in the United States and co-occurrence with the southern pine beetle—*Dendroctonus frontalis* (Coleoptera: Scolytidae or Curculionidae: Scolytidae). *Entomological News, 116*, 235–243.

Myers, J. H., Simberloff, D., Kuris, A. M., & Carey, J. R. (2000). Eradication revisited: Dealing with exotic species. *Trends in Ecology & Evolution, 15*, 316–320.

National Interagency Coordination Center. (2011). *GACC predictive services intelligence.* http://www.predictiveservices.nifc.gov/intelligence/intelligence.htm. Accessed 25 Jan 2012.

National Invasive Species Council. (2001). *Meeting the invasive species challenge: National invasive species management plan* (p. 80). Washington, DC: U.S. Department of Agriculture.

Nelson, C. R., Halpern, C. B., & Agee, J. K. (2008). Thinning and burning results in low-level invasion by nonnative plants but neutral effects on natives. *Ecological Applications, 18*, 762–770.

Nowacki, G. J., & Abrams, M. D. (2008). The demise of fire and "mesophication" of forests in the eastern United States. *Bioscience, 58*, 112–128.

O'Gorman, P. A., & Schneider, T. (2009). The physical basis for increases in precipitation extremes in simulations of 21st-century climate change. *Proceedings of the National Academy of Sciences, USA, 106*, 14773–14777.

Orwig, D. A., Foster, D. R., & Mausel, D. L. (2002). Landscape patterns of hemlock decline in New England due to the introduced hemlock woolly adelgid. *Journal of Biogeography, 29*, 1475–1487.

Pagano, T., & Garen, D. (2005). A recent increase in Western U.S. streamflow variability and persistence. *Journal of Hydrometeorology, 6*, 173–179.

Paradis, A., Elkinton, J., Hayhoe, K., & Buonaccorsi, J. (2008). Role of winter temperature and climate change on the survival and future range expansion of the hemlock woolly adelgid (*Adelges tsugae*) in eastern North America. *Mitigation and Adaptation Strategies for Global Change, 13*, 541–554.

Parker, T. J., Clancy, K. M., & Mathiasen, R. L. (2006). Interactions among fire, insects and pathogens in coniferous forests of the interior western United States and Canada. *Agricultural and Forest Entomology, 8*, 167–189.

Parks, C. G., Radosevich, S. R., Endress, B. A., et al. (2005). Natural and land-use history of the Northwest mountain ecoregions (USA) in relation to patterns of plant invasions. *Perspectives in Plant Ecology, Evolution and Systematics, 7*, 137–158.

Pattison, R. R., & Mack, R. N. (2008). Potential distribution of the invasive tree *Triadica sebifera* (Euphorbiaceae) in the United States: Evaluating CLIMEX predictions with field trials. *Global Change Biology, 14*, 813–826.

Pauchard, A., Kueffer, C., Dietz, H., et al. (2009). Ain't no mountain high enough: Plant invasions reaching new elevations. *Frontiers in Ecology and the Environment, 7*, 479–486.

Peterson, D. L., Arbaugh, M. J., & Robinson, L. J. (1991). Growth trends of ozone-stressed ponderosa pine (*Pinus ponderosa*) in the Sierra Nevada of California, USA. *The Holocene, 1*, 50–61.

Peterson, D. L., Allen, C. D., Baron, J. S., et al. (2012). Response of Western mountain ecosystems to climatic variability and change: A collaborative research approach. In J. Bellant & E. Beever (Eds.), *Ecological consequences of climate change: Mechanisms, conservation, and management* (pp. 163–190). New York: Taylor & Francis.

Pfeifer, E. M., Hicke, J. A., & Meddens, A. J. H. (2011). Observations and modeling of aboveground tree carbon stocks and fluxes following a bark beetle outbreak in the Western United States. *Global Change Biology, 17*, 339–350.

Pierce, J. L., Meyer, G. A., & Jull, A. J. T. (2004). Fire-induced erosion and millennial-scale climate change in northern ponderosa pine forests. *Nature, 432*, 87–90.

Pimentel, D., Zuniga, R., & Morrison, D. (2005). Update on the environmental and economic costs associated with alien-invasive species in the United States. *Ecological Economics, 52*, 273–288.

Platt, W. J., Beckage, B., Doren, R. F., & Slater, H. H. (2002). Interactions of large-scale disturbances: Prior fire regimes and hurricane mortality of savanna pines. *Ecology, 83*, 1566–1572.

Powell, J. A., Jenkins, J. L., Logan, J. A., & Bentz, B. J. (2000). Seasonal temperature alone can synchronize life cycles. *Bulletin of Mathematical Biology, 62*, 977–998.

Powell, K. I., Chase, J. M., & Knight, T. M. (2011). Synthesis of plant invasion effects on biodiversity across spatial scales. *American Journal of Botany, 98*, 539–548.

Power, M. J., Marlon, J., Ortiz, N., et al. (2008). Changes in fire regimes since the Last Glacial Maximum: An assessment based on a global synthesis and analysis of charcoal data. *Climate Dynamics, 30*, 887–907.

Price, J. I., McCollum, D. W., & Berrens, R. P. (2010). Insect infestation and residential property values: A hedonic analysis of the mountain pine beetle epidemic. *Forest Policy and Economics, 12*, 415–422.

Raffa, K. F., Aukema, B. H., Bentz, B. J., et al. (2008). Cross-scale drivers of natural disturbances prone to anthropogenic amplification: The dynamics of bark beetle eruptions. *BioScience, 58*, 501–517.

Rajagopalan, B., Nowak, K., Prairie, J., et al. (2009). Water supply risk on the Colorado River: Can management mitigate? *Water Resources Research, 45*, W08201.

Ralph, F. M., Neiman, P. J., Wick, G. A., et al. (2006). Flooding on California's Russian River: Role of atmospheric rivers. *Geophysical Research Letters, 33*, L13801.

Reynolds, L. V., Ayres, M. P., Siccama, T. G., & Holmes, R. T. (2007). Climatic effects on caterpillar fluctuations in northern hardwood forests. *Canadian Journal of Forest Research, 37*, 481–491.

Rodenhouse, N. L., Christenson, L. M., Parry, D., & Green, L. E. (2009). Climate change effects on native fauna of northeastern forests. *Canadian Journal of Forest Research, 39*, 249–263.

Rohrs-Richey, J. K., Mulder, C. P. H., Winton, L. M., & Stanosz, G. (2011). Physiological performance of an Alaskan shrub (*Alnus fruticosa*) in response to disease (*Valsa melanodiscus*) and water stress. *New Phytologist, 189*, 295–307.

Ross, M. S., Obrien, J. J., Ford, R. G., et al. (2009). Disturbance and the rising tide: The challenge of biodiversity management on low-island ecosystems. *Frontiers in Ecology and the Environment, 7*, 471–478.

Safranyik, L., Carroll, A. L., Régnière, D. W., et al. (2010). Potential for range expansion of mountain pine beetle into the boreal forest of North America. *Canadian Entomologist, 142*, 415–442.

Sage, R. F., Coiner, H. A., Way, D. A., et al. (2009). Kudzu [*Pueraria montana* (Lour.) Merr. var *lobata*]: A new source of carbohydrate for bioethanol production. *Biomass and Bioenergy, 33*, 57–61.

Salinas-Moreno, Y., Ager, A., Vargas, C. F., et al. (2010). Determining the vulnerability of Mexican pine forests to bark beetles of the genus *Dendroctonus* Erichson (Coleoptera: Curculionidae: Scolytinae). *Forest Ecology and Management, 260*, 52–61.

Sasek, T. W., & Strain, B. R. (1990). Implications of atmospheric CO_2 enrichment and climatic change for the geographical distribution of two introduced vines in the U.S.A. *Climatic Change, 16*, 31–51.

Scheffer, M., Bascompte, J., Brock, W. A., et al. (2009). Early-warning signals for critical transitions. *Nature, 461*, 53–59.

Seabloom, E. W., Harpole, W. S., Reichman, O. J., & Tilman, D. (2003). Invasion, competitive dominance, and resource use by exotic and native California grassland species. *Proceedings of the National Academy of Sciences, USA, 100*, 13384–13389.

Seppälä, R., Buck, A., Katila, P. (Eds.). (2009). *Adaptation of forests and people to climate change: A global assessment report* (224pp). Helsinki: International Union of Forest Research Organizations.

Sexton, J. P., McKay, J. K., & Sala, A. (2002). Plasticity and genetic diversity may allow saltcedar to invade cold climates in North America. *Ecological Applications, 12*, 1652–1660.

Shakesby, R. A., & Doerr, S. H. (2006). Wildfire as a hydrological and geomorphological agent. *Earth-Science Reviews, 74*, 269–307.

Sheppard, S., & Picard, P. (2006). Visual-quality impacts of forest pest activity at the landscape level: A synthesis of published knowledge and research needs. *Landscape and Urban Planning, 77*, 321–342.

Sherriff, R. L., Berg, E. E., & Miller, A. E. (2011). Climate variability and spruce beetle (*Dendroctonus rufipennis*) outbreaks in south-central and southwest Alaska. *Ecology, 92*, 1459–1470.

Siegert, N. W., McCullough, D. G., Venette, R. C., et al. (2009). Assessing the climatic potential for epizootics of the gypsy moth fungal pathogen *Entomophaga maimaiga* in the north central United States. *Canadian Journal of Forest Research, 39*, 1958–1970.

Silander, J. A., Jr., & Klepeis, D. M. (1999). The invasion ecology of Japanese barberry (*Berberis thunbergii*) in the New England landscape. *Biological Invasions, 1*, 189–201.

Silveri, A., Dunwiddie, P. W., & Michaels, H. J. (2001). Logging and edaphic factors in the invasion of an Asian woody vine in a mesic North American forest. *Biological Invasions, 3*, 379–389.

Simard, M., Romme, W. H., Griffin, J. M., & Turner, M. G. (2011). Do mountain pine beetle outbreaks change the probability of active crown fire in lodgepole pine forests? *Ecological Monographs, 81*, 3–24.

Simberloff, D. (2009). The role of propagule pressure in biological invasions. *Annual Review of Ecology, Evolution, and Systematics, 40*, 81–102.

Six, D. L., & Bentz, B. J. (2007). Temperature determines symbiont abundance in a multipartite bark beetle-fungus ectosymbiosis. *Microbial Ecology, 54*, 112–118.

Solomon, S., Plattner, G.-K., Knutti, R., & Friedlingstein, P. (2009). Irreversible climate change due to carbon dioxide emissions. *Proceedings of the National Academy of Sciences, USA, 106*, 1704–1709.

Spracklen, D. V., Mickley, L. J., Logan, J. A., et al. (2009). Impacts of climate change from 2000 to 2050 on wildfire activity and carbonaceous aerosol concentrations in the Western United States. *Journal of Geophysical Research, 114*, D20301.

Standards and Trade Development Facility. (2009). *Climate change and agriculture trade: Risks and responses*. http://www.standardsfacility.org/Climate_change.htm. Accessed 24 Jan 2012.

Stein, B. A., & Flack, S. R. (Eds.). (1996). *America's least wanted: Alien species invasions of U.S. ecosystems* (31pp). Arlington: The Nature Conservancy.

Stone, J. K., Coop, L. B., & Manter, D. K. (2008). Predicting effects of climate change on Swiss needle cast disease severity in Pacific Northwest forests. *Canadian Journal of Plant Pathology, 30*, 169–176.

Sturrock, R. N., Frankel, S. J., Brown, A. V., et al. (2011). Climate change and forest diseases. *Plant Pathology, 60*, 133–149.

Swetnam, T. W., & Betancourt, J. L. (1998). Mesoscale disturbance and ecological response to decadal climatic variability in the American Southwest. *Journal of Climate, 11*, 3128–3147.

Tchakerian, M. D., & Couslon, R. N. (2011). Ecological impacts of southern pine beetle. In R. N. Coulson, & K. D. Klepzig (Eds.), *Southern pine beetle II* (Gen. Tech. Rep. SRS-140, pp. 223–234). Asheville: U.S. Department of Agriculture, Forest Service, Southern Research Station.

Thomson, A. J., & Benton, R. (2007). A 90-year sea warming trend explains outbreak patterns of western spruce budworm on Vancouver Island. *The Forestry Chronicle, 83*, 867–869.

Thomson, A. J., Shepherd, R. F., Harris, J. W. E., & Silversides, R. H. (1984). Relating weather to outbreaks of western spruce budworm, *Choristoneura occidentalis* (Lepidoptera: Tortricidae), in British Columbia. *The Canadian Entomologist, 116*, 375–381.

Tilman, D. (1997). Community invasibility, recruitment limitation, and grassland biodiversity. *Ecology, 78*, 81–92.

Tkacz, B., Brown, H., Daniels, A., et al. (2010). *National roadmap for responding to climate change* (FS-957b). Washington, DC: U.S. Department of Agriculture, Forest Service.

Trân, J. K., Ylioja, T., Billings, R. F., et al. (2007). Impact of minimum winter temperatures on the population dynamics of *Dendroctonus frontalis*. *Ecological Applications, 17*, 882–899.

Trenberth, K. E., Fasullo, J. T., & Kiehl, J. (2009). Earth's global energy budget. *Bulletin of the American Meteorological Society, 90*, 311–323.

Troendle, C. A., MacDonald, L. H., Luce, C. H., & Larsen, I. J. (2010). Fuel management and water yield. In W. J. Elliot, I. S Miller, & L. Audin (Eds.), *Cumulative watershed effects of* fuel management in the Western United States (Gen. Tech. Rep. RMRS-GTR-231, pp. 124–148). Fort Collins: U.S. Department of Agriculture, Forest Service, Rocky Mountain Research Station. Chapter 7.

U.S. Department of Agriculture, Forest Service (USDA FS). (2010). *Major forest insect and disease conditions in the United States: 2009 update* (FS-952, 28pp). Washington, DC: U.S. Forest Service.

van Mantgem, P. J., Stephenson, N. L., Byrne, J. C., et al. (2009). Widespread increase of tree mortality rates in the Western United States. *Science, 323*, 521–524.

Veblen, T. T., Hadley, K. S., Reid, M. S., & Rebertus, A. J. (1991). The response of subalpine forests to spruce beetle outbreak in Colorado. *Ecology, 72*, 213–231.

Wagner, D. L., Defoliart, L., Doak, P., & Schneiderheinze, J. (2008). Impact of epidermal leaf mining by the aspen leaf miner (*Phyllocnistis populiella*) on the growth, physiology, and leaf longevity of quaking aspen. *Oecologia, 157*, 259–267.

Wenger, S. J., Isaak, D. J., Luce, C. H., et al. (2011). Flow regime, temperature, and biotic interactions drive differential declines of trout species under climate change. *Proceedings of the National Academy of Sciences, USA, 108*, 14175–14180.

Werner, R. A., Holsten, E. H., Matsouka, S. M., & Burnside, R. E. (2006). Spruce beetles and forest ecosystems in south-central Alaska: A review of 30 years of research. *Forest Ecology and Management, 227*, 195–206.

Westerling, A. L., Gershunov, A., Brown, T. J., et al. (2003). Climate and wildfire in the Western United States. *Bulletin of the American Meteorological Society, 84*, 595–604.

Westerling, A. L., Hidalgo, H. G., Cayan, D. R., & Swetnam, T. W. (2006). Warming and earlier spring increase western U.S. forest wildfire activity. *Science, 313*, 940–943.

Westerling, A. L., Turner, M. G., Smithwick, E. A. H., et al. (2011). Continued warming could transform Greater Yellowstone fire regimes by mid-21st century. *Proceedings of the National Academy of Sciences, USA, 108*, 13165–13170.

Whitlock, C., Shafer, S. L., & Marlon, J. (2003). The role of climate and vegetation change in shaping past and future fire regimes in the northwestern US and the implications for ecosystem management. *Forest Ecology and Management, 178*, 5–21.

Williams, D. W., & Liebhold, A. M. (1995a). Forest defoliators and climatic change: Potential changes in spatial distribution of outbreaks of western spruce budworm (Lepidoptera: Tortricidae) and gypsy moth (Lepidoptera: Lymantriidae). *Environmental Entomology, 24*, 1–9.

Williams, D. W., & Liebhold, A. M. (1995b). Herbivorous insects and global change: Potential changes in the spatial distribution of forest defoliator outbreaks. *Journal of Biogeography, 22*, 665–671.

Willis, C. G., Ruhfel, B. R., Promack, R. B., et al. (2010). Favorable climate change response explains non-native species' success in Thoreau's woods. *PLoS One, 5*, e8878.

Woods, A. J., Heppner, D., Kope, H. H., et al. (2010). Forest health and climate change: A British Columbia perspective. *The Forestry Chronicle, 86*, 412–422.

Worrall, J. J., Egeland, L., & Eager, T. (2008). Rapid mortality of *Populus tremuloides* in southwestern Colorado, USA. *Forest Ecology and Management, 255*, 686–696.

Ylioja, T., Slone, D. H., & Ayres, M. P. (2005). Mismatch between herbivore behavior and demographics contributes to scale-dependence of host susceptibility in two pine species. *Forest Science, 51*, 522–531.

Zavaleta, E. S., Hobbs, R. J., & Mooney, H. A. (2001). Viewing invasive species removal in a whole-ecosystem context. *Trends in Ecology & Evolution, 16*, 454–459.

Ziska, L. H., & Dukes, J. S. (2011). *Weed biology and climate change* (248pp). Ames: Wiley-Blackwell.

Ziska, L. H., & George, K. (2004). Rising carbon dioxide and invasive, noxious plants: Potential threats and consequences. *World Resource Review, 16*, 427–447.

Ziska, L. H., & Teasdale, J. R. (2000). Sustained growth and increased tolerance to glyphospsate observed in a C_3 perennial weed quackgrass (*Elytrigia repens*), grown at elevated carbon dioxide. *Australian Journal of Plant Physiology, 27*, 159–166.

Chapter 5
Climate Change and Forest Values

David N. Wear, Linda A. Joyce, Brett J. Butler, Cassandra Johnson Gaither, David J. Nowak, and Susan I. Stewart

5.1 Introduction

Human concerns about the effects of climate change on forests are related to the values that forests provide to human populations, that is, to the effects on ecosystem services derived from forests. Forests are valued for the services they provide such as timber products, water resources, aesthetics, and spiritual qualities. Effects of climate change on forest ecosystems will change service flows, people's perception of value, and their decisions regarding land and resource uses. Thus, social systems will need to adapt to climate changes, producing secondary and tertiary effects on the condition of forests in the United States.

D.N. Wear (✉)
Southern Research Station, U.S. Forest Service, Raleigh, NC, USA
e-mail: dwear@fs.fed.us

L.A. Joyce
Rocky Mountain Research Station, U.S. Forest Service, Fort Collins, CO, USA
e-mail: ljoyce@fs.fed.us

B.J. Butler
Northern Research Station, U.S. Forest Service, Amherst, MA, USA
e-mail: bbutler01@fs.fed.us

C. Johnson Gaither
Southern Research Station, U.S. Forest Service, Athens, GA, USA
e-mail: cjohnson09@fs.fed.us

D.J. Nowak
Northern Research Station, U.S. Forest Service, Syracuse, NY, USA
e-mail: dnowak@fs.fed.us

S.I. Stewart
Northern Research Station, U.S. Forest Service, Evanston, IL, USA
e-mail: sistewart@fs.fed.us

Forests and derivative ecosystem services are produced and consumed in (1) rural settings, where human population densities are low and forest cover dominates, (2) urban communities, where forests and trees may be scarce but their relative value, measured as direct ecosystem services, may be high, and (3) transition zones between rural and urban settings (the wildland-urban interface [WUI]), where forest settings comingle with human populations. These three settings pose different challenges for resource management and policy as related to climate change, and each defines a different set of opportunities to affect changes in forest conditions and service flows.

Here we explore how climate change interacts with forest condition, human values, policy, management, and other institutions, and the potential effects of these interactions on human well-being. We examine (1) socioeconomic context (ownership structure, how value is derived, institutional context), (2) interactions between land-use changes and climate change that affect forest ecosystems, and (3) social interactions with forests under climate change (climate factors, community structure, social vulnerability).

5.2 Socioeconomic Context: Ownership, Values, and Institutions

In the United States, forest conditions and the flow of ecosystem services from forest land reflect a long history of intensive, extensive, and passive management, as well as the influence of policy affecting public and private lands (Williams 1989). Future forest management and policy require an understanding of socioeconomic interactions with forests and how they might determine future conditions in a warmer climate. The socioeconomic context of forests in the United States includes (1) ownership patterns that define the institutional context of management, (2) forest contributions to human well-being through provision of various ecosystem services, and (3) institutional settings that affect decision making.

5.2.1 Forest Ownership Patterns

Forest owners comprise the individuals and groups most directly affected by, and most capable of mitigating, the potential impacts of climate change on forests. Working within social and biophysical constraints, owners ultimately decide the fate of the forest: whether it will remain forested and how (or if) it will be managed. Of the 304 million ha of forest land in the United States, 56 % is non-governmental owned by individuals, families, corporations, Native American tribes, and other groups (Butler 2008) (Fig. 5.1). The remaining forest land is publicly owned and controlled by federal, state, and local government agencies. Ownership patterns

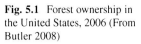

Fig. 5.1 Forest ownership in the United States, 2006 (From Butler 2008)

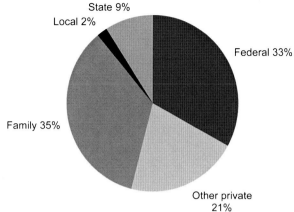

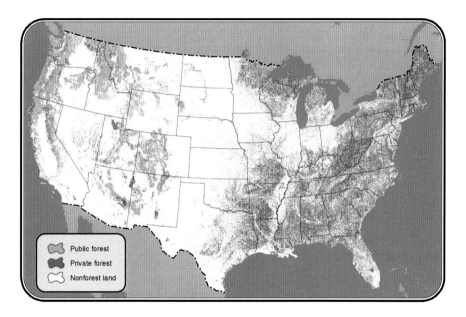

Fig. 5.2 Distribution of public and private forest ownership in the United States

differ across the country (Fig. 5.2). In the East, where 51 % of U.S. forests are located, private ownership is 81 %, and as high as 94 % in some states. Western forests are dominated by public, primarily federal, ownership (70 %), with public ownership in some states as high as 98 % (Butler 2008).

Public forests are often managed for multiple uses, although a single use may dominate at local scales (e.g., water protection, timber production, wildlife habitat, preservation of unique places). The federal government owns 33 % of all forest land, with the greatest percentage being managed by the U.S. Forest Service (59 million ha) and Bureau of Land Management (19 million ha). State agencies own 9 %

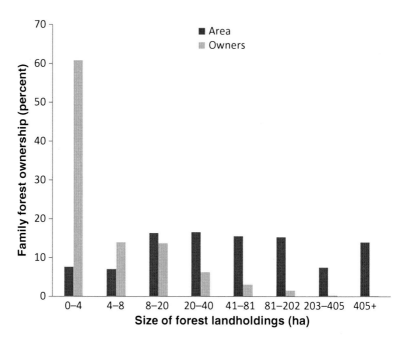

Fig. 5.3 Family forest ownerships in the United States by size of forest holdings, 2006 (From Butler 2008)

of U.S. forest land, and county and municipal governments own 2 %. Most state management objectives are similar to federal uses. Common objectives of many local land management agencies are water protection, recreation, and open space preservation.

Of the major forest ownership categories, families and individuals own a plurality (35 %, 106 million ha) of the forest land in the United States—there are over 10 million of these family ownerships. Although most (61 %) family forest ownerships are small (0.4–3.6 ha), 53 % of the land is owned by those with at least 41 ha (Fig. 5.3). Most family forest ownerships focus on amenity values, although for a significant number of ownerships, especially with larger forest holdings, timber production and land investment are important. Approximately 27 % of family forest ownerships have harvested trees. Although few family forest owners have a written management plan, participated in a cost-share program, green-certified their land, or obtained a conservation easement (Butler 2008), most family forest owners have a strong land and conservation ethic (Butler et al. 2007). In recent years, family forests have been undergoing "parcelization," the dividing of larger parcels of land into smaller ones. When parcelization is accompanied by new houses, roads, or other changes, then forest fragmentation increases, which can harm some ecosystem functions. Twenty percent of current family forest landowners are at least 75 years old, suggesting that a large proportion of forest land will soon change hands, with the potential for increased parcelization and altered ownership objectives.

Most other private forest land is controlled by corporations (56 million ha, 18 % of all forest land), including traditional forest industry and forest management companies, timber investment management organizations (TIMOs), and timberland real estate investment trusts (REITs). Other corporations also own forest land but do not have forest management as their primary business. Native American tribes, nongovernmental organizations, clubs, and unincorporated partnerships control 8.5 million ha (3 %) of the forest land. Some ownerships are explicitly for forest conservation (e.g., land trusts), and others are largely for recreation (e.g., hunting clubs) and other proposes.

From 1977 to 2007, the total area of U.S. forest land increased by 8.9 million ha (4 %) (Smith et al. 2009), with most of the increase occurring in public (especially state) ownership. However, from 1997 to 2007, private forest land decreased a net 0.4 million ha. Over the next 50 years, U.S. forest land is projected to have a net loss of 9.3 million ha (Alig et al. 2003), mostly on private lands subject to urbanization. Since the 1980s, the types of corporations that own forest land underwent a major change. Traditionally, most corporate forest land was owned by forest industry companies, which owned both forest land and facilities to process wood. Beginning in the 1980s most of these companies separated their forest holdings from other assets, and many began to divest themselves of land. This decrease was accompanied by an increase in TIMOs and REITs which often have shorter investment time horizons and no need to supply mills.

Private forest owners will need to play a critical role for successful mitigation of climate change effects, because they own over half of U.S. forest land. Therefore, an ongoing dialogue is needed about the level and types of management necessary to sustain desired ecosystem services in forests and to enhance resilience of existing forest ecosystems in the face of a warmer climate and increased disturbances. Policies that aim to mitigate the effects of climate change on forests must consider the needs, desires, and resources of all forest owners.

5.2.2 Economic Contributions of Forests

Forests deliver many values to private landowners, but also to the public at large. In rural areas, forest cover can generally be equated with forest land use, because forests are a consequence of a decision either to dedicate land to growing trees or to allow land to return to a fallow state. Rural forest ownership can provide direct returns, consumptive values, and monetary returns. Direct returns accrue through extractive activities (mainly commercial timber harvesting) or *in situ* values (e.g., hunting leases, conservation easements). Consumptive values can accrue through direct use of forests for recreation, existence value, and aesthetics. Most monetary returns are from timber production, with some additional returns from recreation leases, conservation easements, and payments for other ecosystem services.

The United States produces more timber by volume than any other nation, and timber represents a significant source of value for forest landowners. Although the volume of roundwood used for industry and fuelwood nearly doubled between 1945 and the late 1980s, production since then has declined. In 2006, the year before the latest recession, total timber production stood at about 90 % of its peak value in 1988. The economic contribution of harvested timber has also declined. Between 1997 and 2006, the value of shipments (the sum of net selling values of freight on board of all products shipped by the sector) fell from a peak of $334 billion to $309 billion (Howard et al. 2010b), a result of declines in the paper products industry.

Although per capita consumption of wood products has been trending downward since the late 1980s, population growth has continued to push total consumption upward, from 0.37 billion m^3 in 1988 to 0.57 billion m^3 in the 1990s and 2000s (Howard et al. 2010a). Between 1957 and 2006, U.S. per capita consumption of wood products averaged 2 m^3 per person, peaking in the late 1980s (2.32 m^3 per person) and falling in the 2000s (1.95 m^3 per person). This reduction was caused mostly by reduced consumption of fuelwood, indicating that consumption of wood and paper products has risen in direct proportion to population growth (Howard et al. 2010b).

The value of U.S. timber production returned to forest landowners was $22 billion in 1997, with 89 % returned to private landowners (USDA FS 2011), roughly 7 % of the value of shipments for wood products. In 2006, the value of all wild-harvested non-timber resources was $0.5 billion, and direct payments to landowners for forest-based ecosystem services (conservation easements, hunting leases, wetland mitigation banks) was $2 billion (USDA FS 2011). In rural forests, ecosystem services such as aesthetics, dispersed recreation, spiritual values, and protection of water quality rarely provide monetary compensation. Current policy initiatives (e.g., the 2008 Farm Bill [Food, Conservation, and Energy Act of 2008]) focus on providing payments, often through constructed markets, to compensate landowners for ecosystem services. It has also been proposed that municipalities compensate landowners in municipal watersheds for protecting water quality (Brauman et al. 2007; Greenwalt and McGrath 2009). In urban settings, trees remove pollution, store C, cool microclimates, and provide recreation. Trees in the WUI typically have little extractive value other than as fuel wood, and these environments are greatly influenced by human activities that occur there.

5.2.3 Policy Context of Forest Management in Response to Climate Change

Forest land ownership, nongovernmental involvement in forest values, and policy instruments (laws and taxes) that influence land management decisions all affect forest management and landowner behavior. Because land-use decisions are the dominant cause of landscape change, human institutions strongly affect future forest conditions and thus responses to climate change. Forest management in the United States derives from the interaction of private and public ownership. Private ownership affords extensive property rights but is constrained by tax and regulatory policy. Although production of nonmarket goods is a primary rationale for public ownership of forest land (e.g., Krutilla and Haigh 1978), nonmarket ecosystem services, which deliver considerable value to society, are not likely to be fully valued in private transactions. Government can direct the actions of individual landowners toward producing other nonmarket benefits by altering incentives (e.g., reforestation subsidies) and selectively restricting property rights (e.g., through forestry practice regulations). Nongovernmental organizations can directly affect changes in land use and resource allocation through outright purchases of land or purchase of development or other rights using conservation easements.

Private forest owners might be expected to alter their management plans more rapidly in response to climate change effects, altered market prices, and policy instruments that affect the provision of non-market ecosystem services. However, management objectives differ greatly between corporate and family forest owners and among subgroups of family forest owners (Butler et al. 2007). The private forest sector has shown high responsiveness to market signals in harvesting timber and investing in future timber production, especially in the southeastern United States where intensively managed pine plantations doubled between 1990 and 2010 (Wear and Prestemon 2004; Wear and Greis 2011). Private forests in the United States have less inventory (reflecting "younger" forest ages), are more accessible, and are more likely to be harvested or actively managed than public forests. On the other hand, public management can seek to maximize public welfare derived from forests and produce benefits not provided by markets, although public (especially federal) management adjusts slowly to changing conditions (Wilkensen and Anderson 1987; Yaffee 1994).

Future responses to climate change, and especially to programs designed to mitigate greenhouse gas (GHG) emissions, will probably be larger on private lands, where market signals and direct policy instruments (incentives and disincentives) are readily translated into management actions. Responses may include more harvesting (a result of new product markets such as biofuels) and altered forest management (responding to demands for forest-based C storage). Responses could also occur as increased or decreased forest area, depending on comparative returns to land from forest and agricultural uses. Therefore, larger policy impacts on GHG mitigation management activities are expected in the eastern United States, where private ownership dominates and transportation and processing infrastructure for wood products are more extensive.

In the forest sector, policy responses to climate change have focused largely on mitigation that reduces the use of fossil fuel (through bioenergy products) or amount of carbon dioxide (CO_2) in the atmosphere (through sequestration). Use of woody biomass for energy can offset fossil fuel consumption, and C storage can be increased through forest growth and conversion of trees to durable wood products. Policies outside the forest sector may have secondary effects on forests, largely through land-use changes; for example, crop price support programs may motivate conversion of some forests to agricultural land.

5.3 Rural Forests, Land-Use Change, and Climate Change

Land-use changes are influenced by choices of landowners, market forces, and economic and environmental policies. In rural environments, market forces influence shifts between agricultural uses and forest uses. Urban expansion converts forest land, with loss of some trees, and intensification of urban areas often leads to the loss of most trees. In the WUI, conversion of large forest tracts to residential areas is driven by home buyers who value the amenities of living in or near forests and are willing to pay more to do so. These changes in land cover can affect local temperature and precipitation (Fall et al. 2009) and interact with climate change to influence forest dynamics.

History demonstrates a tradeoff between agricultural and forest uses in the United States, based on shifting advantages and returns. Climate is a key driver of agricultural productivity, and along with population growth, may influence returns to agriculture and land use switching among cropland, forest, and other uses. Crop productivity is negatively related to non-autumn temperature increases, positively related to non-autumn precipitation (Mendelsohn et al. 1994), and affected by climatic variability (Mendehsohn et al. 2007). Large declines in productivity are projected for important crops in the United States, especially in the latter part of the twenty-first century (Schlenker and Roberts 2009). Assessments of climate change on forest productivity have been less definitive (see Chap. 3), because forest ecosystems are more complex and forest dynamics are not as well understood in relation to climate. Furthermore, disturbances such as wildfire and insect outbreaks can result in immediate changes, including extensive mortality and erosion (see Chap. 4).

Future rural land uses will be affected by a combination of population-driven urbanization, comparative returns to agriculture and forestry, and policies that influence the expression the first two factors. The recent Renewable Resources Planning Assessment Act (RPA) (Wear 2011; USDA FS 2012) forecasts an increase in developed uses from about 30 million ha in 1997 to 54–65 million ha in 2060, based on alternative projections of population and income. Comparative returns to agriculture and forestry could be altered directly and indirectly by climate change, and at the margin, shifts in agricultural productivity could lead to a switch between forests and crops. Shifts in comparative returns to forestry and agriculture would probably result from policy designed to encourage bioenergy production. The degree to which a bioenergy sector favors agricultural feedstocks (e.g., corn) or cellulosic feedstocks from forests will determine the comparative position of forest and agricultural land use. Federal policy to date has subsidized corn ethanol production, but the 2008 Farm Bill and some State-level policies encourage use of wood in electricity generation. In some areas of the United States, policies that mitigate climate change through bioenergy and C sequestration may influence land use and forest area more than climate change itself.

5.4 Trees and Climate in Urban Environments

Trees in developed areas may provide a disproportionately higher value of ecosystem services because of their proximity to human habitation. As the area of urban use expands, the extent and importance of urban trees will increase. Climate change will likewise have important effects on these trees, and urban trees have the potential to moderate climate in urban environments.

In 2000, urban areas occupied 24 million ha (3.1 %) of the conterminous United States and contained over 80 % of the country's population (Nowak et al. 2005), and urban/community lands occupied 41 million ha (5.3 %) (Nowak and Greenfield 2012b). Urban areas are projected to increase to 8.1 % by 2050 (Nowak et al. 2005; Nowak and Walton 2005), resulting in an increase in urban land of 39.2 million ha and a concomitant conversion of 11.8 million ha of forest to urban land (Nowak and Walton 2005).

Percentage of tree cover nationally is projected to decrease by 1.1–1.6 % between 2000 and 2060 (USDA FS 2012). Within urban areas of the conterminous United States, tree cover is declining at a rate of 0.03 % per year, equivalent to 7,900 ha or four million trees per year, or 0.27 % percent of city area per year for densely populated areas (Nowak and Greenfield 2012a). Cities developed in naturally forested regions typically have a higher percentage of tree cover than cities developed in grassland or desert areas (Nowak et al. 1996, 2001; Nowak and Greenfield 2012b).

The structural value of urban trees (cost of replacement or compensation for loss) in the United States is estimated at $2.4 trillion (Nowak et al. 2002b). Urban trees provide additional benefits, such as air pollution removal and C sequestration (Nowak et al. 2006). Thus, as climate changes, urban forests and their associated benefits will be affected even as they help to reduce CO_2 emissions. Urban trees in the United States are estimated to store 643 million Mg C ($50.5 billion value; based on a price of $20.30 per Mg C), with a gross C sequestration rate of $25.6 million Mg C year^{-1} (2.0 billion year^{-1}) (Nowak et al. 2013).

The biggest effects of climate change on urban trees and forests will likely be caused by warmer air temperature, strengthening wind patterns, extreme weather events, and higher concentrations of atmospheric CO_2. In addition, urban surfaces and activities (e.g., buildings, vegetation, emissions) influence local air temperature, precipitation, and windspeed. Urban areas often create an "urban heat island," where air temperatures are higher (1–6 °C) than in surrounding areas (US EPA 2008). In some areas in the southeastern United States, monthly rainfall rates are 28 % higher within 30–60 km downwind of cities, with a 5.6 % increase over the city (Shepherd et al. 2002; Shepherd 2005).

Potential effects of climate change on urban tree populations include changes in tree vigor and physiological condition, species composition (e.g., Iverson and Prasad 2001), insect and disease occurrence, and management and maintenance activities that mitigate tree health and species composition changes (Nowak 2000, 2010). Management activities to sustain healthy tree cover will alter C emissions (because of fossil fuel use), species composition, and urban forest attributes such as biodiversity, wildlife habitat, and human preferences and attitudes toward urban vegetation. Climate change effects on the urban forest may be accelerated or reduced depending on whether urban forests are managed for better adapted species or managed more intensively (e.g., through watering and fertilization) to reduce climate impacts.

Nowak (2000) proposed four ways in which urban forests affect climate change:

- *Removing and storing CO_2*—Trees remove CO_2 from the atmosphere and sequester C in their biomass (McKinley et al. 2011). The net C storage in a given area with a given tree composition cycles through time as the population grows and declines. When forest growth (C accumulation) is larger than decomposition, net C storage increases. Some C from previous generations can also be sequestered in soils. Management activities can enhance long-term C storage by growing large, long-lived species; minimizing use of fossil fuels to manage vegetation; strategically planting vegetation to reduce air temperature and energy use, and using urban tree biomass for energy production (Nowak et al. 2002a).
- *Emitting atmospheric chemicals through vegetation maintenance*—Urban tree management often uses large amounts of energy to maintain vegetation structure (transportation, vehicles, other equipment), and emissions from these activities need to be considered in evaluating the net effect of management.
- *Altering urban microclimates*—Trees are part of the urban structure, and they affect the urban microclimate by cooling the air through transpiration, blocking winds, shading surfaces, and helping to mitigate heat-island effects.
- *Altering energy use in buildings*—Urban trees can reduce energy use in summer through shade and reduced air temperatures, and they can either increase or decrease winter energy use (Heisler 1986), depending on tree location around buildings (e.g., providing shade, blocking winter winds).

5.5 Climate Change and the Wildland-Urban Interface

The WUI encompasses where people live in direct contact with forests and other wildlands, and where development of forested lands for residential and commercial uses has direct, ongoing effects on the forest (Radeloff et al. 2005). Key changes driven by climate change, population growth, and markets for land uses are especially concentrated in this zone. Growth in the WUI has outpaced growth outside the WUI, a trend expected to continue in coming decades, particularly in Western states (Theobald and Romme 2007; Hammer et al. 2009). Area in the WUI is expected to increase 17 % within 50 km of federal lands by 2030 (Radeloff et al. 2010). Proximity to protected areas is an attraction to home buyers because it guarantees that changes to the viewshed will be minimal (Radeloff et al. 2010; Wade and Theobald 2010). These lands have protected status in part to ensure that plant and animal species will be sustained, which makes their attractiveness for adjacent housing troubling from an ecological perspective (Gimmi et al. 2011).

People who live in the WUI are more likely to be aware of forest disturbances that might be exacerbated by climate change, such as drought, wildfire, insect outbreaks, and spread of invasive plants. Awareness and acceptance of the need to prepare for wildland fire has grown in the WUI over the past decade. Since 2002, the Firewise program (http://firewise.org) has assisted communities across the country in developing and maintaining residential areas to minimize fire risk through modification of vegetation and use of fire-resistant building materials (Cohen 2000; NFPA 2011).

Ideally, entire communities would be "fire adapted," where fire passes through and around a community without causing extensive damage. Recent studies of risk perception and response related to wildfire (Cova et al. 2009; McCaffrey 2009) conform to established psychological concepts of risk and actions to reduce risk (Slovik 1987). Climate change, like wildfire, presents many challenges for the ability of people to understand, judge, and act on new information. Specific, observable changes in forest resources, particularly in familiar and local forests, are best able to engage the attention and concern of the general public. Because people attach such great value to forests, it may be more feasible to engage landowners in adapting to climate change than for other resource sectors, especially in the case of fire, insect outbreaks, and other disturbances that have highly visible effects.

Like climate change itself, activity in the WUI affects forest structure and dynamics. Regulations such as zoning ordinances that limit housing density and neighborhood covenants governing property management are intended (in part) to protect the environment. However, multiple small disturbances can overwhelm the ability of forests to adapt by requiring a rapid transition to new conditions, a situation not typically addressed through land use and other residential policies.

Communities in or near forests could be well designed and governed under effective regulations, yet still lack the capacity needed to adapt to climate change. For example, a zoning code that specifies the number of trees retained in a subdivision without accounting for their configuration may result in a fragmented forest and degraded wildlife habitat. However, once known and understood, resource management concerns that are communicated effectively can change human behavior. For example, in Fremont County, Colorado, WUI residents actively learned from each other and engaged in managing several complex WUI resource issues (Larsen et al. 2011). Given expectations for continued WUI growth, together with the effects of climate change, such activities will be essential for maintaining the capacity of forests to adapt to climate-caused changes in the biophysical environment.

5.6 Social Interactions with Forests Under Climate Change

People and the actions they take directly alter the capacity of forests to sequester C and adapt to a changing climate. By modifying the landscape, people alter forest extent, sustainability, and capacity to meet the needs of other species. People and

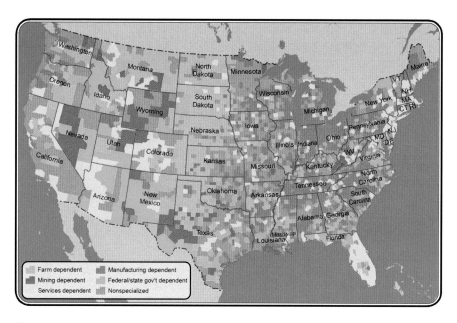

Fig. 5.4 Economic dependence in the United States (From USDA ERS 2012)

societies mediate the relationship between forests and climate change directly (by altering forests) and indirectly (by changing other biophysical conditions that in turn alter forests). Interactions between social relationships with forests and climate change will potentially alter ecosystem services on which people depend.

In natural resource-based communities, socioeconomic relationships based on commodities (e.g., timber) or amenities (e.g., recreation) may be disproportionately affected by climate-forest interactions. On tribal lands, which may be vulnerable to effects of climate change because of strong links among economies, cultures, and natural resources, the ability to adapt by moving to another location is limited because of strong ties to a specific place (often a reservation). Assessing the resilience of natural resource-based and tribal communities to climate change requires understanding ecological, social, and economic vulnerabilities.

5.6.1 Natural Resource-Based Communities

Natural resource-based communities are closely linked with their geographic setting and environmental context. People in these communities derive economic benefits from the surrounding natural resources and withstand their associated natural disturbances, such as wildfires and hurricanes. Natural resource-based communities depend to varying degrees on different resource sectors (USDA ERS 2011). Farm dependency has declined greatly; in 2000, only 20 % of nonmetropolitan counties were considered farming dependent (Dimitri et al. 2005) (Fig. 5.4). Other counties, particularly in the West, depend on federal or state government and mining. Of 368 recreation-dominated counties, 91 % are in rural areas (Lal et al. 2010).

Natural resource-based communities in or near forests often experience the consequences of natural disasters or environmental stresses sooner than do farther-removed populations (Lynn and Gerlitz 2006; Haque and Etkin 2007; Lynn et al. 2011). Although developing regions of the globe present the most glaring examples of forest-dependent communities vulnerable to climate change, these relationships also occur in developed regions of North America. Individual and community vulnerability can be affected by characteristics such as income level, race, ethnicity, health, language, literacy, and land-use patterns. Thus, the social vulnerability of natural resource-based communities can exacerbate biophysical vulnerabilities.

Specific social characteristics associated with forest-based communities can increase climate change risks (Davidson et al. 2003). For example, human capital development is typically lower with respect to educational attainment in these areas, reducing the potential for laborers to transfer skills to other occupations. Politicization of the role of deforestation in climate change can create a larger populace that is unsympathetic to labor dilemmas facing communities dependent on forestry. Uncertainty about climate change effects, coupled with the long-term planning horizon of forest management, elevates risks associated with investments in forest-based industries, leading to under-investment in communities that depend on a single sector economy. As noted in the Fourth Intergovernmental Panel on Climate Change Assessment, social vulnerability may be geographically dispersed: "There are sharp differences across regions and those in the weakest economic position are often the most vulnerable to climate change and are frequently the most susceptible to climate-related damages.... There is increasing evidence of greater vulnerability of specific groups such as the poor and elderly not only in developing but also in developed countries" (Pachauri and Resinger 2007; Solomon et al. 2007). In addition, climate change may not be perceived as a real threat by local residents or key decision makers, resulting in reluctance to devise adaptive strategies to reduce stresses.

5.6.2 Tribal Forests

American Indians and Alaska Natives rely on reservation lands and access to traditional territories outside of reservations for economic, cultural, and spiritual values. Tribes have unique rights, including treaties with the federal government that protect access to water, hunting, fishing, gathering, and cultural practices (Pevar 1992; Lynn et al. 2011). Indian reservations contain 7.2 million ha of forestland, of which 3.1 million ha are classified as timberland and 2.3 million ha as commercial timberland (Gordon et al. 2003), including conifer forest in the Pacific Northwest, dry pine forest and juniper woodland in the Southwest, mixed hardwood-conifer forest in the Lake States, and spruce forest in the southern Appalachian Mountains (Gordon et al. 2003) (Fig. 5.5).

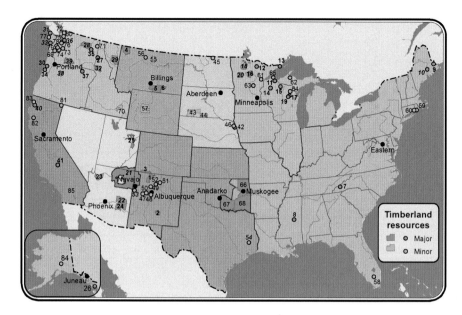

Fig. 5.5 Reservations with significant timberland resources. Numbers 1 through 41 have over 4,000 ha of commercial timberland per reservation. Numbers 41 through 83 have less area in timberland, but what they have is economically viable (Adapted from IFMAT (2003), Intertribal Timber Council, with permission). See Vose et al. (2012, p. 115) for list of reservation names

Tribal forests and woodlands provide jobs and revenue from timber production, non-timber forest products, grazing, fishing, and hunting. They also provide recreation opportunities, energy resources, places for religious ceremonies and solitude, and material for shelter, clothing, medicines, and food. Climate change effects on tribal forests will have implications for treaty rights if the ranges of culturally significant plant and animal species move outside reservation boundaries; water resources and tribal water rights may be especially affected by climate change (Karl et al. 2009; Curry et al. 2011). Current adaptation approaches on tribal lands include watershed management surrounding sacred waters, natural hazard management, and efforts to create green jobs (Lynn et al. 2011), and some tribes have begun to explore options to manage forest lands for C sequestration. The fixed location of tribal lands constrains the adaptive capacity of tribal communities with regard to climate change, especially on reservations that are small or fragmented.

5.6.3 Social Vulnerability and Climate Change

Socially vulnerable populations are often considered as marginal groups in terms of material well-being, which reduces their ability to anticipate, cope with, or recover from environmental stresses (Kelly and Adger 2000). Vulnerability is not just susceptibility or sensitivity to loss arising from hazard exposure, but also a function of sensitivity and adaptive capacity (Brooks 2003; Smit and Wandel 2006; Polsky et al. 2007). *Exposure* is the proximity to a physical hazard or stressor. *Sensitivity* is the susceptibility of humans in sociodemographic terms to physical hazard, which can also include sensitivities of the built environment. *Adaptive capacity* is any mitigation and adaptation to hazard via sociodemographic factors or other means.

Vulnerability can occur after individuals or communities have experienced an environmental stressor, such as incremental changes in climate (Watson 2001), which is why so much emphasis is placed on projecting a "future," an endpoint of vulnerability—sometimes called *outcome vulnerability*—as a consequence of increased GHGs and resultant climate change. Biophysical effects on humans and physical systems are then projected, and adaptation options can then be formulated.

Contextual vulnerability differs from outcome vulnerability in that it analyzes current vulnerabilities within the current social structure of a given place. An analysis of contextual vulnerability (e.g., economic reliance on river-based tourism) focuses on relationships among political actors (elected officials), institutions (rules for concessionaires), socioeconomic well-being (workforce education level), and culture to identify how goods and information are distributed across society. From this evidence, the analysis projects response to a future threat (e.g., whether guides will be able to maintain their concession for river rafting as in-stream flows decline). The ability of human communities to cope with environmental and societal stressors determines how well they will respond to future stressors. This approach first identifies vulnerable communities, then develops management actions that will improve adaptive capacity.

Socioeconomic vulnerability assessment (SEVA) is a promising approach for linking social and biophysical vulnerabilities to climate change. A SEVA first summarizes secondary data from the U.S. Census Bureau (U.S. Census Bureau 2007) and similar sources. Then, a SEVA (1) briefly discusses the social history of the forest and its human geography, including communities of place and communities of interest, (2) links current and expected biophysical changes to community-relevant outcomes, (3) determines stakeholder perceptions of values at risk, and (4) prioritizes threats to vulnerable communities and identifies those for which adaptation is feasible. This basic approach is flexible enough to accommodate a range of different conditions and levels of detail, depending on the goals of the assessment team.

5.7 Conclusions

Interactions between changes in biophysical environments (climate, disturbance, and ecological function) and human responses to those changes (management and policy) will determine the effects of climate change on human communities. The ultimate effects on people are measured in terms of changes in ecosystem services provided by forested landscapes, including traditional timber products and new extractive uses, rural and urban recreation, cultural resources, contributions of urban forests to human health, and protection of water quality. Climate change has been linked to bioenergy and C sequestration policy options, emphasizing the effect of climate-human interactions on forests as well as the role of forests in mitigating climate change. Any effect of climate change on forests will result in a ripple effect of policy and economic response affecting economic sectors and human communities.

The key mechanism of change in human-dominated landscapes is choice. Where private ownership dominates, choices regarding land use and resource production directly and indirectly affect changes in forest conditions and the flow of ecosystem services. The choices are directly influenced by shifts in land productivity, prices of various products, and ultimately economic returns for different land uses. Land-use shifts in rural areas under climate change could include conversion between forests and agricultural uses, depending on market conditions. Climate changes are expected to alter productivity (local scale) and prices (market scale). Land-use patterns dictate the availability of ecosystem services from forests and from trees within other land types. Both WUI and urban areas are projected to increase, often at the expense of rural forests. Anticipated climate changes, coupled with population growth, strongly increase the extent and value of urban trees in providing ecosystem services and for mitigating climate change impacts at fine scales. However, climate change also increases the challenge of keeping trees healthy in urban environments.

Collective choice, in the form of various policies, also affects land use and forest conditions relative to climate change. Policies targeting climate mitigation, especially for bioenergy production and C sequestration, directly influence forest extent and use. Implemented through markets, these policies yield secondary and tertiary effects on forest composition and structure through direct action and through resource input and product substitutions in related sectors. These and other policies (forest management regulations, land-use restrictions, property taxes) provide a context for and potentially constrain adaptive choices by private landowners.

Human communities in urban, WUI, and rural environments will experience changes to forests as a result of climate change. Those communities dependent on forests for economic or cultural values are likely to see the effects of climate change first. The potential for human communities to adapt to potential climate changes is linked to their exposure to climate change, which differs along the rural-to-urban gradient, and also to the nature of the social and institutional structures in each environment. We can prepare for or mitigate future climate stresses in these environments by ensuring that present-day human communities are resilient and therefore better able to respond to future stresses.

References

Alig, R. J., Plantinga, A. J., Ahn, S., & Kline, J. D. (2003). *Land use changes involving forestry in the United States: 1952 to 1997, with projections to 2050* (Gen. Tech. Rep. PNW-GTR-587, 92pp). Portland: U.S. Department of Agriculture, Forest Service, Pacific Northwest Research Station.

Brauman, K. A., Dailey, G. C., Duartel, T. K., & Mooney, H. A. (2007). The nature and value of ecosystem services: An overview highlighting hydrologic services. *Annual Review of Environmental Resources, 32*, 67–98.

Brooks, N. (2003). *Vulnerability, risk and adaptation: A conceptual framework* (Working Paper 38, 16pp). Norwich: University of East Anglia, School of Environmental Sciences, Tyndall Centre for Climate Change Research.

Butler, B. J. (2008). *Family forest owners of the United States, 2006* (Gen. Tech. Rep. NRS-27, 73pp). Newtown Square: U.S. Department of Agriculture, Forest Service, Northern Research Station.

Butler, B. J., Tyrrell, M., Feinberg, G., et al. (2007). Understanding and reaching family forest owners: Lessons from social marketing research. *Journal of Forestry, 105*, 348–357.

Cohen, J. D. (2000). Preventing disaster: Home ignitability in the wildland-urban interface. *Journal of Forestry, 98*, 15–21.

Cova, T. J., Drews, F. A., Siebeneck, L. K., & Musters, A. (2009). Protective actions in wildfires: Evacuate or shelter-in-place? *Natural Hazards Review, 10*, 151–162.

Curry, R., Eichman, C., Staudt, A., et al. (2011). *Facing the storm: Indian tribes, climate-induced weather extremes, and the future for Indian country* (28pp). Boulder: National Wildlife Federation, Rocky Mountain Regional Center.

Davidson, D. J., Williamson, T., & Parkins, J. (2003). Understanding climate change risk and vulnerability in Northern forest-based communities. *Canadian Journal of Forest Research, 33*, 2252–2261.

Dimitri, C., Effland, A., & Conklin, N. (2005). *The 20th-century transformation of U.S. agriculture and farm policy* (Economic Information Bulletin Number EIB3, 17pp). Washington, DC: U.S. Department of Agriculture, Economic Research Service.

Fall, S., Niyogi, D., Gluhovsky, A., et al. (2009). Impacts of land use land cover on temperature trends over the continental United States: Assessment using the North American regional reanalysis. *International Journal of Climatology, 30*, 1980–1993.

Food, Conservation, and Energy Act. (2008). *Public Law 110–234*. Washington, DC: V.S. Government.

Gimmi, U., Schmidt, S. M., Hawbaker, T. J., et al. (2011). Decreasing effectiveness of protected areas due to increasing development in the surroundings of U.S. National Park Service holdings after park establishment. *Journal of Environmental Management, 92*, 229–239.

Gordon, J., Berry, J., Ferrucci, M., et al. (2003). *An assessment of Indian forests and forest management in the United States: By the Second Indian Forest Management Assessment Team for the Intertribal Timber Council* (134pp). Portland: Clear Water Printing.

Greenwalt, T., & McGrath, D. (2009). Protecting the city's water: Designing a payment for ecosystem services program. *Natural Resources and Environment, 24*, 9–13.

Hammer, R. B., Stewart, S. I., & Radeloff, V. C. (2009). Demographic trends, the wildland-urban interface, and wildfire management. *Society and Natural Resources, 22*, 777–782.

Haque, C. E., & Etkin, D. (2007). People and community as constituent parts of hazards: The significance of societal dimensions in hazards analysis. *Natural Hazards, 41*, 271–282.

Heisler, G. M. (1986). Energy savings with trees. *Journal of Arboriculture, 12*, 113–125.

Howard, J. L., Westby, R., & Skog, K. E. (2010a). *Criterion 6 indicator 25: Value and volume of wood and wood products production, including primary and secondary processing* (Res. Note FPL-RN-0316, 14pp). Madison: U.S. Department of Agriculture, Forest Service, Forest Products Laboratory.

Howard, J. L., Westby, R., & Skog, K. E. (2010b). *Criterion 6 indicator 28: Total and per capita consumption of wood and wood products in roundwood equivalents* (Res. Note FPL-RN-0317, 20pp). Madison: U.S. Department of Agriculture, Forest Service, Forest Products Laboratory.

Indian Forest Management Assessment Team [IFMAT]. (2003). *An assessment of Indian forests and forest management in the United States* (134pp). Portland: Intertribal Timber Council.

Iverson, L. R., & Prasad, A. M. (2001). Potential changes in tree species richness and forest community types following climate change. *Ecosystems, 4*, 186–199.

Karl, T. R., Melillo, J. M., Peterson, T. C. (Eds.). (2009). *Global climate change impacts in the United States* (196pp). Cambridge: Cambridge University Press.

Kelly, P. M., & Adger, W. N. (2000). Theory and practice in assessing vulnerability to climate change and facilitating adaptation. *Climatic Change, 47*, 325–352.

Krutilla, J., & Haigh, J. A. (1978). An integrated approach to National Forest management. *Environmental Law, 8*, 373–415.

Lal, P., Alavalapati, J., & Mercer, D. E. (2010). Socioeconomic impacts of climate change on rural communities in the United States. In R. J. Alig (Tech. Coord.), *Effects of climate change on natural resources and communities: A compendium of briefing papers* (Gen. Tech. Rep. PNW-GTR-837, pp. 73–118). Portland: U.S. Department of Agriculture, Forest Service, Pacific Northwest Research Station. Chapter 3.

Larsen, S. C., Foulkes, M., Sorenson, C. J., & Thomas, A. (2011). Environmental learning and the social construction of an exurban landscape in Fremont County, Colorado. *Geoforum, 42*, 83–93.

Lynn, K., & Gerlitz W. (2006). Mapping the relationship between wildfire and poverty. In P. L. Andrews & B.W. Butler (Comps.), *Fuels management—How to measure success* (Proceedings RMRS-P-41, pp. 401–415). Fort Collins: U.S. Department of Agriculture, Forest Service, Rocky Mountain Research Station.

Lynn, K., MacKendrick, K., & Donoghue, E. M. (2011). *Social vulnerability and climate change: Synthesis of literature* (Gen. Tech. Rep. PNW-GTR-838, 70pp). Portland: U.S. Department of Agriculture, Forest Service, Pacific Northwest Research Station.

McCaffrey, S. (2009). Crucial factors influencing public acceptance of fuels treatments. *Fire Management Today, 69*, 9–12.

McKinley, D. C., Ryan, M. G., Birdsey, R. A., et al. (2011). A synthesis of current knowledge on forests and carbon storage in the United States. *Ecological Applications, 21*, 1902–1924.

Mendehsohn, R., Basist, A., Dinar, A., et al. (2007). What explains agricultural performance: Climate normals or climate variance? *Climatic Change, 81*, 85–99.

Mendelsohn, R., Nordhaus, W. D., & Shaw, D. (1994). The impact of global warming on agriculture: A Ricardian analysis. *American Economic Review, 84*, 753–771.

National Fire Protection Association [NFPA]. (2011). *Firewise at NFPA: A brief history*. http://www.firewise.org/About/History.aspx. Accessed 10 Jan 2012.

Nowak, D. J. (2000). The interactions between urban forests and global climate change. In K. K. Abdollahi, Z. H. Ning, & A. Appeaning (Eds.), *Global climate change and the urban forest* (pp. 31–44). Baton Rouge: GCRCC and Franklin Press.

Nowak, D. J. (2010). Urban biodiversity and climate change. In N. Muller, P. Werner, & J. G. Kelcey (Eds.), *Urban biodiversity and design* (pp. 101–117). Hoboken: Wiley-Blackwell Publishing.

Nowak, D. J., & Greenfield, E. J. (2012a). Tree and impervious cover change in U.S. cities. *Urban Forestry and Urban Greening, 11*, 21–30.

Nowak, D. J., & Greenfield, E. J. (2012b). Tree and impervious cover in the United States. *Landscape and Urban Planning, 107*, 21–30.

Nowak, D. J., Greenfield, E. J., Hoehn, R., & La Point, E. (2013). Carbon storage and sequestration by trees in urban and community areas of the United States. *Environmental Pollution, 178*, 229–236.

Nowak, D. J., & Walton, T. (2005). Projected urban growth and its estimated impact on the U.S. forest resource (2000–2050). *Journal of Forestry, 103*, 383–389.

Nowak, D. J., Rowntree, R. A., McPherson, E. G., et al. (1996). Measuring and analyzing urban tree cover. *Landscape and Urban Planning, 36*, 49–57.

Nowak, D. J., Noble, M. H., Sisinni, S. M., & Dwyer, J. F. (2001). Assessing the US urban forest resource. *Journal of Forestry, 99*, 37–42.

Nowak, D. J., Crane, D. E., & Dwyer, J. F. (2002a). Compensatory value of urban trees in the United States. *Journal of Arboriculture, 28*, 194–199.

Nowak, D. J., Stevens, J. C., Sisinni, S. M., & Luley, C. J. (2002b). Effects of urban tree management and species selection on atmospheric carbon dioxide. *Journal of Arboriculture, 28*, 113–122.

Nowak, D. J., Walton, J. T., Dwyer, J. F., et al. (2005). The increasing influence of urban environments on U.S. forest management. *Journal of Forestry, 103*, 377–382.

Nowak, D. J., Crane, D. E., & Stevens, J. C. (2006). Air pollution removal by urban trees and shrubs in the United States. *Urban Forestry and Urban Greening, 4*, 115–123.

Pachauri, R. K., & Reisinger, A. (Eds.). 2007. *Climate change 2007: Synthesis report* (104pp). Cambridge: Cambridge University Press.

Pevar, S. L. (1992). *The rights of Indians and tribes: The basic ACLU guide to Indian and tribal rights* (2nd ed., 315pp). Carbondale: Southern Illinois University Press.

Polsky, C., Neff, R., & Yarnal, B. (2007). Building comparable global change vulnerability assessments: The vulnerability coping diagram. *Global Environmental Change, 17*, 472–485.

Radeloff, V. C., Hammer, R. B., & Stewart, S. I. (2005). The wildland urban interface in the United States. *Ecological Applications, 15*, 799–805.

Radeloff, V. C., Stewart, S. I., Hawbaker, T. J., et al. (2010). Housing growth in and near United States' protected areas limits their conservation value. *Proceedings of the National Academy of Sciences, USA, 107*, 940–945.

Schlenker, W., & Roberts, M. J. (2009). Nonlinear temperature effects indicate sever damages to US crop yields under climate change. *Proceedings of the National Academy of Sciences, USA, 106*, 15594–15598.

Shepherd, J. M. (2005). A review of current investigations of urban-induced rainfall and recommendations for the future. *Earth Interactions, 9*, 1–6.

Shepherd, J. M., Pierce, H., & Negri, A. J. (2002). Rainfall modification by major urban areas: Observations from spaceborne rain radar on the TRMM satellite. *Journal of Applied Meteorology, 41*, 689–701.

Slovik, P. (1987). Perception of risk. *Science, 236*, 280–285.

Smit, B., & Wandel, J. (2006). Adaptation, adaptive capacity and vulnerability. *Global Environmental Change, 16*, 282–292.

Smith, W. B., Miles, P. D., Perry, C. H., & Pugh, S. A. (2009). *Forest resources of the United States, 2007* (Gen. Tech. Rep. WO-78, 336pp). Washington, DC: U.S. Department of Agriculture, Forest Service.

Solomon, S., Qin, D., Manning, M., et al. (2007). *Climate change 2007: The physical science basis—Contribution of Working Group I to the fourth assessment report of the Intergovernmental Panel on Climate Change* (996pp). Cambridge: Cambridge University Press.

Theobald, D. M., & Romme, W. H. (2007). Expansion of the US wildland–urban interface. *Landscape and Urban Planning., 83*, 340–354.

U.S. Census Bureau. (2007). *U.S. census data*. www.census.gov. Accessed Jan 2007.

U.S. Department of Agriculture, Economic Research Service [USDA ERS]. (2011). *Atlas of rural and small-town America*. http://www.ers.usda.gov/data/ruralatlas

U.S. Department of Agriculture, Forest Service [USDA FS]. (2011). *National report on sustainable forests—2010* (FS-979). Washington, DC.

U.S. Department of Agriculture, Forest Service [USDA FS]. (2012). *Future of Americas forest and rangelands: Forest Service 2010 Resources Planning Act assessment* (Gen. Tech. Rep. WO-87, 198pp). Washington, DC.

U.S. Environmental Protection Agency. (2008). *Heat island effect*. http://www.epa.gov/hiri/. Accessed July 2008.

Vose, J. M., Peterson, D. L., & Patel-Weynand, T. (2012). Effects of climatic variability and change on forest ecosystems: a comprehensive science synthesis for the U.S. forest sector. (Gen. Tech. Rep. PNW-GTR-870, 265 pp). Portland, OR: U.S. Department of Agriculture, Forest Service, Pacific Northwest Research Station.

Wade, A. A., & Theobald, D. M. (2010). Residential development encroachment on U.S. protected areas. *Conservation Biology, 24*, 151–161.

Watson, R. T. (Ed.). (2001). *Climate change 2001: Synthesis report* (398pp). Cambridge/New York: Cambridge University Press.

Wear, D. N. (2011). *Forecasts of county-level land uses under three future scenarios: A technical document supporting the Forest Service 2010 RPA assessment* (Gen. Tech. Rep. SRS-141, 41pp). Asheville: U.S. Department of Agriculture, Forest Service, Southern Research Station.

Wear, D. N., & Greis, J. G. (Eds.). (2011). *The Southern forest futures project: Technical report.* http://www.srs.fs.usda.gov/futures/reports/draft/pdf/technicalreport.pdf. Accessed 27 June 2012.

Wear, D. N., & Prestemon, J. (2004). Timber market research, private forests, and policy rhetoric. In M. Rauschner, & K. Johnsen (Eds.), *Southern forest science: Past, present and future* (Gen. Tech. Rep. SRS-75, pp. 289–300). Asheville: U.S. Department of Agriculture, Forest Service, Southern Research Station.

Williams, M. (1989). *Americans and their forests: A historical geography* (599pp). New York: Cambridge University Press.

Willkinson, C. F., & Anderson, H. M. (1987). *Land and resource planning in the national forests* (396pp). Washington, DC: The Island Press.

Yaffee, S. L. (1994). *The wisdom of the spotted owl: policy lessons for a new century* (430pp). Washington, DC: Island Press.

Chapter 6
Regional Highlights of Climate Change

David L. Peterson, Jane M. Wolken, Teresa N. Hollingsworth,
Christian P. Giardina, Jeremy S. Littell, Linda A. Joyce,
Christopher W. Swanston, Stephen D. Handler, Lindsey E. Rustad,
and Steven G. McNulty

6.1 Introduction

The U.S. Global Change Research Program provides a national framework for research and communication of scientific information and periodically develops syntheses through the National Climate Assessment (NCA) (http://globalchange. gov). Although the assessment is comprised of national syntheses for many

D.L. Peterson (✉)
Pacific Northwest Research Station, U.S. Forest Service, Seattle, WA, USA
e-mail: peterson@fs.fed.us

J.M. Wolken
School of Natural Resources and Agricultural Sciences, University of Alaska,
Fairbanks, AK, USA
e-mail: jmwolken@alaska.edu

T.N. Hollingsworth
Pacific Northwest Research Station, U.S. Forest Service, Fairbanks, AK, USA
e-mail: thollingsworth@fs.fed.us

C.P. Giardina
Pacific Southwest Research Station, U.S. Forest Service, Hilo, HI, USA
e-mail: cgiardina@fs.fed.us

J.S. Littell
Alaska Climate Science Center, U.S. Geological Survey, Anchorage, AK, USA
e-mail: jlittell@usgs.gov

L.A. Joyce
Rocky Mountain Research Station, U.S. Forest Service, Fort Collins, CO, USA
e-mail: ljoyce@fs.fed.us

C.W. Swanston
Northern Research Station, U.S. Forest Service, Houghton, MI, USA
e-mail: cswanston@fs.fed.us

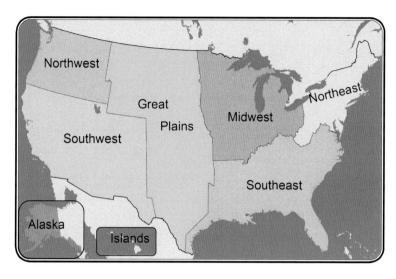

Fig. 6.1 Regions of the United States as defined by the U.S. Global Change Research Program National Climate Assessment

different sectors (forests, agriculture, water resources, transportation, etc.), it also summarizes climate change effects in eight regions of the United States, each of which is represented by a separate regional assessment. Most of the information in this book is derived from a report for the forest sector of the NCA. Although the report and this book contain many specific examples, here we summarize and highlight some of the most important climate change issues facing each region of the United States, exploring a diversity of issues beyond the information contained in the preceding chapters.

The geographic domains of the regions as defined by the USGCRP are shown in Fig. 6.1. These regions may differ from other physical, biological, or political definitions of U.S. regions, but are used here for consistency with other components of the NCA. Each section provides different perspectives and a variable mix of biophysical and social issues. Despite the diversity of themes presented here, we note that climatic extremes, ecological disturbance, and their interactions are expected to have the biggest effects on ecosystems and social systems in most regions in the coming decades.

S.D. Handler
Northern Research Station, U.S. Forest Service, Houghton, MI, USA
e-mail: sdhandler@fs.fed.us

L.E. Rustad
Northern Research Station, U.S. Forest Service, Durham, NH, USA
e-mail: lrustad@fs.fed.us

S.G. McNulty
Southern Research Station, U.S. Forest Service, Raleigh, NC, USA
e-mail: smcnulty@fs.fed.us

6.2 Alaska

J.M. Wolken (✉)
School of Natural Resources and Agricultural Sciences, University of Alaska,
Fairbanks, AK, USA
e-mail: jmwolken@alaska.edu

T.N. Hollingsworth
Pacific Northwest Research Station, U.S. Forest Service, Fairbanks, AK, USA
e-mail: thollingsworth@fs.fed.us

Alaskan forests cover one-third of the state's 52 million ha of land (Parson et al. 2001), and are regionally and globally significant. Ninety percent of Alaskan forests are classified as boreal, representing 4 % of the world's boreal forests, and are located throughout interior and south-central Alaska. The remaining 10 % of Alaskan forests are classified as coastal-temperate, representing 19 % of the world's coastal-temperate forests (National Synthesis Assessment Team 2003), and are located in southeastern Alaska. Regional changes in disturbance regimes of Alaskan forests (Wolken et al. 2011) directly affect the global climate system through greenhouse gas (GHG) emissions (Tan et al. 2007) and altered surface energy budgets (Chapin et al. 2000; Randerson et al. 2006). Climate-related changes in Alaskan forests also have regional societal consequences, because some forests are in proximity to communities (both urban and rural) and provide a diversity of ecosystem services (Reid et al. 2005; Wolken et al. 2011).

In interior Alaska, the most important biophysical factors responding to changes in climate are permafrost thaw and changes in fire regime. The region is characterized by discontinuous permafrost, defined as ground (soil or rock) that remains at or below 0 °C for at least 2 years (Harris et al. 1988). Thawing permafrost may substantially alter surface hydrology, resulting in poorly drained wetlands and thaw lakes (Smith et al. 2005) or well-drained ecosystems on substrates with better drainage. Permafrost thaw may occur directly as a result of changes in regional and global climate, but it is particularly significant following disturbance to the organic soil layer by wildfire (Fig. 6.2). As permafrost thaws, large pools of stored carbon (C) in frozen ground are susceptible to increased decomposition, which will have not only regional effects on gross primary productivity (Vogel et al. 2009) and species composition (Schuur et al. 2007) but also feedbacks to the global C system (Schuur et al. 2008).

Recent changes in the fire regime in interior Alaska are linked to climate. The annual area burned in the Interior has doubled in the last decade compared to any decade since 1970, with three of the largest wildfire years on record also occurring during this time (Kasischke et al. 2010). Black spruce (*Picea mariana* [Mill.] Britton, Sterns & Poggenb.) forests, the dominant forest type in the Interior, historically burned in low-severity, stand-replacing fires every 70–130 years (Johnstone et al. 2010a). However, postfire succession of black spruce forests has recently shifted toward deciduous-dominated forests with the increase in wildfire

Fig. 6.2 In 2004, Alaska's largest wildfire season on record, the Boundary Fire, burned 217,000 ha of forest in interior Alaska (Photo by State of Alaska, Division of Forestry)

severity (Johnstone and Kasischke 2005; Kasischke and Johnstone 2005; Johnstone and Chapin 2006) (Fig. 6.2) and the reduction in fire-return interval (Johnstone et al. 2010a, b; Bernhardt et al. 2011) (see Chap. 4). With continued warming, changes in the fire regime will increase the risk to life and property for Interior Alaskan residents (Chapin et al. 2008).

South-central Alaska may be particularly sensitive to climate changes because of its confluence of human population growth and changing disturbance regimes (e.g., insects, wildfire, invasive species). Warmer temperatures have contributed to recent spruce beetle (*Dendroctonus rufipennis* Kirby) outbreaks in this region by reducing the beetle life cycle from 2 years to 1 year (Berg et al. 2006; Werner et al. 2006). Higher fuel loads resulting from beetle-caused tree mortality are expected to increase the frequency and severity of wildfires (Berg et al. 2006), which raises societal concerns of increased risks to life and property (Flint 2006). Most goods are shipped to Alaska via ports in south-central Alaska, so invasive plant species will probably become an increasingly important risk factor. Several invasive plant species in Alaska have already spread into burned areas (e.g., Siberian peashrub [*Caragana arborescens* Lam.], narrowleaf hawksbeard [*Crepis tectorum* L.], and white sweetclover [*Melilotus alba* Medik.]) (Lapina and Carlson 2004; Cortés-Burns et al. 2008).

Changes in surface hydrology in south-central Alaska have also been linked to warmer temperatures. In the Kenai lowlands in south-central Alaska, many water bodies have shrunk in response to warming since the 1950s and have subsequently been invaded by woody vegetation (Klein et al. 2005). Recently, the rate of woody invasion has accelerated as a result of a 56 % decline in water balance since 1968 (Berg et al. 2009). As a result of these combined effects of wetland drying and vegetation succession, wetlands are becoming weak C sources rather than strong C sinks, which has important consequences for the global climate system.

In southeastern Alaska, climatic warming has affected forest ecosystems primarily through effects on precipitation. Historically, this region has average winter temperatures close to 0 °C and long growing seasons, so even moderate warming could increase rain and reduce snow. Many glaciers extending from Glacier Bay and the Juneau ice field have receded since 1750, with observed reductions in snow (Motyka et al. 2002; Larsen et al. 2005). Continued warming and corresponding reductions in snow precipitation will influence the hydrologic cycle and thus alter fish and mammal habitat, organic matter decomposition, and the C cycle.

For the past 100 years, the culturally and economically important Alaska cedar (*Callitropsis nootkatensis* [D. Don] Oerst. ex. D.P. Little), also known as yellow-cedar, has been dying throughout southeastern Alaska (Hennon et al. 2006). The onset of this decline in 1880 (Hennon et al. 1990) is attributed to warmer winters and reduced snow, combined with early spring freezing events (Beier et al. 2008). The decline in Alaska cedar also has societal consequences because it is the highest valued commercial timber species exported from the region (Robertson and Brooks 2001). Native Alaskans also value this tree for ceremonial carvings; subsistence uses include fuel, clothing, baskets, bows, tea, and medicine (Schroeder and Kookesh 1990; Pojar and MacKinnon 1994). If cedar decline continues, it will alter the structure and function of forest ecosystems, as well as the lifeways of people in this region.

6.3 Hawaii and the U.S.-Affiliated Pacific Islands

C.P. Giardina (✉)
Pacific Southwest Research Station, U.S. Forest Service, Hilo, HI, USA
e-mail: cgiardina@fs.fed.us

Hawaii and the U.S.-affiliated Pacific islands, including Guam, American Samoa, Commonwealth of Northern Marianas Islands, Federated States of Micronesia, Republic of Palau, and the Marshall Islands, contain a high diversity of flora, fauna, ecosystems, geographies, and cultures, with climates ranging from lowland tropical to alpine desert. Forest ecosystems range from equatorial mangrove swamps to subalpine dry forests on high islands, with most other forest life zones between. As a result, associated climate change effects and potential management strategies vary across the region (Mimura et al. 2007). The vulnerability of Pacific islands is caused by the (1) fast rate at which climate change is occurring; (2) diversity of climate-related threats and drivers of change (sea level rise, precipitation changes, invasive species); (3) low financial, technological, and human resource capacities to adapt to or mitigate projected effects; (4) pressing economic concerns affecting island communities; and (5) uncertainty about the relevance of large-scale projections for local scales. However, island societies may be somewhat resilient to climate change, because cultures are based on traditional knowledge, tools, and institutions that have allowed small island communities to persist during historical periods of biosocial change. Resilience is also provided by strong, locally based land and shore ownerships, subsistence economies, opportunities for human migration, and tight linkages among decision makers, state-level managers, and land owners (Barnett 2001; Mimura et al. 2007).

The distribution and persistence of most forest species are largely determined by temperature and precipitation, and coastal forests are also affected by sea level. Based on known historical climate-vegetation relationships, many forests are expected to experience significant changes in distribution and abundance by the end of the twenty-first century. Over the past 30 years, air temperature for mid-elevation ecosystems in Hawaii increased by 0.3 °C per decade, exceeding the global average rate (Giambelluca et al. 2008a). Stream flow decreased by 10 % during the period 1973–2002 compared to 1913–1972 (Oki 2004), which is similar to what is suggested by simulation modeling for a warmer climate (Safeeq and Fares 2011). Preliminary climatic downscaling for the Hawaiian Islands projects that continued warming and drying will be coupled with more intense rain events separated by more dry days (Chu and Chen 2005; Chu et al. 2010; Norton et al. 2011).

The direct effects of climate change on forests will be variable and strongly dependent on interactions with other disturbances, especially novel fire regimes that are expanding into new areas because of invasion by fire-prone exotic grass and shrub species (Fig. 6.3), such as fountain grass (*Cenchrus setaceus* [Forssk.] Morrone) and common gorse (*Ulex europaeus* L.) in Hawaii and guinea grass (*Urochloa maxima* [Jacq.] R.D. Webster) across the region (D'Antonio and Vitousek 1992). These invasions have the potential to alter or even eliminate native forests through conversion of forested systems to open, exotic-dominated grass and shrub lands. In wet forests, invasive plants can alter hydrologic processes by increasing water use by vegetation (Cavaleri and Sack 2010), and these effects may be more severe under warmer or drier conditions (Giambelluca et al. 2008b). Because invasive species have invaded most native-dominated ecosystems (Asner et al. 2005, 2008), anticipated direct (higher evapotranspiration) and indirect (increased competitive advantage of high water use plants) effects of climate change will modify stream flows and populations of stream organisms. Higher temperature will facilitate expansion of pathogens into cooler, high-elevation areas and potentially reduce native bird populations of Hawaii (Benning et al. 2002).

Most forests have at least some stimulatory effects from carbon dioxide CO_2 (Norby et al. 2005), especially in younger, faster-growing species. Therefore,

Fig. 6.3 In Hawaii's high-elevation forests (shown here) and in forests across the Pacific, projected warming and drying will increase invasive plants such as fire-prone grasses, resulting in novel fire regimes and conversion of native forests to exotic grasslands. For areas already affected in this way, climate change will increase the frequency and in some cases intensity of wildfire (Photo by Christian Giardina, U.S. Forest Service)

the effects of climate on fire regimes and stream flow described above may be accentuated by rising CO_2 through increased fuel accumulation and increased competitiveness of invasive species; higher water use across the landscape may be partially offset by higher water use efficiency in some species. For strand, mangrove, and other coastal forests, anticipated sea level rise for the region (about 2 mm year^{-1}) (Mimura et al. 2007) will have moderate (initial or enhanced inundation with expansion to higher elevation) to very large (extirpation of forest species in the absence of upland refugia) effects on the distribution and persistence of these systems. Enhanced storm activity and intensity in the region during some large-scale climatic events (e.g., El Niño Southern Oscillation) will increase the effects of storm surges on these coastal systems and increase salt water intrusions into the freshwater lens that human and natural systems require for existence (Mimura et al. 2007). A combination of sea level rise and increased frequency and severity of storm surges could result in extensive loss of forest habitat in low-lying islands.

Climate change effects on island ecosystems (Table 6.1) will extend across federal, state, tribal, and private lands, the most vulnerable being coastal systems and human communities. Sea level rise, apparent trajectories for storm intensity and frequency in the region, and warming and drying trends (for Hawaii) are based on robust measurements that suggest high confidence in projected ecological changes. Vulnerabilities and risks are most relevant in coastal zone forests, but all forests

Table 6.1 Potential climate change related risks, and confidence in projections (From Mimura et al. 2007)

Risk	Confidence level
Small islands have characteristics that make them especially vulnerable to the effects of climate change, sea level rise, and extreme events	Very high
Sea level rise is expected to exacerbate inundation, storm surges, erosion, and other coastal hazards, thus threatening infrastructure, settlements, and facilities that support the livelihood of island communities	Very high
Strong evidence exists that under most climate change scenarios, water resources in small islands will be seriously compromised	Very high
On some islands, especially those at higher latitudes, warming has already led to the replacement of some local plant species	High
It is very likely that subsistence and commercial agriculture on small islands will be adversely affected by climate change	High
Changes in tropical cyclone tracks are closely associated with the El Niño Southern Oscillation, so warming will increase the risk of more persistent and severe tropical cyclones	Moderate

of the region are at greater risk of degradation from secondary drivers of change, especially fire, invasive species, insects, and pathogens.

Island systems of the Pacific are home to some of the most intact traditional cultures on earth and communities that generally are strongly linked to forest resources. Sea level rise, increased storm frequency and intensity, and more severe droughts will reduce the habitability of atolls, representing a major potential impact in Pacific island countries (Barnett and Adger 2003). For low-lying Islands of the Pacific, enhanced storm activity and severity and sea level rise will cause the relocation of entire communities and even nations; the first climate refugees have already had to relocate from homelands in the region (Mimura et al. 2007). For high islands, warming and drying in combination with expanded cover of invasive species, and in some cases increased fire frequency and severity, will alter the hydrological function of forested watersheds, with cascading effects on groundwater recharge as well as downstream agriculture, urban development, and tourism (Mimura et al. 2007).

Few options are available for managing climate change effects in Pacific island ecosystems. For some very low-lying islands and island systems, such as the Marshall Islands where much of the land mass is below anticipated future sea levels, climate change will reduce fresh water supply and community viability. When fresh water becomes contaminated with salt water, the options for persisting in a location are logistically challenging and often unsustainable. For higher islands, adaptation practices include shoreline stabilization through tree planting, reduced tree harvest, facilitated upward or inward migration of forest species, and shoreline development planning (Mimura 1999). Because many Pacific island lands are owned and managed traditionally, adaptation and mitigation can be enhanced at the community level through education and outreach focused on coastal management and protection, mitigation of sea level rise, forest watershed protection, and restoration actions.

Because Hawaii has significant topographic relief, as well as moderately sophisticated management infrastructure, anticipatory planning and facilitation of inward species migration are already being practiced in some coastal wetlands. The spread of invasive species can be slowed by multifaceted management strategies (biocontrol, physical and chemical control) and restoration of areas with fire-prone invasives (green break planting, native species planting, physical and chemical control of weed species). To this end, management prescriptions for simultaneously addressing conservation objectives and climate change effects are being addressed by the Hawaii Department of Land and Natural Resources Watershed Initiative, U.S. Fish and Wildlife Service (USFWS) Pacific Island Climate Change Cooperative, Hawaii Restoration and Conservation Initiative, and Hawaii Conservation Alliance Effective Conservation Program, as well as individual climate change management plans (e.g., USFWS Hakalau Forest National Wildlife Refuge Comprehensive Conservation Plan). Because minimal scientific information is available for the U.S.-affiliated Pacific islands, research is needed to identify thresholds beyond which social-ecological systems in atolls will be permanently compromised, and the contributions of resource management, behavior, and biophysical factors to pushing systems across thresholds (Barnett and Adger 2003).

6.4 Northwest

J.S. Littell (✉)
Alaska Climate Science Center, U.S. Geological Survey, Anchorage, AK, USA
e-mail: jlittell@usgs.gov

The state of knowledge about climatic effects on forests of the Northwest region was recently summarized in a peer-reviewed assessment of these effects in Washington (Littell et al. 2009, 2010) and a white paper on climatic effects on Oregon vegetation (Shafer et al. 2010). Recent modeling studies provide additional scenarios for effects of climate change on wildfire, insects, and dynamic vegetation in the Northwest. This summary describes evidence for such effects on climate-sensitive forest species and vegetation distribution, fire, insect outbreaks, and tree growth.

Based on projections of direct effects of climate change on the distribution of Northwest tree species and forest biomes, widespread changes in equilibrium vegetation are expected. Statistical models of tree species-climate relationships show that each tree species has a unique relationship with limiting climatic factors (McKenzie et al. 2003; Rehfeldt et al. 2006, 2008; McKenney et al. 2011). These relationships have been used to project future climate suitability for species in western North America (Rehfeldt et al. 2006, 2009; McKenney et al. 2007, 2011) and in Washington in particular (e.g., Littell et al. 2010 after Rehfeldt et al. 2006). Climate is projected to become unfavorable for Douglas-fir (*Pseudotsuga menziesii* [Mirb.] Franco) over 32 % of its current range in Washington, and up to 85 % of the range of some pine species may be outside the current climatically suitable range (Rehfeldt et al. 2006; Littell et al. 2010). Based on preliminary projections from the global climate model (GCM) CCSM2 and the process model 3PG, Coops and Waring (2010) projected that the range of lodgepole pine (*Pinus contorta* var. *latifolia* Engelm. ex S. Watson) will decrease in the Northwest. Using similar methods, Coops and Waring (2011) projected a decline in current climatically suitable area for 15 tree species in the Northwest by the 2080s; 5 of these species would lose less than 20 % of this range, and the range of the other 10 species would decline up to 70 %.

Various modeling studies project significant changes in species distribution in the Northwest, but with considerable variation within and between studies. McKenney et al. (2011) summarized responses of tree species to climate change across western North America for three emission scenarios. Projected changes in suitable climates for Northwest tree species ranged from near balanced (−5 to +10) to greatly altered species distribution at the subregional scale (−21 to −38 species), depending on the emissions scenario. Modeling results by Shafer et al. (2010) indicate either relatively little change over the twenty-first century under a moderate warming, wetter climate (CSIRO Mk3, B1), or, in western Oregon, a nearly complete conversion from maritime to evergreen needleleaf forest and subtropical mixed forest under a warmer, drier climate (HadCM3, A2). Lenihan et al. (2008) concluded that shrublands would be converted to woodlands, and woodlands to forest in response to elevated CO_2, a trend that would be facilitated by fire suppression.

Potential changes in fire regimes and area burned have major implications for ecosystem function, resource values in the wildland-urban interface (WUI), and expenditures and policy for fire suppression and fuels management. The projected effects of climate change on fire in the Northwest generally suggest increases in both fire area burned and biomass consumed in forests (McKenzie et al. 2004; Littell et al. 2009, 2010). Littell et al. (2010) used statistical climate-fire models to project future area burned for the combined area of Idaho, Montana, Oregon, and Washington. Median regional area burned per year is projected to increase from the current 0.2 million ha, to 0.3 million ha in the 2020s, 0.5 million ha in the 2040s, and 0.8 million ha in the 2080s. Furthermore, the area burned compared to the period 1980 through 2006 is expected to increase, on average, by a factor of 3.8 in forested ecosystems (western and eastern Cascades, Okanogan Highlands, Blue Mountains). Rogers et al. (2011) used the MC1 dynamic vegetation model to project fire area burned, given climate and dynamic vegetation under three GCMs. Compared to 1971–2000, large increases are predicted by 2100 in both area burned (76–310 %), and burn severities (29–41 %).

Tree vigor and insect populations are both affected by temperature: host trees can be more vulnerable because of water deficit, and bark beetle outbreaks are correlated with high temperature (Powell and Logan 2005) and low precipitation (Berg et al. 2006). Littell et al. (2010) projected relationships between climate (vapor pressure deficit) and mountain pine beetle (*Dendroctonus ponderosae* Hopkins) (MPB) attack in the late twenty-first century. They also projected potential changes in MPB adaptive seasonality, which suggested that the region of climatic suitability will move higher in elevation, eventually reducing the total area of suitability. Using future temperature scenarios for the Northwest, Bentz et al. (2010) simulated changes in adaptive seasonality for MPB and single-year offspring survival for the spruce beetle (*D. rufipennis* Kirby) (SBB). The probability of MPB adaptive seasonality increases in higher elevation areas, particularly in the southern and central Cascade Range for the early twenty-first century and in the north Cascades and central Idaho for the late twenty-first century. Single-year development of SBB offspring also increases at high elevations across the region in both the early and late twenty-first century.

Response of tree growth to climate change will depend on subregional-to-local characteristics that change the sensitivity of species along the climatic gradients of their ranges (e.g., Littell et al. 2008; Chen et al. 2010; Peterson and Peterson 2001). Douglas-fir is expected to grow more slowly in much of the drier part of its range (Chen et al. 2010) but may currently be growing faster in many locations in the Northwest (Littell et al. 2008). Although no regional synthesis of expected trends in tree growth exists, the projected trend toward warmer and possibly drier summers in the Northwest (Mote and Salathé 2010) is likely to increase growth where trees are energy limited (at higher elevations) and decrease growth where trees are water limited (at lowest elevations and in driest areas) (Case and Peterson 2005; Holman and Peterson 2006; Littell et al. 2008). Growth at middle elevations will depend on summer precipitation (Littell et al. 2008).

The effects of climate change on forest processes in the Northwest are expected to be diverse, because the mountainous landscape of the region is complex, and species distribution and growth can differ at small spatial scales. Forest cover will change faster via disturbance and subsequent regeneration over decades, rather than via gradual readjustment of vegetation to a new climate over a century or more. Additional data are needed on interactions between disturbances and on connections between climate-induced changes in forests and ecosystem services, including water supply and quality, air quality, and wildlife habitat.

6.5 Southwest

D.L. Peterson (✉)
Pacific Northwest Research Station, U.S. Forest Service, Seattle, WA, USA
e-mail: peterson@fs.fed.us

Dying pinyon pines (*Pinus edulis* Engelm.) in New Mexico and adjacent states in the early 2000s became an iconic image of the effects of a warming climate in U.S. forests. Several consecutive years of drought reduced the vigor of pines, allowing pinyon ips (*Ips confusus* LeConte) to successfully attack and kill pines across more than one million ha (Breshears et al. 2005). The pinyon pine dieback was one of the most important manifestations of extreme climate in North America during the past decade, an indicator that a physiological threshold was exceeded because of the effects of low soil moisture (Floyd et al. 2009). Although this is not direct evidence of the effects of climate change, it demonstrates the effects of severe drought, a phenomenon expected more frequently in the future, on large-scale forest structure and function in arid environments.

Disturbance processes that are facilitated by climatic extremes, primarily multiyear droughts, dominate the potential effects of climatic variability and change on both short-term and long-term forest dynamics in the Southwest (Allen and Breshears 1998). Although diebacks in species other than pinyon pine have not been widespread, large fires and insect outbreaks appear to be increasing in both frequency and spatial extent throughout the Southwest. In Arizona and New Mexico, 14–18 % of the forested area was killed by wildfire and bark beetles between 1997 and 2008 (Williams et al. 2010). This forest mortality appears to be related to the current trend of increasing temperature and decreasing precipitation, at least in the southern portion of the region, since the mid 1970s (Weiss et al. 2009; Cayan et al. 2010).

In late spring 2011, following a winter with extremely low precipitation and a warm spring, the Wallow Fire burned 217,000 ha of forest and woodland in eastern Arizona and western New Mexico, receiving national attention for its size and intensity (Incident Information System 2011). The Wallow Fire was the largest recorded fire in the conterminous United States, and forced the evacuation of eight communities, cost $109 million to suppress (4,700 firefighters involved) and $48 million to implement rehabilitation measures, and resulted in high consumption of organic material and extensive overstory mortality across much of the burned landscape. A total of 880,000 ha burned in Arizona and New Mexico in 2011 (National Interagency Fire Center 2011). Large, intense fires illustrate how extreme drought can cause rapid, widespread change in forest ecosystems.

Recent large fires may portend future increases in wildfire. Using an empirical analysis of historical fire data on federal lands, McKenzie et al. (2004) projected the following increases in annual area burned for these Southwestern States, given a temperature increase of 1.5 °C: Arizona, 150 %; Colorado, 80 %; New Mexico, 350 %; and Utah, 300 %. California and Nevada were projected to be relatively insensitive to temperature, but their data included extensive non-forest area. In a more recent analysis, Littell et al. (n.d.) project the following increases for a 1 °C temperature increase: Arizona. 380–470 %; California, 310 %; Colorado, 280–660 %; Nevada, 280 %; New Mexico, 320–380 %; and Utah, 280–470 %. Applying the Parallel Climate Model to California, Lenihan et al. (2003) projected that area burned will increase at least 10 % per year (compared to historical level) by around 2100 (temperature increase of 2.0 °C).

The general increase in fire that is expected in the future, and that may already be occurring, will result in younger forests, more open structure, increased dominance of early successional plant species, and perhaps some invasive species. Because annual accretion of biomass is relatively low in this region, production of live and dead fuels in the understory in 1 year affects the likelihood of fire in the next year (Littell et al. 2009). The interaction of climate, fuel loading, and fuel moisture will contribute to both future area burned and fire severity.

The ongoing expansion of bark beetle outbreaks in western North America has been especially prominent in Colorado. Since 1996, multiple beetle species have caused high forest mortality on 2.7 million ha, of which 1.4 million ha were infested with mountain pine beetle (USDA FS 2011). Facilitated by extended drought and warmer winters, mountain pine beetle outbreaks have focused primarily on older (stressed) lodgepole pine forest. In Arizona and New Mexico, 7.6–11.3 % of forest and woodland area was affected by extensive tree mortality owing to bark beetles from 1997 through 2008 (Williams et al. 2010). As in other areas of the West, bark beetles appear to be attacking trees at higher elevations than in the past (Gibson et al. 2008).

In a detailed analysis of tree growth data for the United States, Williams et al. (2010) found that growth in the Southwest was positively correlated with interannual variability in total precipitation and negatively correlated with daily maximum temperature during spring through summer, which suggests that increased future drought will have a profound effect on growth and productivity. Projecting an A2 emission scenario on these growth-climate relationships produced significant growth reductions for forests in Arizona, Colorado, and New Mexico after 2050, affecting primarily ponderosa pine (*Pinus ponderosa* Lawson & C. Lawson), Douglas-fir, and pinyon pine. Projected growth decreases were larger than for any other region of the United States (Williams et al. 2010).

Simulation modeling of potential changes in vegetation in California suggests that significant changes can be expected by 2100 (Lenihan et al. 2003). Modeling results show that mixed-evergreen forest will replace evergreen conifer forest throughout much of the latter's historical range. This process may include gradual replacement of Douglas-fir–white fir (*Abies concolor* [Gordon & Glend.] Lindl. ex Hildebr.) forest by Douglas-fir–tanoak (*Lithocarpus densiflorus* [Hook. & Arn.] Rehd.) forest and the replacement of white fir-ponderosa pine forest by ponderosa pine-California black oak (*Quercus kelloggii* Newberry) forest in the Sierra Nevada. Tanoak-Pacific madrone (*Arbutus menziesii* Pursh)-canyon live oak (*Q. chrysolepis* Liebm.) woodland may replace blue oak (*Q. douglasii* Hook. & Arn.) woodlands, chaparral, and perennial grassland. In general, shrubland will replace oak woodland, and grassland will replace shrubland throughout the state. Evergreen conifer forest will advance into the high elevation subalpine forest in the Sierra Nevada, and species such as Shasta red fir (*Abies magnifica* A. Murray) and lodgepole pine may become more common in subalpine parklands and meadows. A high degree of regional variability in species changes can be expected, and large-scale transitions will need to be facilitated through fire disturbance that enables regeneration.

Increased disturbance from fire and insects, combined with lower forest productivity at most lower elevation locations because of a warmer climate, will probably result in lower C storage in most forest ecosystems. The fire-insect stress complex may keep many low-elevation forests in younger age classes in perpetuity. The normal cycle of fire followed by high precipitation (in winter in California, in early summer in much of the rest of the Southwest) may result in increased erosion and downstream sediment delivery (Allen 2007). In a warmer climate, it may be possible to reduce fire severity and protect WUI areas through assertive use

Fig. 6.4 The effectiveness of fuel treatments is seen in this portion of the 2011 Wallow Fire near Alpine, Arizona. High-intensity crown fire was common in this area, but forest that had been thinned and had surface fuels removed experienced lower fire intensity, and structures in the residential area were protected (Photo by U.S. Forest Service)

of fuel treatments (Peterson et al. 2011), as shown recently in the Wallow Fire (Bostwick et al. 2011) (Fig. 6.4). It may also be possible to reduce large-scale beetle epidemics by maintaining multiple forest age classes across the landscape (Li et al. 2005). Significant financial resources and collaboration across different agencies and landowners will be necessary to successfully implement these adaptive strategies.

6.6 Great Plains

L.A. Joyce (✉)
Rocky Mountain Research Station, U.S. Forest Service, Fort Collins, CO, USA
e-mail: ljoyce@fs.fed.us

Natural vegetation of the Great Plains is primarily grassland and shrubland ecosystems with trees occurring in scattered areas along streams and rivers, on planted woodlots, as isolated forests such as the Black Hills of South Dakota, and near the biogeographic contact with Rocky Mountains and eastern deciduous forests.

Trees are used in windbreaks and shelterbelts for crops and within agroforestry systems, extending the tree-covered area considerably (e.g., over 160,000 ha in Nebraska) (Meneguzzo et al. 2008). Urban areas in the Great Plains benefit from trees providing wildlife habitat, water storage, recreation, and aesthetic value.

Forests in the northern Great Plains (North Dakota, South Dakota, Kansas, Nebraska) comprise less than 3 % of the total land area within each state (Smith et al. 2009). More than half of the forest land in South Dakota is in public land ownership in contrast to the other three states. Dominant forest species are ponderosa pine (*Pinus ponderosa* Lawson and C. Lawson var. *scopulorum* Engelm.), fir-spruce, and western hardwoods. Eastern cottonwood (*Populus deltoides* Bartr. ex Marsh.) forests are an important source of timber in North Dakota (Haugen et al. 2009) and Nebraska (Meneguzzo et al. 2008). Many cottonwood stands in this region are quite old, and regeneration has been minimal owing to infrequent disturbance (South Dakota Resource Conservation and Forestry Division 2007; Meneguzzo et al. 2008; Moser et al. 2008; Haugen et al. 2009). The decline of this species often leads to establishment of non-native species (Haugen et al. 2009) or expansion of natives such as green ash (*Fraxinus pennsylvanica* Marsh.), which is susceptible to the invasive emerald ash borer (*Agrilus planipennis* Fairmaire). In North Dakota, quaking aspen (*Populus tremuloides* Michx.) forests are generally in poor health and have minimal regeneration because of fire exclusion (Haugen et al. 2009). In South Dakota, forest land is dominated by ponderosa pine forest, which supports a local timber industry in the Black Hills area. Management concerns include densely stocked stands, high fuel loadings and fire hazard, and mountain pine beetle outbreaks. Eastern redcedar (*Juniperus virginiana* L.) is expanding in many states, the result of fire exclusion and prolonged drought conditions (South Dakota Resource Conservation and Forestry Division 2007; Meneguzzo et al. 2008). This presents opportunities for using redcedar for wood products, but also raises concerns about trees encroaching into grasslands and altering wildlife habitat (Moser et al. 2008). Land-use activities that support biofuel development, particularly on marginal agricultural land, may affect forests in this area (Meneguzzo et al. 2008; Haugen et al. 2009).

Forests in the southern Great Plains (Oklahoma, Texas) comprise less than 17 % of the land area (Smith et al. 2009), are often fragmented across large areas, and are mostly privately owned. In Texas, the forest products industry is one of the top ten manufacturing sectors in the state, with a fiscal impact of $33.6 billion on the state economy (Xu 2002). Loss of forest to urbanization, oil and gas development, and conversion to cropland and grassland has led to a permanent reduction in forest cover (Barron 2006; Johnson et al. 2010).

Forests in the western Great Plains (Montana, Wyoming) comprise less than 27 % of the land area (Smith et al. 2009), and most of this land is in public ownership. Montana has large contiguous areas of forest, particularly in the western part of the state where public land, forest industry, and private land intermingle. Both Montana and Wyoming have forested areas on mountains where the surrounding ecosystems are grassland and shrubland. The three major forest types in Montana are also the most commercially important species: Douglas-fir, lodgepole pine, and ponderosa pine (Montana Department of Natural Resources and Conservation 2010). Fire exclusion has caused higher fire hazard and more mountain pine beetle outbreaks. In recent years, the forest industry has been adversely affected by reduced timber supply and general economic trends. Wyoming forests are dominated by lodgepole pine, followed by spruce-fir and ponderosa pine, and land ownership is a mosaic of public, private, and industrial. Similar to Montana, the forest industry in Wyoming has faced several challenges but continues to be a significant component of the state economy (Wyoming State Forestry Division 2009). Both Montana and Wyoming have urban forests, riparian forests, and windbreaks and shelterbelts associated with agriculture. Tree species used in windbreaks and shelterbelts, including ponderosa pine and the nonnatives Scots pine (*Pinus sylvestris* L.) and Austrian pine (*P. nigra* Arnold) are being attacked by mountain pine beetles, and green ash is susceptible to the emerald ash borer. Similar to other parts of the Great Plains, some lower elevation riparian forests are in decline, because regeneration has been reduced by fire exclusion, water diversions, drought, agricultural activities, and urban development.

Little information is available on the potential effects of climate change on Great Plains forests. However, this area has been part of continental and national studies (Bachelet et al. 2008), and areas such as the Greater Yellowstone Ecosystem have a long history of research that has recently included climate change. Tree species in the Yellowstone area are expected to move to higher elevation in a warmer climate (Bartlein et al. 1997; Koteen 2002; Whitlock et al. 2003). However, projecting future vegetation distribution is complicated by the complex topography of Wyoming, which influences the microclimatic environment that controls vegetation distribution. Forests in this area and Montana are currently affected by insect outbreaks and wildfire, and changes in these disturbances under climate change could potentially disrupt ecosystems across large landscapes. A recent modeling study suggests that a warmer climate will increase the frequency and spatial extent of wildfire in the Yellowstone area (Westerling et al. 2011).

In a review of the literature on the effects of climate change in semiarid riparian ecosystems, Perry et al. (2012) noted that climate-driven changes in streamflow are expected to reduce the abundance of dominant, native, early-successional tree species and increase herbaceous, drought-tolerant, and late-successional woody species (including nonnative species), leading to reduced habitat quality for riparian fauna. Riparian systems will be especially important locations on which to focus monitoring for the early effects of climate change.

Reduced tree distribution in the Great Plains will likely have a negative effect on agricultural systems, given the important role of shelterbelts and windbreaks in reducing soil erosion. In these "linear forests," warmer temperatures are expected to reduce aboveground tree biomass and spatial variation in biomass at lower elevations, but may increase biomass on upland habitats (Guo et al. 2004). Whereas most studies in this region have explored the potential influence of elevated CO_2 on grassland, Wyckoff and Bowers (2010) analyzed the relationship between historical climate and tree growth and suggest that the interaction of climate change and elevated CO_2 could be a potential factor in the expansion of forests from the eastern United States into the Great Plains. Carbon sequestration through agroforestry has been suggested as a potential mitigation activity (Morgan et al. 2010).

Across the Great Plains, forests are currently exposed to many stressors. Common to all states in this region is a concern about land-use changes that would reduce the total area of forests, fragment intact forests, and alter forest dynamics. Current stressors such as insects, fungal pathogens, and altered hydrologic dynamics may be exacerbated by a warmer climate. The potential for stress complexes that include wildfire, longer droughts, and increased risk of insect outbreaks could significantly modify Great Plains forest environments (see Chap. 4).

6.7 Midwest

C.W. Swanston (✉)
Northern Research Station, U.S. Forest Service, Houghton, MI, USA
e-mail: cswanston@fs.fed.us

S.D. Handler
Northern Research Station, U.S. Forest Service, Houghton, MI, USA
e-mail: sdhandler@fs.fed.us

Forests are a defining landscape feature for much of the Midwest, from boreal forests surrounding the northern Great Lakes to oak-hickory (*Quercus* spp., *Carya* spp.) forests blanketing the Ozark Highlands. Forests cover approximately 28 % of the area in the eight-state Midwest region and help sustain human communities ecologically, economically, and culturally. Most of the Midwest is contained within the Laurentian Mixed Forest, Eastern Broadleaf Forest (Continental and Oceanic), and Prairie Parklands ecoregions (Bailey 1995) (Fig. 6.5).

The broad diversity in species composition and structure across the Midwest will likely engender higher resilience to a changing climate than less diverse biogeographic regions, but each ecoregion might be best characterized by a few strong vulnerabilities. With this in mind, key vulnerabilities related to climate change are summarized below according to ecoregions. The term "vulnerability" refers to a decline in vigor and productivity, in addition to more severely altered community composition or ecosystem function, and a species or ecosystem may be considered vulnerable to climate change by virtue of significantly decreased well-being, even if it is not projected to disappear completely (Swanston et al. 2011).

The *Laurentian Mixed Forest* spans the northern areas of the Great Lakes states (Fig. 6.5), typified by a glaciated landscape with low relief covered with mesic broadleaf deciduous forests, sometimes mixed with conifers, and often grading to pure conifers on poor soils. Winters are cold and of long duration, often with heavy snowfall, and summers are warm and provide much of the annual precipitation. As a transitional zone between boreal forests in the north and broadleaf forests to the

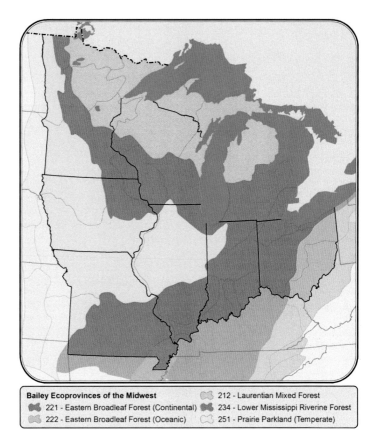

Fig. 6.5 Ecoregions in the Midwest, according to Bailey (1995)

south, the Laurentian forests are often dominated by boreal species at the southern edge of their suitable habitat range. Many of these species, such as black spruce, balsam fir (*Abies balsamea* [L.] Mill.), paper birch (*Betula papyrifera* Marsh.), and northern whitecedar (*Thuja occidentalis* L.), are projected to lose suitable habitat through much of their current range (Iverson et al. 2008a, b; Walker et al. 2002). Associated ecosystems may thus be more likely to experience stress and undergo more distinct community transitions (Swanston et al. 2011; Xu et al. 2012).

Forested wetlands, including peatlands, may be especially susceptible to a combination of range shifts and changes in hydrologic regimes (e.g., Swanston et al. 2011). These systems store a large amount of belowground C (Johnson and Kern 2003) that could be at risk if fire increases in drier conditions. Sub-boreal species such as sugar maple (*Acer saccharum* Marsh.) may be less affected than boreal species, but any effects may be more apparent aesthetically and economically owing to their prevalence on the landscape (Iverson et al. 2008b).

The *Eastern Broadleaf Forest* (Fig. 6.5) mostly consists of the Continental ecoregion, with low rolling hills, some glaciation in the north, and the Ozark

Highlands to the south. Precipitation generally comes during the growing season but decreases in the western ecoregion. Oak-hickory forest is dominant, grading to maple (*Acer* spp.), American beech (*Fagus grandifolia* Ehrh.), and American basswood (*Tilia americana* L.) in the north. Oak decline is increasing the mortality of oak species throughout the southern half of the Midwest and is correlated with drought periods (Wang et al. 2007). Species in the red oak (*Quercus rubra* L., *Q. coccinea* Münchh., *Q. velutina* Lam.) group are particularly susceptible to decline and make up a large proportion of upland forests in this ecoregion. White oak (*Q. alba* L.) may also be declining on the western margins of its range (Goldblum 2010), which may be further amplified by higher summer temperatures in the future (Iverson et al. 2008b). Oak decline could worsen if droughts become more frequent or severe, and elevated fine and coarse fuels could result from tree mortality, thereby increasing wildfire hazard.

Wildfire suppression has gradually favored more mesic species such as maple, leaving fire-adapted species like oaks and shortleaf pine (*Pinus echinata* Mill.) at a competitive disadvantage (Nowacki and Abrams 2008). With adequate moisture and continued fire suppression, these forests are likely to persist but may become increasingly susceptible to wildfire in a drier climate (Lenihan et al. 2008). A general decline in resilience, in combination with increased disturbances such as fire, could make these forests more susceptible to invasive species such as kudzu (*Pueraria lobata* [Lour.] Merr., an aggressive vine) and Chinese and European privet (*Ligustrum sinense* Lour. and *L. vulgare* L., highly invasive shrubs), that may to expand into the Midwest as winter minimum temperatures increase (Bradley et al. 2010).

The *Prairie Parklands* (Fig. 6.5) are predominantly covered by agriculture and prairie, with interspersed upland forests of oak and hickory. Forest stands are also found near streams and on north-facing slopes. Fragmentation and parcelization of forest ecosystems are more extreme in the Prairie Parklands than in other Midwest ecoregions. For example, over 90 % of forest land in Iowa is currently divided into private holdings averaging less than 7 ha (Flickinger 2010). Combined with extensive conversion of available land to agricultural monocultures, this ecoregion currently exists as a highly fragmented landscape for forest ecosystems, effectively impeding the natural migration of tree species. Model simulations indicate that factors such as increasing summer temperatures and dryness, coupled with inadequate fire suppression, could lead to loss of ecosystem function and transition to grasslands or woodland/savanna even under low emissions scenarios (Lenihan et al. 2008).

Human communities are an integral part of the landscape in the Midwest and have greatly shaped current forests and prairie-forest boundaries (Abrams 1992; Mladenoff and Pastor 1993). Contemporary land use and ownership patterns provide critical input to policy responses to ecological issues, including climate change. In the Midwest, 68 % of forests are in private ownership (Butler 2008; Nelson et al. 2010). Stewardship of private lands reflects diverse values and motivations (Bengston et al. 2011), providing a challenge to effective outreach (Kittredge 2004). Likewise, a coordinated response to forest ecosystem threats is further challenged by parcelization (DeCoster 1998; Mehmood and Zhang 2001). Fostering climate preparedness as a component of sustainable land stewardship will require significantly increased outreach and coordination to communicate relevant and credible information to private forest landowners. Conversely, inadequate attention to land stewardship will place this forest sector at greater risk of avoidable impacts of climate change.

6.8 Northeast

L.E. Rustad (✉)
Northern Research Station, U.S. Forest Service, Durham, NH, USA
e-mail: lrustad@fs.fed.us

Climate is a key regulator of terrestrial biogeochemical processes. A recent synthesis of climate-change effects on forests of the Northeast region concluded that changes in climate that are already underway will result in changes in forest species composition, length of growing season, and forest hydrology, which together exert significant controls on forest productivity and sustainability (Rustad et al. 2009). Since 1900, mean annual temperature in the region has risen by an average of 0.8 °C, precipitation has increased by 100 mm, the onset of spring (based on phenologic indicators) has advanced by approximately 4 days, streamflows have generally increased, and dates of river and lake ice melt have advanced by 1–2 weeks (Huntington et al. 2009). Projections for the twenty-first century suggest that temperature will increase by 2.9–5.3 °C, precipitation will increase by 7–14 % (with minimal change in summer precipitation), the onset of spring will advance by 10–14 days, riverflows will increase during winter and spring but decrease in summer because of increased frequency of short-term droughts, and winter ice and snow will diminish. Variability and intensity of weather are also expected to increase, with more precipitation during large events with longer intervening dry spells, and more frequent and severe extreme events.

Forests cover large areas of the land surface in the northeastern United States, from 59 % in Rhode Island to 89 % in Maine (National Land Cover Database 2001). These forests are currently dominated by (1) southern hardwoods (oak, hickory) and pines in the southernmost region; (2) northern hardwoods (American beech, paper birch, and yellow birch [*Betula alleghaniensis* Britt.], sugar maple, and red maple [*Acer rubrum* L.]) in the central part and at lower elevations throughout; and (3) boreal-conifer forests in the north and at higher elevations (red spruce [*Picea rubens* Sarg.], black spruce, and balsam fir. eastern hemlock (*Tsuga canadensis* [L.] Carrière), an important shade tolerant, late successional species, is found throughout the Northeast.

Paleoecological data reveal a strong climate signal in current species assemblages and show that tree species have shifted in response to a gradually changing climate over the past 12,000 years since deglaciation. Projecting how the distribution and abundance of species will shift in the future in response to climate change is com-

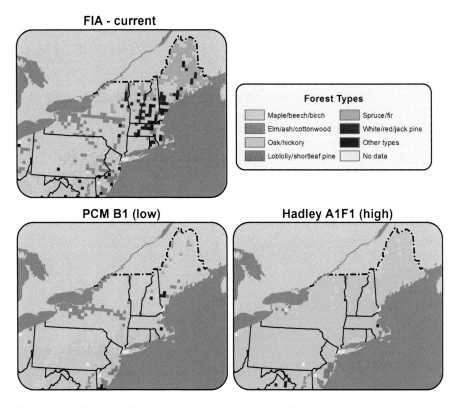

Fig. 6.6 Suitable habitat for forest vegetation in New England is expected to shift with changes in climate (year 2100) associated with different emissions scenarios (From Mohan et al. 2009, with permission)

plicated by the longevity of current individuals in the existing forest, robustness of the genetic pool to accommodate adaptation to new climatic conditions, limitations on regeneration and dispersal, and interactions with factors such as elevated nitrogen (N) deposition, elevated tropospheric ozone, land use change, habitat fragmentation, and changes in disturbance regimes caused by invasive species, pathogens, and fire.

In lieu of projecting future forest composition, some researchers have used "climatic envelopes," which combine information on current species distributions with climatic projections for the future, based on an ensemble of earth system models and emissions scenarios, to generate maps of "suitable habitat" for individual species and assemblages of species as forest types. For example, Iverson et al. (2008b) projected that a warming climate will result in a large contraction of suitable habitat for spruce-fir forest, moderate decline in suitable habitat for the maple-birch-beech forest, and expansion of suitable habitat for oak-dominated forest (Fig. 6.6). Projections of change in suitable habitat for individual tree species indicate that, of the 84 most common species, 23–33 will lose suitable habitat, 48–50 will gain habitat, and 1–10 will experience no change. The tree species predicted to have the most affected habitat include balsam fir, quaking aspen, paper birch (80–87 % decrease in suitable habitat), and black and white oak (greater than 100 % increase in suitable habitat) (Iverson et al. 2008b).

As climate and species composition change, so will forest productivity and C sequestration. More favorable climatic conditions for growth, particularly longer growing seasons, are correlated with higher productivity, whereas climatic extremes such as droughts, extreme cold or heat, and windstorms have been linked with tree diebacks and periods of lower productivity (Mohan et al. 2009). At Hubbard Brook Experimental Forest (New Hampshire), green canopy duration increased by 10 days over a 47-year period for a northern hardwood forest, suggesting a future longer period for growth and higher productivity (Richardson et al. 2006).

Model projections indicate that forest productivity for hardwood species is likely to be enhanced by future warmer temperatures, longer growing seasons, and increased concentrations of atmospheric CO_2. For example, Ollinger et al. (2008) used the model PnET-CN to project that net primary productivity in deciduous forests would increase by 52–250 % by 2100, depending on the global climate model and emission scenario used. The same model projected that current-day spruce forests are likely to show a climate-driven decrease in productivity along with a contraction of range. The effects of changing tree species assemblages and concurrent stress associated with forest fragmentation, atmospheric pollution, and invasive plant and animal species complicate these projections.

Changes in climate, hydrology, and forest tree species composition will have cascading effects on associated biogeochemical processes. Warmer temperatures and extended growing seasons will probably increase rates of microbial decomposition, N mineralization, nitrification, and denitrification, which will provide increased short-term availability of nutrients such as calcium, magnesium, and N for forest growth, as well as the potential for elevated losses of these same nutrients to surface waters (Campbell et al. 2009). Forests may respond to climate change with significant increases in nitrate leaching from soils to surface waters, with consequences for downstream water quality and eutrophication (Campbell et al. 2009). Potential accelerated loss of calcium and magnesium, especially from areas that have already experienced loss of these nutrients owing to decades of acidic deposition, may increase soil acidification. Warmer temperatures will also probably increase rates of root and microbial respiration, with an increased release of CO_2 from the soil to the atmosphere (Rustad et al. 2000).

Climate change will affect the distribution and abundance of many wildlife species in the region through changes in habitat, food availability, thermal tolerances, species interactions such as competition, and susceptibility to parasites and pathogens (Rodenhouse et al. 2009). Decades of survey data show that migratory birds are arriving earlier and breeding later in response to recent warming, with consequences for the annual production of young and survival (Rodenhouse et al. 2009). Among 25 species of resident birds studied, 15 are increasing in abundance, which is consistent with the observation that ranges of these species are limited by winter climate. Of the remaining species, 5 are decreasing in abundance and 5 show no change. Significant range expansions have also been observed, with 27 of 38

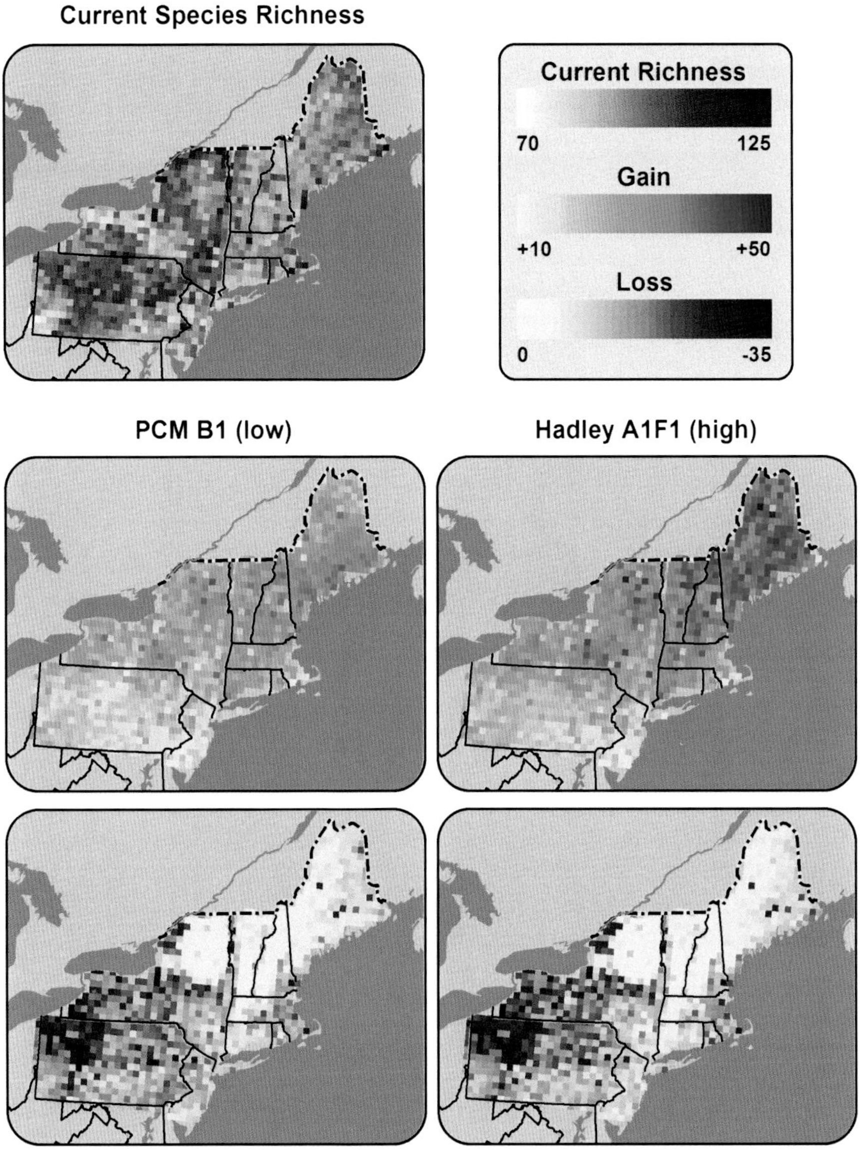

Fig. 6.7 Climate change (year 2100) is expected to affect bird species richness more intensely in some areas of the northeastern United States than in others (From Rodenhouse et al. 2008, with permission)

species studied expanding their ranges in a northward direction (Fig. 6.7). Using a climatic envelope approach, Rodenhouse et al. (2009) projected that twice as many resident bird species are expected to increase in abundance as decrease; for migrants (which comprise more than 85 % of the avifauna), an equal number are expected to increase as decrease.

Climate-related historical and future projected changes in native and introduced insects deserve special mention because these species contribute heavily to disturbance in Northeastern forests, and some species are particularly adept at adjusting to changing climatic conditions. Direct effects of climate change are likely to include summer warming-induced acceleration of reproductive and development rates, winter warming-induced increase in the ability to overwinter, and moisture-related changes in survival and fecundity. If minimum winter temperature increases as projected, this may allow the northward migration of many unwanted species, such as the hemlock wooly adelgid (*Adelges tsugae* Annand) (Skinner et al. 2003). Based on recent projections, climatic warming could allow the adelgid to spread throughout the range of eastern hemlock, potentially altering forest composition, nutrient cycling, and surface water quality (Dukes et al. 2009).

An increase in extreme weather events may have a larger effect on natural and managed systems than the more gradual change in mean climatic conditions. Legacies of past extreme windstorms and ice storms are apparent across the forested landscape of the region. It is imperative for the scientific and land management communities to better understand and anticipate the future occurrence and effects of these extreme events on forest composition, productivity, biogeochemistry, and fauna.

The twentieth century climate of the northeastern United States changed more rapidly than at any time since the last glaciation, and this rate of change is expected to continue throughout the twenty-first century. The direct and indirect effects of climate change on Northeastern forests, individually and in combination with other stressors such as acidic deposition, N and mercury deposition, tropospheric ozone, and various land uses, have the potential to cause significant changes in ecosystem structure and function (see Chap. 4). Additional research on indirect and interacting effects of these changes on forest ecosystems will be especially valuable for understanding potential effects of climate change, and for developing adaptation options that will enhance the sustainability of the diverse forests of this region.

6.9 Southeast

S.G. McNulty (✉)
Southern Research Station, U.S. Forest Service, Raleigh, NC, USA
e-mail: smcnulty@fs.fed.us

Forests of the southeastern United States are a complex mixture of private and public land, interspersed with rapidly urbanizing areas and agriculture. A long history of active forest management, including intensive management such as forest plantations, fertilization, and prescribed fires, has created stand conditions and management regimes that differ from other areas of the United States. For example, relative to forests of the western United States, smaller tracts of accessible forest land may be more amenable to management actions that can be used to mitigate

C emissions or help forests adapt to climate change. On the other hand, the large private ownership of relatively small forest land holdings makes it challenging to implement uniform or coordinated large-scale management activities.

The majority of U.S. wood and fiber is produced in the Southeast, but climate change could significantly alter productive capacity in the region (Wertin et al. 2010). Loblolly pine is the most important commercial species, and although current air temperature is near optimal for growth across much of its range, as temperature continues to increase, conditions for pine growth may begin to deteriorate (McNulty et al. 1998b). Even if regional forest productivity remains high, the center of forest productivity could shift farther north into North Carolina and Virginia, causing economic and social effects in areas gaining and losing timber industry jobs (Sohngen et al. 2001).

Carbon sequestration is an increasingly valued component of forest productivity, and a large portion of the C stored in U.S. forests occurs in the Southeast (Pan et al. 2011). In addition to potentially reducing forest productivity (and therefore C uptake), climate change could increase decomposition of soil organic matter and CO_2 release (Boddy 1983). When added to the potential for increased wildfires, the potential for ecosystem C sequestration may decrease in the future, and the ecosystem value of sequestered forest C may shift from the southern to northern United States (Hurteau et al. 2008).

Wildfires, hurricanes, drought, insect outbreaks, and pathogen outbreaks have been a driving force for millennia in Southeastern forests. However, during the past two centuries, the type and magnitude of ecosystem stress and disturbance have changed and will likely continue to change as the climate warms (Dale et al. 2001). Wind and extreme precipitation events associated with hurricanes can have major effects on Southeastern forests. A single hurricane can reduce total forest C sequestration by 10 % in the year in which it occurs (McNulty 2002), although not all forest species are equally susceptible to wind damage. Longleaf pine (*Pinus palustris* Mill.) shows less damage than does loblolly pine (*P. taeda* L.) when exposed to an equal level of wind stress (Johnsen et al. 2009), suggesting that the former species would be more resistant to an increase in windstorms. Extreme precipitation events that accompany hurricanes can cause extended submersion of low-lying forests, which can kill tree roots by causing anaerobic soil conditions (Whitlow and Harris 1979).

Wildfires are a natural component of ecosystem maintenance and renewal in the Southeast, which has more area burn annually, with wildfire and prescribed fire, than any other region of the United States (except Alaska in some years) (Andreu and Hermansen-Baez 2008). However, decades of fire exclusion coupled with increasing air temperatures have increased the potential for crown fire in some Southeastern forests. Future fire potential is expected to increase from low to moderate in summer and autumn in eastern sections in the South, and from moderate to high in western portions (Liu et al. 2010). As fire seasons lengthen in the future, the window for prescribed burning may decrease because of increased fuel flammability, thus potentially affecting the management of fuels and C dynamics; fuel treatments with prescribed fire emit 20 % less CO_2 than wildfires, at least in the short term

(Wiedinmyer and Hurteau 2010). Historically, longleaf pine was a dominant species across the region. It is well adapted to drought, with thick bark and fast seedling growth, allowing it to thrive in habitats subjected to periodic wildfire (Brockway and Lewis 1997). Most of the longleaf pine was cut during the twentieth century, followed by replanting with the faster growing loblolly pine, which is preferred by the timber industry but is less resistant to wildfire. Land managers are reassessing the preferential use of loblolly pine, because longleaf pine would be more resistant to the increased fire, drought, and wind expected with climate change.

Insect and pathogen outbreaks are increasing in Southeastern forests (Pye et al. 2011), potentially threatening the long-term productivity and structure of forest ecosystems. Higher temperature has caused a longer growing season of at least 2 weeks compared to historical lengths, allowing additional time for insects and pathogens to find and colonize susceptible trees (Ayres and Lombardero 2000). In addition, timing of the predator–prey cycle may be changing. For example, when the growing season begins earlier, insects may be hatching and maturing before migratory insectivorous bird species return, allowing more insects to reach maturity, speed up the reproductive cycle, and locate susceptible host trees. Finally, higher temperature and subsequent soil drying increases stress in trees, reducing their physiological capacity to resist attack (McNulty et al. 1998a).

Some aspects of the high biodiversity in the Southeast may be susceptible to climate change (Thompson et al. 2009), particularly species that are near the environmental limit of their range. Red spruce and eastern hemlock are well adapted to the cool climates of the last glacial age. However, the extent and dominance of these two species have decreased greatly as a result of stress complexes that include warmer temperature, air pollution, and insects (McNulty and Boggs 2010; Elliott and Vose 2011). With further warming, red spruce and eastern hemlock are projected to be extirpated from the southern United States before 2100 (Prasad et al. 2007), and small populations of balsam fir will also be at risk. Altered tree species dominance will affect birds and other terrestrial vertebrate species that depend on forest habitat.

Cold water fish species, which are generally confined to northern and mountainous areas of the Southeast where cooler water (and air) temperatures allow dissolved oxygen contents to remain at sufficient levels, will likely face increased stress from higher temperature at the southern limit of their range. In addition, rainfall intensity has been increasing for over a century (Karl et al. 1995), which can in turn increase soil erosion and stream turbidity (Trimble 2008). A combination of higher air temperature and lower water quality may significantly reduce trout abundance across the Southeast during the coming decades (Flebbe et al. 2006).

Abundant, year-round rainfall has historically provided a sufficient supply of water for industrial, commercial, residential, agricultural, and hydro-electric use in the Southeast, but several factors may contribute to a shift in water abundance. The population of the Southeast is increasing and much of this increase is centered in

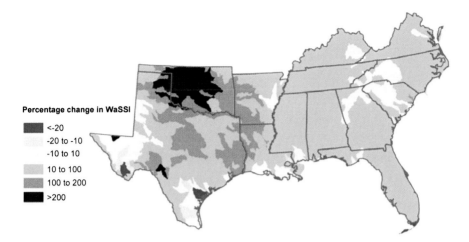

Fig. 6.8 Percentage change in water supply stress owing to climate change, as defined by the water supply stress index (*WaSSI*) for 2050 using the CSIROMK2 B2 climate scenario. WaSSI is calculated by dividing water demand by supply, where higher values indicate higher stress on watersheds and water systems. From Lockaby et al. (2011)

metropolitan areas, whereas much of the water originates in forested headwaters, often long distances from urban areas. On an annual basis, average water supply is approximately 20 times higher than demand, although short-term (1–3 years) drought can significantly increase pressure on available water (Lockaby et al. 2011) (Fig. 6.8). A combination of increased population, changing land-use patterns, and shifts in rainfall patterns could further amplify water shortages, and even if precipitation rates remain unchanged, higher tree water use in response to higher temperature, or shifting management regimes for new products such as biofuels, could contribute to water shortfalls (Sun et al. 2008; Lockaby et al. 2011). Seasonal timing of precipitation within the year could also affect water supply. If precipitation occurs in fewer, more intense events, then proportionally less water will be retained by forest ecosystems, and more will be lost as runoff, potentially causing flooding, soil erosion, and stream sedimentation (Trimble 2008).

The Southeast has diverse year-round recreational opportunities, some of which could be severely affected by climate change. Ski areas in the region are marginally profitable, and increased winter warming may increase the proportion of rain to snow and prevent snow making (Millsaps and Groothuis 2003). Reduced quality or quantity of the ski season could force most of the marginal ski areas to close. Similarly, cold water fisheries are a major recreational attraction, and revenues from lodging, food, and secondary activities are a major economic boost to local mountain economies. Therefore, extirpation of trout from these areas could significantly harm the recreation industry.

References

Abrams, M. D. (1992). Fire and the development of oak forests. *BioScience, 42*, 346–353.
Allen, C. D. (2007). Interactions across spatial scales among forest dieback, fire, and erosion in northern New Mexico landscapes. *Ecosystems, 10*, 797–808.
Allen, C. D., & Breshears, D. D. (1998). Drought-induced shift of a forest-woodland ecotone: Rapid landscape response to climate variation. *Proceedings of the National Academy of Sciences, USA, 95*, 14839–14842.
Andreu, A., & Hermansen-Baez, L. A. (2008). *Fire in the South 2: The Southern wildfire risk assessment* (32pp). A report by the Southern Group of State Foresters. http://www.southernwildfirerisk.com/reports/FireInTheSouth2.pdf
Asner, G. P., Elmore, A. J., Hughes, R. F., et al. (2005). Ecosystem structure along bioclimatic gradients in Hawai'i from imaging spectroscopy. *Remote Sensing of Environment, 96*, 497–508.
Asner, G. P., Hughes, R. F., Vitousek, P. M., et al. (2008). Invasive plants transform the three-dimensional structure of rain forests. *Proceedings of the National Academy of Sciences, USA, 105*, 4519–4523.
Ayres, M. P., & Lombardero, M. J. (2000). Assessing the consequences of global change for forest disturbance from herbivores and pathogens. *The Science of the Total Environment, 262*, 263–286.
Bachelet, D., Lenihan, J., Drapek, R., & Neilson, R. P. (2008). VEMAP vs VINCERA: A DGVM sensitivity to differences in climate scenarios. *Global and Planetary Change, 64*, 38–48.
Bailey, R. G. (1995). *Description of the ecoregions of the United States* (108pp). Washington, DC: U.S. Department of Agriculture, Forest Service.
Barnett, J. (2001). Adapting to climate change in Pacific island countries: The problem of uncertainty. *World Development, 29*, 977–993.
Barnett, J., & Adger, W. (2003). Climate dangers and atoll countries. *Climatic Change, 61*, 321–337.
Barron, E. (2006). *State of the Texas forest 2005* (37pp). College Station: Texas Forest Service.
Bartlein, P., Whitlock, C., & Shafer, S. (1997). Future climate in the Yellowstone National Park region and its potential impact on vegetation. *Conservation Biology, 11*, 782–792.
Beier, C. M., Sink, S. E., Hennon, P. E., et al. (2008). Twentieth-century warming and the dendroclimatology of declining yellow-cedar forests in southeastern Alaska. *Canadian Journal of Forest Research, 38*, 1319–1334.
Bengston, D. N., Asah, S. T., & Butler, B. J. (2011). The diverse values and motivations of family forest owners in the United States: An analysis of an openended question in the National Woodland Owner Survey. *Small-Scale Forestry, 10*, 339–355.
Benning, T. L., LaPointe, D., Atkinson, C. T., & Vitousek, P. M. (2002). Interactions of climate change with biological invasions and land use in the Hawaiian Islands: Modeling the fate of endemic birds using a geographic information system. *Proceedings of the National Academy of Sciences, USA, 99*, 14246–14249.
Bentz, B. J., Régnière, J., Fettig, C. J., et al. (2010). Climate change and bark beetles of the western United States and Canada: Direct and indirect effects. *BioScience, 60*, 602–613.
Berg, E. E., Henry, J. D., Fastie, C. L., et al. (2006). Spruce beetle outbreaks on the Kenai Peninsula, Alaska, and Kluane National Park and Reserve, Yukon Territory: Relationship to summer temperatures and regional differences in disturbance regimes. *Forest Ecology and Management, 227*, 219–232.
Berg, E. E., Hillman, K. M., Dial, R., & DeRuwe, A. (2009). Recent woody invasion of wetlands on the Kenai Peninsula Lowlands, south-central Alaska: A major regime shift after 18,000 years of wet Sphagnum-sedge peat recruitment. *Canadian Journal of Forest Research, 39*, 2033–2046.
Bernhardt, E. L., Hollingsworth, T. N., & Chapin, F. S. (2011). Fire mediates climate-driven shifts in understorey community composition of black spruce stands of interior Alaska. *Journal of Vegetation Science, 22*, 32–44.

Boddy, L. (1983). Carbon dioxide release from decomposing wood: Effect of water content and temperature. *Soil Biology and Biochemistry, 15*, 501–510.

Bostwick, P., Menakis, J., & Sexton, T. (2011). *How fuel treatments saved homes from the 2011 Wallow fire*. http://www.fs.fed.us/fire/management/fuel_treatments.pdf. 19 Dec 2011.

Bradley, B. A., Wilcove, D. S., & Oppenheimer, M. (2010). Climate change increases risk of plant invasion in the eastern United States. *Biological Invasions, 12*, 1855–1872.

Breshears, D. D., Cobb, N. S., Rich, P. M., et al. (2005). Regional vegetation die-off in response to global-change type drought. *Proceedings of the National Academy of Sciences, USA, 102*, 15144–15148.

Brockway, D. G., & Lewis, C. E. (1997). Long-term effects of dormant-season prescribed fire on plant community diversity, structure and productivity in a longleaf pine wiregrass ecosystem. *Forest Ecology and Management, 97*, 167–183.

Butler, B. J. (2008). *Family forest owners of the United States, 2006* (General Technical Report NRS-27, 72pp). Newtown Square: U.S. Department of Agriculture, Forest Service, Northern Research Station.

Campbell, J. L., Rustad, L. E., Boyer, E. B., et al. (2009). Consequences of climate change for biogeochemical cycling in forests of northeastern North America. *Canadian Journal of Forest Research, 39*, 264–284.

Case, M. J., & Peterson, D. L. (2005). Fine-scale variability in growth-climate relationships of Douglas-fir, North Cascade Range, Washington. *Canadian Journal of Forest Research, 35*, 2743–2755.

Cavaleri, M. A., & Sack, L. (2010). Comparative water use of native and invasive plants at multiple scales: A global meta-analysis. *Ecology, 91*, 2705–2715.

Cayan, D. R., Das, T., & Pierce, T. P. (2010). Future dryness in the southwest US and the hydrology of the early 21st century drought. *Proceedings of the National Academy of Sciences, USA, 107*, 21271–21276.

Chapin, F. S., McGuire, A. D., Randerson, J., et al. (2000). Arctic and boreal ecosystems of western North America as components of the climate system. *Global Change Biology, 6*, 211–223.

Chapin, F. S., III, Trainor, S. F., Huntington, O., et al. (2008). Increasing wildfire in Alaska's boreal forest: Pathways to the potential solutions of a wicked problem. *BioScience, 58*, 531–540.

Chen, P.-Y., Welsh, C., & Hamann, A. (2010). Geographic variation in growth response of Douglas-fir to interannual climate variability and projected climate change. *Global Change Biology, 16*, 3374–3385.

Chu, P.-S., & Chen, H. (2005). Interannual and interdecadal rainfall variations in the Hawaiian Islands. *Journal of Climate, 18*, 4796–4813.

Chu, P.-S., Chen, Y. R., & Schroeder, T. A. (2010). Changes in precipitation extremes in the Hawaiian Islands in a warming climate. *Journal of Climate, 23*, 4881–4900.

Coops, N. C., & Waring, R. H. (2010). A process-based approach to estimate lodgepole pine (*Pinus contorta* Dougl.) distribution in the Pacific Northwest under climate change. *Climatic Change, 105*, 313–328.

Coops, N. C., & Waring, R. H. (2011). Estimating the vulnerability of fifteen tree species under changing climate in Northwest North America. *Ecological Modelling, 222*, 2119–2129.

Cortés-Burns, H., Lapina, I., Klein, S. C. et al. (2008). *Invasive plant species monitoring and control-areas impacted by 2004 and 2005 fires in interior Alaska: A survey of Alaska BLM lands along the Dalton, Steese, and Taylor highways* (BLM-BAER Final Report, 162pp). Anchorage: Bureau of Land Management, Alaska State Office.

Dale, V. H., Joyce, L. A., McNulty, S. G., et al. (2001). Climate change and forest disturbances. *BioScience, 59*, 723–734.

D'Antonio, C. M., & Vitousek, P. M. (1992). Biological invasions by exotic grasses, the grass/fire cycle, and global change. *Annual Review of Ecology and Systematics, 23*, 63–87.

DeCoster, L. A. (1998). The boom in forest owners—A bust for forestry? *Journal of Forestry, 96*, 25–28. doi: http://dx.doi.org/10.1890/ES11-00288.1

Dukes, J. S., Pontius, J., & Orwig, D. (2009). Responses of insect pests, pathogens, and invasive plant species to climate change in the forests of northeastern North America: What can we predict? *Canadian Journal of Forest Research, 39*, 231–248.

Elliott, K. J., & Vose, J. M. (2011). The contribution of the Coweeta Hydrologic Laboratory to developing an understanding of long-term (1934–2008) changes in managed and unmanaged forests. *Forest Ecology and Management, 261*, 900–910.

Flebbe, P. A., Roghair, L. D., & Bruggink, J. L. (2006). Spatial modeling to project southern Appalachian trout distribution in a warmer climate. *Transactions of the American Fisheries Society, 135*, 1371–1382.

Flickinger, A. (2010). In E. Miller (Ed.). *Iowa's forests today: An assessment of the issues and strategies for conserving and maintaining Iowa's forests*. Des Moines: Iowa Department of Natural Resources. http://www.iowadnr.gov/Environment/Forestry/ForestryLinksPublications/IowaForestActionPlan.aspx

Flint, C. G. (2006). Community perspectives on spruce beetle impacts on the Kenai Peninsula, Alaska. *Forest Ecology and Management, 227*, 207–218.

Floyd, M. L., Clifford, M., Cobb, N. S., et al. (2009). Relationship of stand characteristics to drought-induced mortality in three southwestern piñion-juniper woodlands. *Ecological Applications, 19*, 1223–1230.

Giambelluca, T., Delay, J., Asner, G. et al. (2008a). *Stand structural controls on evapotranspiration in native and invaded tropical montane cloud forest in Hawai'i*. [Abstract]. http://adsabs.harvard.edu/abs/2008AGUFM.B43A0422G. 11 Dec 2011.

Giambelluca, T. W., Diaz, H. F., & Luke, S. A. (2008b). Secular temperature changes in Hawai'i. *Geophysical Research Letters, 35*, L12702. doi:10.1029/2008GL034377.

Gibson, K., Skov, K., Kegley, S. et al. (2008). *Mountain pine beetle impacts in high-elevation five-needle pines: current trends and challenges* (R1-08-020, 32pp). Missoula: U.S. Department of Agriculture, Forest Service, Forest Health Protection.

Goldblum, D. (2010). The geography of white oak's (*Quercus alba* L.) response to climatic variables in North America and speculation on its sensitivity to climate change across its range. *Dendrochronologia, 28*, 73–83.

Guo, Q., Brandle, J., Schoeneberger, M., & Buettner, D. (2004). Simulating the dynamics of linear forests in Great Plains agroecosystems under changing climates. *Canadian Journal of Forest Research, 34*, 2564–2572.

Harris, S. A., French, H. M., Heginbottom, J. A., et al. (1988). *Glossary of permafrost and related ground-ice terms* (Technical Memo No. 142, 156pp). Ottawa: National Research of Council Canada, Associate Committee on Geotechnical Research, Permafrost Subcommittee.

Haugen, D. E., Kangas, M., Crocker, S. J., et al. (2009). *North Dakota's forests 2005* (Resource Bulletin NRS-31, 82pp). Newtown Square: U.S. Department of Agriculture, Forest Service, Northern Research Station.

Hennon, P. E., Shaw, C. G., & Hansen, E. M. (1990). Dating decline and mortality of *Chamaecyparis nootkatensis* in southeast Alaska. *Forest Science, 36*, 502–515.

Hennon, P. E., D'Amore, D., Wittwer, D., et al. (2006). Climate warming, reduced snow, and freezing injury could explain the demise of yellow-cedar in southeast Alaska, USA. *World Resource Review, 18*, 427–450.

Holman, M. L., & Peterson, D. L. (2006). Spatial and temporal variability in forest growth in the Olympic Mountains, Washington: Sensitivity to climatic variability. *Canadian Journal of Forest Research, 36*, 92–104.

Huntington, T. G., Richardson, A. D., McGuire, K. J., & Hayhoe, K. (2009). Climate and hydrological changes in the northeastern United States: Recent trends and implications for forested and aquatic ecosystems. *Canadian Journal of Forest Research, 39*, 199–212.

Hurteau, M. D., Koch, G. W., & Hungate, B. A. (2008). Carbon protection and fire risk reduction: Toward a full accounting of forest carbon offsets. *Frontiers in Ecology and the Environment, 6*, 493–498.

Incident Information System. (2011). *Wallow*. http://www.inciweb.org/incident/2262. 19 Dec 2011.

Iverson, L. R., Prasad, A. M., & Matthews, S. (2008a). Modeling potential climate change impacts on the trees of the northeastern United States. *Mitigation and Adaptation Strategies for Global Change, 13*, 517–540.

Iverson, L. R., Prasad, A. M., Matthews, S. N., & Peters, M. (2008b). Estimating potential habitat for 134 eastern US tree species under six climate scenarios. *Forest Ecology and Management, 254*, 390–406.

Johnsen, K. H., Butnor, J. R., Kush, J. S., et al. (2009). Hurricane Katrina winds damaged longleaf pine less than loblolly pine. *Southern Journal of Applied Forestry, 3*, 178–181.

Johnson, M. G., & Kern, J. S. (2003). Quantifying the organic carbon held in forested sols of the United States and Puerto Rico. In J. M. Kimble, L. S. Heath, R. A. Birdsey, R. Lal, J. M. Kimble, L. S. Heath, R. A. Birdsey, & R. Lal (Eds.), *The potential of U.S. forest soils to sequester carbon and mitigate the greehouse effect* (pp. 47–72). New York: CRC Press.

Johnson, E., Geissler, G., & Murray, D. (2010). *The Oklahoma forest resource assessment, 2010: A comprehensive analysis of forest-related conditions, trends, threats and opportunities* (163pp). Oklahoma City: Oklahoma Department of Agriculture, Food, and Forestry.

Johnstone, J. F., & Chapin, F. S. (2006). Effects of soil burn severity on post-fire tree recruitment in boreal forest. *Ecosystems, 9*, 14–31.

Johnstone, J. F., & Kasischke, E. S. (2005). Stand-level effects of soil burn severity on post-fire regeneration in a recently-burned black spruce forest. *Canadian Journal of Forest Research, 35*, 2151–2163.

Johnstone, J. F., Chapin, F. S., Hollingsworth, T. N., et al. (2010a). Fire, climate change, and forest resilience in interior Alaska. *Canadian Journal of Forest Research, 40*, 1302–1312.

Johnstone, J. F., Hollingsworth, T. N., Chapin, F. S., & Mack, M. C. (2010b). Changes in fire regime break the legacy lock on successional trajectories in Alaskan boreal forest. *Global Change Biology, 16*, 1281–1295.

Karl, T. R., Knight, R. W., & Plummer, N. (1995). Trends in high-frequency climate variability in the twentieth century. *Nature, 377*, 217–220.

Kasischke, E. S., & Johnstone, J. F. (2005). Variation in ground-layer surface fuel consumption and its effects on site characteristics in a *Picea mariana* forest complex in Interior Alaska. *Canadian Journal of Forest Research, 35*, 2164–2177.

Kasischke, E. S., Verbyla, D. L., Rupp, T. S., et al. (2010). Alaska's changing fire regime—Implications for the vulnerability of its boreal forests. *Canadian Journal of Forest Research, 40*, 1313–1324.

Kittredge, D. B. (2004). Extension/outreach implications for America's family forest owners. *Journal of Forestry, 102*, 15–18.

Klein, E., Berg, E. E., & Dial, R. (2005). Wetland drying and succession across the Kenai Peninsula lowlands, south-central Alaska. *Canadian Journal of Forest Research, 35*, 1931–1941.

Koteen, L. (2002). Climate change, whitebark pine, and grizzly bears in the Greater Yellowstone Ecosystem. In S. Schneider & T. Root (Eds.), *Wildlife response to climate change: North American case studies* (pp. 343–414). Washington, DC: Island Press.

Lapina, I., & Carlson, M. L. (2004). *Non-native plant species of Susitna, Matanuska, and Copper River basins: Summary of survey findings and recommendations for control actions* (Final Report, 64pp). Anchorage: U.S. Department of Agriculture, Forest Service, State and Private Forestry.

Larsen, C. F., Motyka, R. J., Freymueller, J. T., et al. (2005). Rapid viscoelastic uplift in southeast Alaska caused by post-Little Ice Age glacial retreat. *Earth and Planetary Science Letters, 237*, 548–560.

Lenihan, J. M., Drapek, R., Bachelet, D., & Neilson, R. P. (2003). Climate change effects on vegetation distribution, carbon, and fire in California. *Ecological Applications, 13*, 1667–1681.

Lenihan, J., Bachelet, D., Neilson, R., & Drapek, R. (2008). Simulated response of conterminous United States ecosystems to climate change at different levels of fire suppression, CO_2 emission rate, and growth response to CO_2. *Global and Planetary Change, 64*, 16–25.

Li, C., Barclay, H. J., Hawkes, B. C., & Taylor, S. W. (2005). Lodgepole pine forest age class dynamics and susceptibility to mountain pine beetle attack. *Ecological Complexity, 2*, 232–239.

Littell, J. S. (n.d). *Relationships between area burned and climate in the Western United States: Vegetation-specific historical and future fire*. Manuscript in preparation. On file with: U.S. Geological Survey, Alaska Climate Science Center, 4210 University Drive, Anchorage, AK 99508.

Littell, J. S., Peterson, D. L., & Tjoelker, M. (2008). Douglas-fir growth in mountain ecosystems: Water limits tree growth from stand to region. *Ecological Monographs, 78*, 349–368.

Littell, J. S., McKenzie, D., Peterson, D. L., & Westerling, A. L. (2009). Climate and wildfire area burned in western U.S. ecoprovinces, 1916–2003. *Ecological Applications, 19*, 1003–1021.

Littell, J. S., Oneil, E. E., McKenzie, D., et al. (2010). Forest ecosystems, disturbance, and climatic change in Washington State, USA. *Climatic Change, 102*, 129–158.

Liu, Y., Stanturf, J., & Goodrick, S. (2010). Trends in global wildfire potential in a changing climate. *Forest Ecology and Management, 259*, 685–697.

Lockaby, B. G, Nagy, C., Vose, J. M., et al. (2011). Water and forests. In *Southern forest futures technical report*. Asheville: U.S. Department of Agriculture, Forest Service, Southern Research Station. Chapter 13. http://www.srs.fs.usda.gov/futures/reports/draft/pdf/Chapter%2013.pdf

McKenney, D. W., Pedlar, J. H., Lawrence, K., et al. (2007). Potential impacts of climate change on the distribution of North American trees. *BioScience, 57*, 939–948.

McKenney, D. W., Pedlar, J. H., Rood, R. B., & Price, D. (2011). Revisiting projected shifts in the climate envelopes of North American trees using updated general circulation models. *Global Change Biology, 17*(8), 2720–2730.

McKenzie, D., Peterson, D. W., Peterson, D. L., & Thornton, P. E. (2003). Climatic and biophysical controls on conifer species distributions in mountain forests of Washington State, USA. *Journal of Biogeography, 30*, 1093–1108.

McKenzie, D., Gedalof, Z., Peterson, D. L., & Mote, P. (2004). Climatic change, wildfire, and conservation. *Conservation Biology, 18*, 890–902.

McNulty, S. G. (2002). Hurricane impacts on U.S. forest carbon sequestration. *Environmental Pollution, 116*, s17–s24.

McNulty, S. G., & Boggs, J. L. (2010). A conceptual framework: Redefining forest soil's critical acid loads under a changing climate. *Environmental Pollution, 158*, 2053–2058.

McNulty, S. G., Lorio, P. L., Ayres, M. P., & Reeve, J. D. (1998a). Predictions of southern pine beetle populations using a forest ecosystem model. In R. A. Mickler & S. A. Fox (Eds.), *The productivity and sustainability of Southern forest ecosystems in a changing environment* (pp. 617–634). New York: Springer.

McNulty, S. G., Vose, J. M., & Swank, W. T. (1998b). Predictions and projections of pine productivity and hydrology in response to climate change across the southern United States. In R. A. Mickler & S. A. Fox (Eds.), *The productivity and sustainability of Southern forest ecosystems in a changing environment* (pp. 391–406). New York: Springer.

Mehmood, S. R., & Zhang, D. (2001). Forest parcelization in the United States: A study of contributing factors. *Journal of Forestry, 99*, 30–34.

Meneguzzo, D. M., Butler, B. J., Crocker, S. J. et al. (2008). *Nebraska's forests, 2005* (Resource Bulletin NRS-27, 94pp). Newtown Square: U.S. Department of Agriculture, Forest Service, Northern Research Station.

Millsaps, W., & Groothuis, P. A. (2003). *The economic impact of North Carolina ski areas on the economy of North Carolina: 2002–2003 season*. North Carolina Ski Areas Association. http://www.goskinc.com.

Mimura, N. (1999). Vulnerability of island countries in the South Pacific to sea level rise and climate change. *Climate Research, 12*, 137–143.

Mimura, N., Nurse, L., & McLean, R. F. (2007). Small islands. In M. L. Parry, O. F. Canziani, & J. P. Palutikof (Eds.), *Climate change 2007: Impacts, adaptation and vulnerability. Contribution of Working Group II to the fourth assessment report of the Intergovernmental Panel on Climate Change* (pp. 687–716). Cambridge: Cambridge University Press.

Mladenoff, D. J., & Pastor, J. (1993). Sustainable forest ecosystems in the northern hardwood and conifer region: concepts and management. In G. H. Aplet, J. T. Olsen, N. Johnson, & V. A. Sample (Eds.), *Defining sustainable forestry* (pp. 145–180). Washington, DC: Island Press.

Mohan, J. E., Cox, R. M., & Iverson, L. R. (2009). Composition and carbon dynamics of forests in northeastern North America in a future, warmer world. *Canadian Journal of Forest Research, 39*, 213–230.

Montana Department of Natural Resources and Conservation. (2010). *Montana's state assessment of forest resources: Base findings and GIS methodology* (29pp). Missoula: Montana Department of Natural Resources and Conservation.

Morgan, J. A., Follett, R. F., Allen, L. H., et al. (2010). Carbon sequestration in agricultural lands in the United States. *Journal of Soil and Water Conservation, 65*, 6A–13A.

Moser, W. K., Hansen, M. H., Atchison, R. L., et al. (2008). Kansas forests, 2005 (Resource Bulletin NRS-26, 125pp). Newtown Square: U.S. Department of Agriculture, Forest Service, Northern Research Station.

Mote, P. W., & Salathé, E. P. (2010). Future climate in the Pacific Northwest. *Climatic Change, 102*, 29–50.

Motyka, R. J., O'Neel, S., Conner, C. L., & Echelmeyer, K. A. (2002). Twentieth century thinning of Mendenhall Glacier, Alaska, and its relationship to climate, lake calving, and glacier run-off. *Global and Planetary Change, 35*, 93–112.

National Interagency Fire Center. (2011). *Statistics: National year-to-date report on fires and acres burned by state*. http://www.nifc.gov/fireInfo/fireInfo_stats_YTD2011.html. 19 Dec 2011.

National Land Cover Database. (2001). *Multi-resolution land characteristics consortium: National land cover database*. http://www.mrlc.gov/nlcd01_data.php. 23 Dec 2011.

National Synthesis Assessment Team. (2003). *U.S. national assessment of the potential consequences of climate variability and change: educational resources: Regional paper, Alaska*. http://www.usgcrp.gov/usgcrp/nacc/education/alaska/default.htm. 3 Sept 2010.

Nelson, M. D., Liknes, G. C., & Butler, B. J. (2010). *Map of forest ownership in the conterminous United States [Scale 1:7,500,000]* (Research Map NRS-2). Newtown Square: U.S. Department of Agriculture, Forest Service, Northern Research Station.

Norby, R., DeLucia, E., Gielen, B., et al. (2005). Forest response to elevated CO_2 is conserved across a broad range of productivity. *Proceedings of the National Academy of Sciences, USA, 102*, 18052–18056.

Norton, C. W., Chu, P. S., & Schroeder, T. A. (2011). Projecting changes in future heavy rainfall events for Oahu, Hawaii: A statistical downscaling approach [Abstract]. *Journal of Geophysical Research, 116*, D17110.

Nowacki, G. J., & Abrams, M. D. (2008). The demise of fire and "mesophication" of forests in the eastern United States. *BioScience, 58*, 123–138.

Oki, D. S. (2004). *Trends in streamflow characteristics at long-term gaging stations, Hawaii*. (Scientific Investigations Report 2004–5080, 120pp). Denver: U.S. Geological Survey.

Ollinger, S. V., Goodale, C. L., Hayhoe, K., & Jenkins, J. P. (2008). Potential effects of climate change and rising CO_2 on ecosystem processes in northeastern U.S. forests. *Mitigating Adaptation Strategies for Global Change, 13*, 467–485.

Pan, Y., Birdsey, R. A., Fang, J., et al. (2011). A large and persistent carbon sink in the world's forests, 1990–2007. *Science, 333*, 988–993.

Parson, E. A., Carter, L., Anderson, P., et al. (2001). Potential consequences of climate variability and change for Alaska. In N. A. S. Team (Ed.), *Climate change impacts on the United States– Foundation report* (pp. 283–312). Cambridge: Cambridge University Press.

Perry, L. G., Andersen, D. C., Reynolds, L. V., et al. (2012). Vulnerability of riparian ecosystems to elevated CO_2 and climate change in arid and semiarid western North America. *Global Change Biology, 18*, 821–842.

Peterson, D. W., & Peterson, D. L. (2001). Mountain hemlock growth responds to climatic variability at annual and decadal time scales. *Ecology, 82*, 3330–3345.

Peterson, D. L., Halofsky, J. L., & Johnson, M. C. (2011). Managing and adapting to changing fire regimes in a warmer climate. In D. McKenzie, C. Miller, & D. Falk (Eds.), *The landscape ecology of fires* (pp. 249–267). New York: Springer.

Pojar, J., & MacKinnon, A. (1994). *Plants of the Pacific Northwest coast: Washington, Oregon, British Columbia, and Alaska* (p. 528). Redmond: Lone Pine Publishing.

Powell, J., & Logan, J. (2005). Insect seasonality: Circle map analysis of temperature-driven life cycles. *Theoretical Population Biology, 67*, 161–179.

Prasad, A. M., Iverson, L. R., Matthews, S., & Peters, M. (2007-ongoing). *A climate change atlas for 134 forest tree species of the eastern United States*. Delaware: U.S. Department of Agriculture, Forest Service, Northern Research Station. Database at http://www.nrs.fs.fed.us/atlas/tree.

Pye, J. M., Holmes, T. P., Prestemon, J. P., & Wear, D. N. (2011). Economic impacts of the southern pine beetle. In R. N. Coulson, & K. D. Klepzig (Eds.), *Southern pine beetle II* (General Technical Report SRS-140, pp. 213–222). Asheville: U.S. Department of Agriculture, Forest Service, Southern Research Station.

Randerson, J. T., Liu, H., Flanner, M. G., et al. (2006). The impact of boreal forest fire on climate warming. *Science, 314*, 1130–1132.

Rehfeldt, G. E., Crookston, N. L., Warwell, M. V., & Evans, J. S. (2006). Empirical analyses of plant-climate relationships for the western United States. *International Journal of Plant Sciences, 167*, 1123–1150.

Rehfeldt, G. E., Ferguson, D. E., & Crookston, N. L. (2008). Quantifying the abundance of co-occurring conifers along Inland Northwest (USA) climate gradients. *Ecology, 89*, 2127–2139.

Rehfeldt, G. E., Ferguson, D. E., & Crookston, N. L. (2009). Aspen, climate, and sudden decline in western USA. *Forest Ecology and Management, 258*, 2353–2364.

Reid, W. V., Mooney, H. A., Cropper, A., et al. (2005). *Ecosystems and human well-being: Synthesis* (155pp). Washington, DC: Island Press.

Richardson, A. D., Bailey, A. S., Denny, E. G., et al. (2006). Phenology of a northern hardwood forest canopy. *Global Change Biology, 12*, 1174–1188.

Robertson, G., & Brooks, D. (2001). *Assessment of the competitive position of the forest products sector in southeast Alaska, 1985–94* (General Technical Report PNW-GTR-504, 29pp). Portland: U.S. Department of Agriculture, Forest Service, Pacific Northwest Research Station.

Rodenhouse, N. L., Matthews, S. N., McFarland, K. P., et al. (2008). Potential effects of climate change on birds of the Northeast. *Mitigation and Adaption Strategies for Global Change, 13*, 517–540.

Rodenhouse, N. L., Christenson, L. M., Parry, D., & Green, L. E. (2009). Climate change effects on native fauna of northeastern forests. *Canadian Journal of Forest Research, 39*, 249–263.

Rogers, B. M., Neilson, R. P., Drapek, R., et al. (2011). Impacts of climate change on fire regimes and carbon stocks of the U.S. Pacific Northwest. *Journal of Geophysical Research, 116*, 1–13.

Rustad, L. E., Melillo, J. M., & Mitchell, M. J. (2000). Effects of soil warming on carbon and nitrogen cycling. In R. Mickler, R. Birdsey, & J. Hom (Eds.), *Responses of northern U.S. forests to environmental change* (pp. 357–381). New York: Springer.

Rustad, L. E., Campbell, J. L., Cox, R. M., et al. (2009). NE forests 2100: A synthesis of climate change impacts on forests of the northeastern US and eastern Canada 2009. *Canadian Journal of Forest Research, 39*, iii–iv.

Safeeq, M., & Fares, A. (2011). Hydrologic response of a Hawaiian watershed to future climate change scenarios. *Hydrological Processes*. doi:10.1002/hyp.8328.

Schroeder, R. F., & Kookesh, M. (1990). *Subsistence harvest and use of fish and wildlife resources and the effects of forest management in Hoonah, Alaska* (Technical Paper 142, 326pp). Juneau: Alaska Department of Fish and Game, Division of Subsistence.

Schuur, E. A. G., Crummer, K. G., Vogel, J. G., & Mack, M. C. (2007). Plant species composition and productivity following permafrost thaw and thermokarst in Alaskan tundra. *Ecosystems, 10*, 280–292.

Schuur, E. A. G., Bockheim, J., Canadell, J. G., et al. (2008). Vulnerability of permafrost carbon to climate change: Implications for the global carbon cycle. *Bioscience, 58*, 701–714.

Shafer, S. L., Harmon, M. E., Neilson, R. P., et al. (2010). The potential effects of climate change on Oregon's vegetation. In K. D. Dello & P. W. Mote (Eds.), *Oregon climate assessment report* (pp. 175–210). Corvallis: Oregon State University, College of Oceanic and Atmospheric Sciences, Oregon Climate Change Research Institute.

Skinner, M., Parker, B. L., Gouli, S., & Ashikaga, T. (2003). Regional responses of hemlock woolly adelgid (Homoptera: Adelgidae) to low temperatures. *Environmental Entomology, 32*, 523–528.

Smith, L. C., Sheng, Y., MacDonald, G. M., & Hinzman, L. D. (2005). Disappearing arctic lakes. *Science, 308*, 1429–1429.

Smith, W. B., tech. coord.: Miles, P. D., data coord.: Perry, C. H., Map Coord., Pugh, S. A., data CD coord. (2009). *Forest resources of the United States, 2007* (GeneralTechnical ReportGTR-WO-78, 336pp). Washington, DC: U.S. Department of Agriculture, Forest Service.

Sohngen, B., Mendelsohn, R., & Sedjo, R. (2001). A global model of climate change impacts on timber markets. *Journal of Agricultural and Resource Economics, 26*, 326–343.

South Dakota Resource Conservation and Forestry Division. (2007). *South Dakota forest stewardship plan, 2007 revision* (35pp). Pierre: South Dakota Department of Agriculture, Resource Conservation and Forestry Division.

Sun, G., McNulty, S. G., Moore Myers, J. A., & Cohen, E. C. (2008). Impacts of stresses on water demand and supply across the southeastern United States. *Journal of the American Water Resources Association, 44*, 1441–145.

Swanston, C. W., Janowiak, M., Iverson, L., et al. (2011). *Ecosystem vulnerability assessment and synthesis: A report from the Climate Change Response Framework Project in northern Wisconsin, Version 1* (General Technical Report NRS-82, 142pp). Newtown Square: U.S. Department of Agriculture, Forest Service, Northern Research Station.

Tan, Z., Tieszen, L. L., Zhu, Z., et al. (2007). An estimate of carbon emissions from 2004 wildfires across Alaskan Yukon River Basin. *Carbon Balance and Management, 2*, 12. doi:10.1186/1750-0680-2-12.

Thompson, I., Mackey, B., McNulty, S., & Mosseler, A. (2009). *Forest resilience, biodiversity, and climate change* (Technical Series no. 43, 67pp). Montreal: Secretariat of the Convention on Biological Diversity.

Trimble, S. W. (2008). *Man-induced soil erosion on the Southern Piedmont: 1700–1970* (2nd ed., 80pp, p. 80). Ankeny: Soil and Water Conservation Society.

U.S. Department of Agriculture, Forest Service. (2011). *Rocky Mountain bark beetle*. http://www.fs.usda.gov/main/barkbeetle/home. 19 Dec 2011.

Vogel, J., Schuur, E. A. G., Trucco, C., & Lee, H. (2009). Response of CO_2 exchange in a tussock tundra ecosystem to permafrost thaw and thermokarst development. *Journal of Geophysical Research, 114*, G04018. doi:10.1029/2008JG000901.

Walker, K. V., Davis, M. B., & Sugita, S. (2002). Climate change and shifts in potential tree species range limits in the Great Lakes region. *Journal of Great Lakes Research., 28*, 555–567.

Wang, C., Lu, Z., & Haithcoat, T. L. (2007). Using Landsat images to detect oak decline in the Mark Twain National Forest, Ozark Highlands. *Forest Ecology and Management, 24*, 70–78.

Weiss, J. L., Castro, C. L., & Overpeck, J. T. (2009). Distinguishing pronounced droughts in the southwestern United States: Seasonality and effects of warmer temperatures. *Journal of Climate, 22*, 5918–5932.

Werner, R. A., Holsten, E. H., Matsuoka, S. M., & Burnside, R. E. (2006). Spruce beetles and forest ecosystems in south-central Alaska: A review of 30 years of research. *Forest Ecology and Management, 227*, 195–206.

Wertin, T. M., McGuire, M. A., & Teskey, R. O. (2010). The influence of elevated temperature, elevated atmospheric CO_2 concentration and water stress on net photosynthesis of loblolly pine (*Pinus taeda* L.) at northern, central and southern sites in its native range. *Global Change Biology, 16*, 2089–2103.

Westerling, A. L., Turner, M. G., Smithwick, E. A. H., et al. (2011). Continued warming could transform Greater Yellowstone fire regimes by mid-21st century. *Proceedings of the National Academy of Sciences, USA, 108*, 13165–13170.

Whitlock, C., Shafer, S., Marlon, J., et al. (2003). The role of climate and vegetation change in shaping past and future fire regimes in the northwestern U.S. and the implication for ecosystem management. *Forest Ecology and Management, 178*, 5–21.

Whitlow, T. H., & Harris, R. W. (1979). *Flood tolerance in plants: A state-of-the-art review* (pp. 1–161). Washington, DC: U.S. Department of Commerce, National Technical Information Service.

Wiedinmyer, C., & Hurteau, M. D. (2010). Prescribed fire as a means of reducing forest carbon emissions in the western United States. *Environmental Science and Technology, 44*, 1926–1932.

Williams, A. P., Allen, C. D., Millar, C. I., et al. (2010). Forest responses to increasing aridity and warmth in southwestern North America. *Proceedings of the National Academy of Sciences, USA, 107*, 21289–21294.

Wolken, J. M., Hollingsworth, T. N., Rupp, T. S., et al. (2011). Evidence and implications of recent and projected climate change in Alaska's forest ecosystems. *Ecosphere, 2*(11), art124.

Wyckoff, P. H., & Bowers, R. (2010). Response of the prairie–forest border to climate change: Impacts of increasing drought may be mitigated by increasing CO_2. *Journal of Ecology, 98*, 197–208.

Wyoming State Forestry Division. (2009). *Wyoming statewide assessment* (74pp). Cheyenne: Wyoming State Forestry.

Xu, W. (2002). *Economic impact of the Texas forest sector* (Texas Agricultural Extension Publication 161, 19pp). College Station: Texas AgriLife Extension Service.

Xu, C., Gertner, G., & Scheller, R. (2012). Importance of colonization and competition in forest landscape response to global climatic change. *Climatic Change, 110*, 53–83.

Part III
Responding to Climate Change

Chapter 7
Managing Carbon

Kenneth E. Skog, Duncan C. McKinley, Richard A. Birdsey, Sarah J. Hines, Christopher W. Woodall, Elizabeth D. Reinhardt, and James M. Vose

7.1 Introduction

Storing carbon (C) and offsetting carbon dioxide (CO_2) emissions with the use of wood for energy, both of which slow emissions of CO_2 into the atmosphere, present significant challenges for forest management (IPCC 2001). In the United States, there has been a net increase in C in forests and in harvested wood products stocks (Tables 7.1 and 7.2), a result of historical and recent ecological conditions, management practices, and use of forest products (Birdsey et al. 2006). However,

---

K.E. Skog (✉)
Forest Products Laboratory, U.S. Forest Service, Madison, WI, USA
e-mail: kskog@fs.fed.us

D.C. McKinley
National Research and Development Staff, U.S. Forest Service, Washington, DC, USA
e-mail: dcmckinley@fs.fed.us

R.A. Birdsey
Northern Research Station, U.S. Forest Service, Newtown Square, PA, USA
e-mail: rbirdsey@fs.fed.us

S.J. Hines
Rocky Mountain Research Station, U.S. Forest Service, Fort Collins, CO, USA
e-mail: shines@fs.fed.us

C.W. Woodall
Northern Research Station, U.S. Forest Service, Saint Paul, MN, USA
e-mail: cwoodall@fs.fed.us

E.D. Reinhardt
Fire and Aviation Management, U.S. Forest Service, Washington, DC, USA
e-mail: ereinhardt@fs.fed.us

J.M. Vose
Southern Research Station, U.S. Forest Service, Raleigh, NC, USA
e-mail: jvose@fs.fed.us

Table 7.1 Net annual changes in carbon (C) stocks in forest and harvested wood pools, 1990–2009

C pool	1990	2000	2005	2009
	(Tg C year^{-1})			
Forest				
Live, aboveground	−98.2	−78.3	−122.1	−122.1
Live, belowground	−19.3	−15.7	−24.1	−24.1
Dead wood	−8.6	−3.5	−8.4	−9.1
Litter	−8.8	7.5	−11.4	−11.4
Soil organic C	−14.9	17.6	−53.8	−53.8
Total forest	−149.8	−72.4	−219.9	−220.6
Harvested wood products				
Products in use	−17.7	−12.8	−12.4	1.9
Products in solid waste disposal sites	−18.3	−18.0	−16.3	−16.7
Total harvested wood products	−35.9	−30.8	−28.7	−14.8
Total net flux	−185.7	−103.2	−248.6	−235.4

From USEPA (2011)

Table 7.2 Carbon (C) stocks in forest and harvested wood pools, 1990–2010

C pool	1990	2000	2005	2009
	(Tg C)			
Forest				
Live, aboveground	15,072	16,024	16,536	17,147
Live, belowground	2,995	3,183	3,285	3,405
Dead wood	2,960	3,031	3,060	3,105
Litter	4,791	4,845	4,862	4,919
Soil organic C	16,965	17,025	17,143	17,412
Total forest	42,783	44,108	44,886	45,988
Harvested wood products				
Products in use	1,231	1,382	1,436	1,474
Products in solid waste disposal sites	628	805	890	974
Total harvested wood products	1,859	2,187	2,325	2,449
Total C stock	44,643	46,296	47,211	48,437

From USEPA (2011)

recent projections for the forest sector suggest that annual C storage could begin to decline, and U.S. forests could become a net C emitter of tens to hundreds of Tg C year^{-1} within a few decades (USDA FS 2012a). It is therefore urgent to identify effective C management strategies, given the complexity of factors that drive the forest C cycle and the multiple objectives for which forests are managed. An ideal C management activity contributes benefits beyond increasing C storage by achieving other management objectives and providing ecosystem services in a sustainable manner.

Strategies for effectively managing forest C stocks and offsetting C emissions requires a thorough understanding of biophysical and social influences on the forest

C cycle (Birdsey et al. 1993). Successful policies and incentives may be chosen to support strategies if sufficient knowledge of social processes (e.g., landowner or wood-user response to incentives and markets) is available. For example, if C stocks are expected to decrease owing to decreasing forest land area caused by exurban development, policies or incentives to avoid deforestation in those areas may be effective. If C stocks are expected to decrease owing to the effects of a warmer climate, reducing stand densities may retain C over the long term by increasing resilience to drought and other stressors and by reducing crown fire hazard (Jackson et al. 2005; Reinhardt et al. 2008). Protecting old forests and other forests that have high C stocks may be more effective than seeking C offsets associated with wood use, especially if those forests would recover C more slowly in an altered climate.

If climate change increases productivity in a given area over a long period of time, increasing forest C stocks through intensive management and forest products, including biomass energy, may be especially effective. It is equally important to know which strategies might make some management practices unacceptable (e.g., reducing biodiversity). However, no standard evaluation framework exists to aid decision making on alternative management strategies for maximizing C storage while minimizing risks and tradeoffs.

Here we discuss (1) where forest C is stored in the United States, (2) how to measure forest C through space and time, (3) effectiveness of various management strategies in reducing atmospheric greenhouse gases (GHG), and (4) effectiveness of incentives, regulations, and institutional arrangements for implementing C management.

7.2 Status and Trends in Forest-Related C

Net annual C additions to forests (84 %) and harvested wood products used in human settlements and infrastructure (10 %) account for most of the total annual GHG sequestration in the United States. The two largest C components in forests are aboveground biomass (37 %) and soil organic C (38 %), with the rest distributed among belowground biomass (8 %), litter (11 %), and dead wood (6 %). Because aboveground biomass accumulates, then shifts to dead wood, litter, or wood products in a matter of decades, forest management and land use activities can affect aboveground biomass distribution over decades. In other words, management modifications can potentially increase C accumulation and emission offsets.

Change in forest area and forest C per unit area (C density) determines the change in C stocks over time. Since the 1950s, U.S. forest area has been relatively stable (Fig. 7.1) while C density has been increasing. Large-scale reforestation of the United States since the early 1900s is the primary cause of expansion of forest area over time. Increasing C sequestration is a result of gross growth per year continuing to increase, while mortality has increased slowly and harvest removals have stabilized (Fig. 7.1). Despite national trends of stable forest area and increasing C, it is likely that mortality plus harvest exceeds growth in some areas.

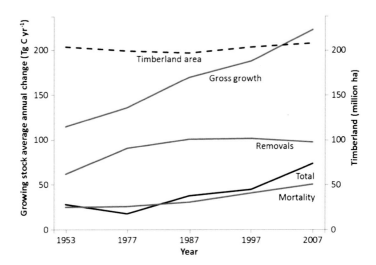

Fig. 7.1 Growing stock carbon change is affected by growth, mortality, and removals, along with timberland area, 1953–2007

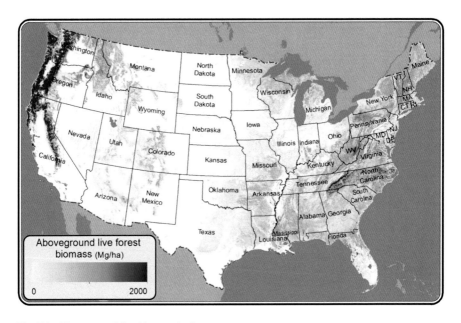

Fig. 7.2 Aboveground live biomass in forests

Aboveground biomass C stocks are largely found in the Pacific coastal region, Appalachian Mountains, Rocky Mountains, Lake States, and central hardwoods (Fig. 7.2). Net annual C sequestration can vary considerably at small spatial scales, and a forest can quickly become a net emitter of C following local disturbances

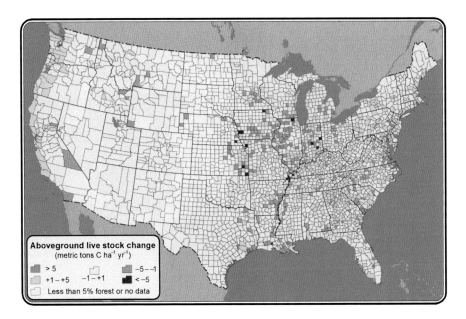

Fig. 7.3 Aboveground live forest carbon change

such as wildfire. Although C stocks have been increasing in most U.S. counties in recent years (Fig. 7.3), uncertainty in annual net sequestration increases greatly as the spatial scale decreases. Given the low density of forest plots that are remeasured each year, estimates of interannual variation in forest C for a local area may be detectable only after major changes in forest structure caused by harvest, wildfire, or other disturbances. Because of inherent variation in C stocks at small spatial scales, it makes more sense to quantify C dynamics at large scales when measuring C sequestration and effectiveness of C management strategies.

7.3 Monitoring and Evaluating Effects of C Management

Effectiveness of C management activities for mitigating GHG emissions is based on forest removal (and retention) of CO_2 from the atmosphere. Figure 7.4 shows C storage and emission processes that can be affected by management of C in forests and wood products. Carbon changes are evaluated by tracking C flows across system boundaries over time. The boundary around the "forest sector" includes forest, wood products, and wood energy processes for a defined forest area. A system can be defined to include only C fluxes to and from forests or wood products, or it may include C fluxes from equipment used to manage forests and manufacture and transport wood products, nonwood products, and fossil fuel feedstocks.

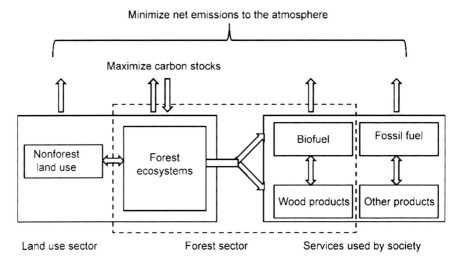

Fig. 7.4 Forest sector and non-forest sector greenhouse gas emissions and stock changes that are influenced by forest management

Forest management can affect GHG emissions beyond the "forest sector." System boundaries can be expanded to include substitution of wood energy for fossil fuels, and substitution of wood products for non-wood products that produce higher levels of GHGs. System boundaries can also be expanded beyond the defined forest area to nonforest areas where actions may cause indirect land use change and associated GHG emissions. System boundaries also include a definition of the time period over which C storage or emissions are evaluated. The effects of altering a C management strategy, storing C, or altering emissions cannot be assessed without clearly defining system boundaries, processes, and time period. Unfortunately, no standard approach exists for evaluating forest biomass as a replacement for fossil fuels.

Evaluating C management strategies associated with forests requires (at a minimum) (1) monitoring C stock changes and emissions over time, and (2) evaluating the effects of altered activities that affect in-forest C (in situ) and associated C storage or emissions outside forests (ex situ). One accounting framework (type A) determines how C fluxes in terrestrial systems and harvested wood products have changed for a current or past period because of management actions and other factors such as natural disturbances. Another accounting framework (type B) determines the degree to which a change in management under various mitigation strategies could increase C storage and decrease emissions.

This accounting approach determines the magnitude of additional C offsets compared to a baseline, where the baseline is the level of C stock, C stock change, level of emissions, or emissions change for a given set of land conditions and activities (e.g., forest management, timber harvest, and disturbances) and off-land activities (e.g., substitution for fossil emissions, as defined by the accounting system and boundaries at a point in time or over a period of time). A baseline can be defined by past conditions or projected future conditions. The effectiveness of a new strategy (e.g., providing an incentive to increase wood use for energy) is determined by changes in landowner behavior. For example, high energy use (high price) may motivate some landowners to convert non-forest land to wood plantations, thus accumulating C and benefiting from wood substitution for fossil fuel. In addition, an increase in wood prices could cause pulpwood to be used for energy rather than for oriented strandboard panel production and associated C storage.

Accounting for the effects of forest management on C must include, explicitly or implicitly, specification of the accounting framework (type A or B) and system boundaries for processes included (e.g., forest sector, service sector, non-forest land use, specific forest area, time period, wood C only, and other GHG emissions). A "common" type A framework defines system boundaries to include current annual C exchange with the atmosphere from forest ecosystems at a given geographic scale, plus C additions and emissions for harvested wood products from those forests (Fig. 7.4). This framework can be used to answer the management question "Are forests and forest products continuing to (collectively) withdraw and store C from the atmosphere?" The framework is also the basis for reporting GHG emissions and sinks in many accounting systems such as that used in annual reports to the United Nations Framework Convention on Climate Change (United Nations 1992).

This framework is not intended to evaluate the full effects on atmospheric CO_2 of a change in strategy, which would require a system boundary that includes changes in non-wood C emissions and C emissions or storage outside the forest. Some excluded changes may include altered fossil fuel use, other land use emissions, and altered no-wood product emissions (Fig. 7.4). Evaluating strategy changes requires a framework that includes all processes that significantly change atmospheric CO_2. If changes in emissions occur over many years, the framework must evaluate CO_2 fluxes over many years. For example, a strategy to increase use of wood for heat, electric power, or biofuels via incentives at a national level would change CO_2 flux estimates compared to a given baseline over an extended time from (1) wood for energy, (2) fossil fuels for energy, (3) land use change (e.g., crops to plantation, or forest to intensive plantation), and (4) flux from forests where wood is removed (including regrowth after removal).

"Leakage" recognizes certain C effects in which the effects of a policy or management change are evaluated with a type B accounting framework. Leakage expressed the C effects of a program change outside the system boundaries defined by a limited set of processes (e.g., C changes for a specific forest area). Leakage includes C changes on land outside of a system boundary (e.g., caused by changes in harvest or land use) (Sohngen et al. 1999; Schwarze et al. 2002; Murray et al. 2004; Gan and McCarl 2007; Pachauri and Reisinger 2007), and differs depending on the mitigation activity (Murray et al. 2004; Gan and McCarl 2007). In the United States, leakage estimates associated with activities on a given land area range from less than 10 % to greater than 90 % (proportion of C benefit lost), depending on the activity and region (Murray et al. 2004). Globally, leakage estimates range between 42 and 95 % (Gan and McCarl 2007). Leakage tends to be highest where programs constrain the supply of forest products (e.g., no harvest is allowed) or constrain land use change (e.g., forest land conversion to agriculture) (Sohngen et al. 1999, 2008; Aukland et al. 2003; Murray et al. 2004; Depro et al. 2008; Sohngen and Brown 2008). In contrast, the indirect effects of a program can increase C benefits outside of a system boundary through "spillover" (Magnani et al. 2009). For example, spillover can occur if an increase in plantation forestry reduces C losses from established forests by increasing C flows in cheaper forest products (Magnani et al. 2009). Defining system boundaries to include indirect effects on C (e.g., multi-national programs) or otherwise accounting for leakage ensures program integrity.

Ineffectiveness of some C storage strategies may be caused by flaws in incentive structures or policies, not by biophysical attributes of the strategy itself. For example, an incentive program might favor harvesting large trees that produce lumber, assuming that lumber would replace building materials that emit more C in manufacturing. If this incentive strategy were implemented, the lumber could go to non-building uses, or an increase in harvest by one landowner could be offset by a decrease by another. This is a flaw of the incentive system, not of the underlying wood substitution strategy. If there were incentives for builders to use wood rather than alternate materials, the strategy could be effective in reducing emissions from manufacturing.

Life cycle assessment (LCA) can be used to evaluate C management strategies by focusing on the change in C storage or emissions associated with producing one unit of wood energy or one unit of wood product. An *attributional LCA*, which is similar to a type A accounting framework, estimates storage and emissions over the life cycle of one unit of product, including specification of forest growth, harvest, manufacturing, end use, disposal, and/or reuse. Attributional LCAs monitor inputs and emissions associated with production and do not include all process that would be affected by a change in production or processes. A *consequential LCA*, which is similar to a type B accounting framework, also estimates storage and emissions over the life cycle of one unit of product, but calculates the *change* in emissions associated with a one-unit change in product production caused by *change* in processes over the life cycle. Consequential LCAs are typically used to analyze the potential response of a change to a system, such as a change in policy, and can include the effects of altered product demand on production and emissions from products across many sectors.

It can be difficult to compare the effectiveness of different C management strategies, because they are often evaluated with different system boundaries, accounting frameworks, models, assumptions, functional units (land area vs. product units), and assumed incentives. However, it is possible to describe the effects of strategies on changing particular processes, uncertainty in attaining specific effects, and timing of the effects.

7.4 Carbon Mitigation Strategies

Carbon mitigation through forest management focuses on (1) increasing forest area (afforestation), avoiding deforestation, or both, (2) C management in existing forests, and (3) use of wood as biomass energy or in wood products for C storage and as a substitute for other building materials. Estimates of CO_2 emissions offset by forests and forest products in the United States (using the type A framework) vary from 10 to 20 % depending assumptions and accounting methods (McKinley et al. 2011), with 13 % (about 221 Tg C year^{-1}) being the estimate as of 2011 (USEPA 2011). The first two mitigation strategies above maintain or increase forest C stocks (using the type B framework with a boundary around forest area and other land capable of growing forests). The third strategy increases C storage or reduces emissions, including C fluxes associated with forests and products removed from the forest (using the type B framework with a boundary around the forest sector, services, and non-forest land processes) (Fig. 7.4). The mitigation potential of these strategies differs in timing and magnitude (Table 7.3).

7.4.1 Land Use Change: Afforestation, Avoiding Deforestation, and Urban Forestry

7.4.1.1 Afforestation

In the United States, estimates of the potential for afforestation (active establishment or planting of forests) to sequester C vary from 1 to 225 Tg C year^{-1} for 2010–2110 (U.S. Climate Change Science Program 2007; USEPA 2005). Afforestation can be done on land that has not been forested for some time (usually more than 20 years). Reforestation refers to establishing forests on land that was previously forested but has been in non-forest use for some time. Mitigation potential, co-benefits, and environmental tradeoffs depend on where afforestation and reforestation efforts are implemented (Table 7.3).

The mitigation potential of afforestation and reforestation is significant and generally has co-benefits, low risk, and few tradeoffs. Forest regrowth on abandoned cropland comprises about half of the additional potential C sink in the United States (Pacala et al. 2001). Sequestering the equivalent of 10 % of U.S. fossil fuel emissions (160 Tg C) would require 44 million ha (or one-third) of U.S. croplands to be converted to tree plantations (Jackson and Schlesinger 2004), with 0.3 to 1.1 million ha needed to sequester 1 Tg C annually (USEPA 2005). Forest establishment on productive, high-value agricultural land is unlikely and may cause leakage (Murray et al. 2004), although establishing forest plantations on less productive, low-value agricultural land is more feasible. Where climatic and soil conditions favor forest growth (over crops), irrigation and fertilization inputs would be low

Table 7.3 Mitigation strategies, timing of impacts, uncertainty in attaining carbon (C) effects, co-benefits and tradeoffs[a]

Mitigation strategy	Timing of maximum impact	Uncertainty about strategy (biophysical risks)	Uncertainty about strategy (structural risks)[b]	Co-benefits	Tradeoffs
Land use change					
Afforestation (on former forest land)	Delayed	Low	Leakage	Erosion control, improved water quality, increased biodiversity and wildlife habitat	Lost revenue from agriculture
Afforestation (on non-forest land)	Delayed	Moderate	Leakage	Biodiversity	Erosion, lower streamflow, decreased biodiversity and wildlife habitat, increased nitrous oxide emissions, competition for agricultural water, local warming from lower albedo
Avoided deforestation	Immediate	Low	Leakage	Watershed protection; maintenance of biodiversity and wildlife habitat, some recreation	Lost economic opportunities affecting farmers or developer directly
Urban forestry	Delayed	High		Reduced energy use for cooling; increased wildlife habitat, possible recreational opportunities	High maintenance might be required in terms of water, energy, and nutrients; possible damage to infrastructure
In situ forest C management					
Decreasing C outputs	Immediate	Moderate	Leakage	Increased old growth; increased structural and species diversity, and wildlife habitat; benefits depend on landscape condition, forest type, and wildlife species (e.g., may not benefit species requiring early sucessional habitat)	Displaced economic opportunities affecting forest owners, forest industry, and employees

7 Managing Carbon

Strategy	Timing	Uncertainty	Risk	Co-benefits	Tradeoffs
Increasing forest growth	Delayed	Low	Leakage	Higher wood production, potential for quicker adaptation to climate change	Lower streamflow, loss of biodiversity, release of nitrous oxide, greater impact of disturbance on C storage
Fuel treatments	Delayed	High		Lower risk from fire and insects, increased economic activity, possible additional offsets from use of wood, climate change adaptation tool	Lost economic opportunities to firefighting business and employees, lower carbon on site, site damage caused by treatment
Ex situ forest C management					
Product substitution	Part immediate, part delayed	Moderate	Leakage	Increased economic activity in forest product industries	Active forest management on larger area, lower C storage in forests
Biomass energy	Immediate to delayed, depending on source	Moderate/high	Leakage	Increased economic activity in forest product industries, possible lower cost of forest restoration	Intensive management on larger area; lower C storage in forests

[a] Uncertainty is defined here as the extent to which an outcome is unknown. Most mitigation strategies have a risk of leakage and reversal, which could compromise C benefits. Timing of maximum impact is adapted from Solomon et al. (2007) and uncertainty, co-benefits, and tradeoffs from McKinley et al. (2011)

[b] The potential degree of leakage or other structural risk for a strategy depends on the incentives, regulations, or policy used to implement it. For example, if an incentive program to increase forest growth occurs in only one region, then growth may be decreased in other regions. If the incentive is nationwide, there is little leakage within the United States, but there may be leakage to other countries. Other structural risks can result from improper selection of locations to implement the strategy. For example, fire hazard reduction treatments could be done on land areas where the removals of forest C are larger and of longer duration than the expected avoided emissions from fire. There can also be risk in selecting wood for fuel (e.g., from older forests) where C recovery will be slow

relative to gains in C storage, creating co-benefits such as erosion control, improved water quality, higher species diversity, and wildlife habitat.

Afforestation on lands that do not naturally support forests may require human intervention and environmental tradeoffs. Carbon storage in tree and shrub encroachment in grasslands, rangelands, and savannas could potentially be 120 Tg C year^{-1}, a C sink that could be equivalent to more than half of what existing U.S. forests sequester annually (U.S. Climate Change Science Program 2007), demonstrating the potential (unintentional) effects of land use change and other human activities (Van Auken 2000). However, planting trees (especially non-native species) where they were not present historically may alter the water table, cause soil erosion on hill slopes, and absorb more solar energy compared with a native ecosystem (Jobbágy and Jackson 2004; Farley et al. 2008; Jackson et al. 2008; McKinley and Blair 2008). Irrigation, where necessary, may compete with agricultural water supply and other uses, and water-demanding tree species can reduce streamflow (Farley et al. 2005; Jackson et al. 2005). Use of nitrogen (N) fertilizers may increase emissions of nitrous oxide, a GHG with 300 times greater radiative effect than CO_2.

7.4.1.2 Avoiding Deforestation

Avoiding the loss of forested land can prevent loss of C to the atmosphere. Estimates of potential C mitigation through avoided deforestation are not available for the United States; however on a global scale, deforestation results in the gross annual loss of 90,000 km^2, or 0.2 % of all forests (FAO 2007; Pachauri and Reisinger 2007), which releases 1,400–2,000 Tg C year^{-1}; two-thirds of the deforestation occurs in tropical forests in South America, Africa, and Southeast Asia (Houghton 2005; Pachauri and Reisinger 2007). Over a recent 150-year period, global land use change released 156,000 Tg C to the atmosphere, mostly from deforestation (Houghton 2005). In contrast, forest area in the United States increased at a net rate of 340,000 ha year^{-1} between 2002 and 2007. Increased forest area and regrowth are responsible for most of the current U.S. sink (USEPA 2011). However, land development and conversion of forest to agricultural land is expected to decrease forest area in the United States by 9 million ha by 2050 (Alig et al. 2003). In addition, increased area burned by fire may result in the conversion of some forests to non-forest (McKenzie et al. 2009), or a permanent reduction in C stocks on existing forests if fire-return intervals are reduced (Harden et al. 2000; Balshi et al. 2009). Successful regeneration after wildfires will help avoid conversion of forest to vegetation that retains less C (Keyser et al. 2008; Donato et al. 2009).

Avoided deforestation protects existing forest C stocks and has many co-benefits (Table 7.3), including maintaining the functionality of watersheds, plant and animal habitat, and some recreational activities (McKinley et al. 2011). However, incentives to avoid deforestation in one area may increase forest harvest elsewhere, deriving minimal reduction in atmospheric CO_2. Avoided deforestation may also decrease economic opportunities for timber, agriculture, and urban development in some areas (Meyfroidt et al. 2010). Leakage can be large for avoided deforestation, particularly if harvest is not allowed (Murray et al. 2004).

7.4.1.3 Urban Forestry

Planting and managing trees in and around human settlements offers limited potential to store additional C, but urban trees provide indirect reductions of fossil fuel emissions and have many co-benefits. Although urban C stocks in the United States are surprisingly large (Churkina et al. 2010), the potential for urban forestry to help offset GHG emissions is limited because urban areas comprise only 3.5 % of the landscape (Nowak and Crane 2002), and urban trees require intensive management. Urban forests affect local climate by cooling with shading and transpiration, potentially reducing fossil fuel emissions associated with air conditioning (Akbari 2002). In urban forests planted over very large areas, trees have both warming effects and cooling effects, resulting in complex patterns of convection that can alter air circulation and cloud formation (Jackson et al. 2008). Mortality of urban trees is generally high (Nowak et al. 2004), and they require ongoing maintenance, particularly in cities in arid regions. Risks increase when irrigation, fertilization, and other maintenance are necessary to maintain tree vigor (Pataki et al. 2006).

7.5 *In Situ* Forest Carbon Management

Carbon mitigation through forest management focuses on efforts to increase forest C stock by either decreasing C outputs in the form of harvest and disturbance, or increasing C inputs through active management. Carbon mitigation for a combined effort including increased harvest intervals, increased growth, and preserved establishment could remove 105 Tg C year^{-1}. Achieving these results would require large land areas, because 500,000–700,000 ha of manageable forest land are needed to store 1 Tg C year^{-1} (USEPA 2005).

7.5.1 Increasing Forest C by Decreasing Harvest and Protecting Large C Stocks

Forest management can increase forest C by increasing the interval between harvests or decreasing harvest intensity (Thornley and Cannell 2000; Liski et al. 2001; Harmon and Marks 2002; Jiang et al. 2002; Seely et al. 2002; Kaipainen et al. 2004; Balboa-Murias et al. 2006; Harmon et al. 2009). Increasing harvest intervals have the biggest effect on forests harvested at ages before peak rates of growth begin to decline (culmination of mean annual increment [CMAI]), such as some Douglas-fir

(*Pseudotsuga menziesii* [Mirb.] Franco) forests in the northwestern United States. Increasing rotation age for forests with low CMAI, such as Southern pine species that are already harvested near CMAI, would yield a decreasing benefit per year of extended rotation.

Harvesting forests with high biomass and planting a new forest reduce overall C stocks more in the near term than if the forest were retained, even counting the C storage in harvested wood products (Harmon et al. 1996, 2009). For example, some old-growth forests in Oregon store as much as 1,100 Mg C ha^{-1} (Smithwick et al. 2002), which would require centuries to regain if these stocks were liquidated and replaced, even with fast-growing trees (McKinley et al. 2011). Partial harvests, including leaving dead wood on site, maintain higher C stocks compared to clearcuts (Harmon et al. 2009) while concurrently allowing forests to be used for wood products or biomass energy. Although thinning increases the growth rate and vigor of residual trees, it generally reduces net C storage rates and C storage at the stand scale (Schonau and Coetzee 1989; Dore et al. 2010). Studies on the effects of harvest on soil C provide mixed results (Johnson and Curtis 2001; Nave et al. 2010). Benefits of decreased wood (and C) outputs from forests include an increase in structural and species diversity (Table 7.3). Risks include C loss from disturbance and reduced substitution of wood for more C-intensive materials.

7.5.2 Managing Forest Carbon with Fuel Treatments

Since 1990, CO_2 emissions from wildland forest fires in the conterminous United States have averaged 67 Tg C year^{-1} (USEPA 2009a, 2010). The possibility that fuel treatments, although reducing onsite C stocks, may contribute to mitigation by providing a source for biomass energy and avoiding future wildfire emissions is attractive, especially because fuel treatments have many co-benefits. Fuel treatments are a widespread forest management practice in the western United States (Battaglia et al. 2010) and are designed to alter fuel conditions to reduce wildfire intensity, crown fires, tree mortality, and suppression difficulty (Reinhardt et al. 2008; Scott and Reinhardt 2001). Fuel treatment to reduce crown fire hazard can be done by reducing surface fuels, ladder fuels (small trees), and canopy fuels (Peterson et al. 2005), all of which remove C from the site (Stephens et al. 2009; Reinhardt et al. 2010) and alter subsequent forest C dynamics.

Crown fires often result in extensive tree mortality, whereas many tree species can survive surface fires. This contrast in survival has led to the notion that fuel treatments may offer a C benefit by removing some C from the forest to protect the remaining C (Finkral and Evans 2008; Hurteau et al. 2008; Mitchell et al. 2009; Stephens et al. 2009; Dore et al. 2010). Thinned stands that burn in a surface fire typically have much higher tree survival and lower C losses than similar, unthinned stands that burn in a crown fire, although the net effect of fuel treatment C removal and surface fire emissions may exceed that from crown fire alone, even when materials from fuel treatments are used for wood products (Reinhardt et al.

2010). Because fuel treatment benefits are transient, they may lapse before a wildfire occurs, in which case the C removed by the fuel treatment is not offset by reduced wildfire emissions.

Modeling studies suggest that fuel treatments in most landscapes will result in a net decrease in landscape C over time (Harmon et al. 2009; Mitchell et al. 2009; Ager et al. 2010), because the savings in wildfire emissions is gained only on the small fraction of the landscape where fire occurs each year. The following conditions would be required to yield a substantial C benefit: (1) relatively light C removal would substantially reduce emissions, (2) fire occurrence is high in the near term (while fuel treatments are still effective), and (3) thinnings can provide wood for energy or long-lived products that yield substitution benefits. If fuel treatments are implemented, it is preferable from a C management standpoint to use removed fuels for energy production or wood products, rather than burning them onsite (Coleman et al. 2010; Jones et al. 2010). Feasibility and energy implications depend in part on hauling distance (Jones et al. 2010). An alternative to hauling biomass to conversion facilities is *in situ* pyrolysis to produce energy-dense liquid fuel and biochar which can remain onsite to enhance soil productivity and sequester C (Coleman et al. 2010). Even if thinning and fuel treatments reduce overall forest C, they may have the benefit of providing small C emissions every few decades, rather than large pulses from wildfire (Restaino and Peterson 2013).

7.5.3 Increasing Forest C Stocks by Increasing Forest Growth

Increasing growth rates in existing or new forests can increase C storage and the supply of forest products or biomass energy. Practices that increase forest growth include fertilization, irrigation, use of fast-growing planting stock, and control of weeds, pathogens, and insects (Albaugh et al. 1998, 2003, 2004; Nilsson and Allen 2003; Borders et al. 2004; Amishev and Fox 2006; Allen 2008). The potential for increasing forest growth differs by site and depends on specific climate, soil, tree species, and management.

Increased yields from these practices can be impressive. In pine forests in the southern United States, tree breeding has improved wood growth by 10–30 % (Fox et al. 2007b) and has increased resistance to insects and other stressors (McKeand et al. 2006). In this region, pine plantations using improved seedlings, control of competing vegetation, and fertilizer grow wood four times faster than naturally regenerated second-growth pine forests without competition control (Carter and Foster 2006). Tree breeding and intensive management also provide an opportunity to plant species and genotypes better adapted to future climates.

Many U.S. forests are N limited and would likely respond to fertilization (Reich et al. 1997). Nitrogen and phosphorus fertilizers have been used in about 6.5 million ha of managed forests in the southern United States to increase wood production (Liski et al. 2001; Seely et al. 2002; Albaugh et al. 2007; Fox et al. 2007a). Fertilization can produce 100 % gains for wood growth (Albaugh et al. 1998,

2004), although the benefits of fertilization for growth and C increase need to be balanced by the high GHG emissions associated with fertilizer production and from eutrophication in aquatic systems (Carpenter et al. 1998) (Table 7.3). Other risks include reduced water yield (faster growth uses more water), which is more pronounced in arid and semiarid forests, and potential loss of biodiversity if faster growth relies on monocultures (limited diversity can make some forests vulnerable to insects and pathogens). Increasing the genetic and species diversity of trees and increasing C stocks could be compatible goals in some areas (Woodall et al. 2011).

Markets for forest products can provide revenue to invest in accelerating forest growth. For example, expectation of revenue from the eventual sale of high value timber products would support investment in treatments or tree planting to increase growth rate. Taxation or other government incentives may also support growth-enhancing management. To the extent that incentives to alter growth alter timber harvest and wood product use, evaluation will require a type B accounting with system boundaries that include the forest sector, services sector, and non-forest land.

7.6 *Ex Situ* Forest C Management

Wood is removed from the forest for a variety of uses, each of which has different effects on C balance. Carbon can be stored in wood products for a variable length of time, oxidized to produce heat or electrical energy, or converted to liquid transportation fuels and chemicals that would otherwise come from fossil fuels (Fig. 7.5). In addition, a substitution effect occurs when wood products are used in place of other products that emit more GHG in manufacturing (Lippke et al. 2011).

Strategies that increase use life, use of wood products in place of higher emitting alternate products, and storage in long-lived wood products can complement strategies aimed at increasing forest C stocks. Risk and uncertainty in attaining benefits need to be considered when comparing strategies for increasing forest C with strategies for attaining wood product C offsets—successful strategies need to ensure energy offsets are attained in an acceptable period of time and that substitution effects are attained.

7.6.1 *Carbon in Forest Products*

Wood and paper store C when in use and also in landfills (Fig. 7.5). Rates of net C accumulation depend on rates of additions, disposal, combustion, and landfill decay. The half-life for single-family homes made of wood built after 1920 is about 80 years (Skog 2008; USEPA 2008), whereas the half-life of paper and paperboard products is less than 3 years (Skog 2008). About two-thirds of discarded wood and one-third of discarded paper go into landfills (Skog 2008). Decay in landfills is typically anaerobic and very slow (Barlaz 1998), and 77 % of the C in solid wood products and 44 % in paper products remain in landfills for decades (Chen et al. 2008; Skog 2008). However, current rates of methane release and capture can eliminate this C storage benefit for some low-lignin paper products (Skog 2008). About 2,500 Tg C was accumulated in wood products and landfills in the United States from 1910 to 2005 (Skog 2008), with about 700 Tg C (in 2001) in single- and multi-family homes. In 2007, net additions to products in use and those in landfills combined were 27 Tg C year^{-1} (USEPA 2009b), with about 19 Tg C year^{-1} from products in use (Skog 2008).

7.6.2 Product Substitution

Net C emissions associated with production and use of forest products is typically much less than with steel and concrete. Use of 1 Mg of C in wood materials in construction in place of steel or concrete can result in 2 Mg less C emissions (Schlamadinger and Marland 1996; Sathre and O'Connor 2008). Using wood from faster-growing forests for substitution can sometimes be more effective in lowering atmospheric CO_2 than storing C in the forest (Marland and Marland 1992; Marland et al. 1997; Baral and Guha 2004) (Fig. 7.5a). On the other hand, harvesting forests with very high C stocks that have accumulated over many decades may result in a large deficit of biological C storage that could take many decades to restore (McKinley et al. 2011) (Fig. 7.5b). Opportunities for substitution are largely in non-residential buildings (McKeever et al. 2006; Upton et al. 2008) because most houses are already built with wood, although some building practices, such as using wood for walls, can create a substitution effect in residential buildings (Lippke and Edmonds 2006). Attaining the substitution effect requires incentives that encourage increased use of wood.

7.6.3 Biomass Energy

Biomass energy could prevent the release of an estimated 130–190 Tg C year^{-1} from fossil fuels (Perlack et al. 2005; Zerbe 2006). Biomass energy comprises 28 % of renewable energy supply and 2 % of total energy use in the United States; the latter amount has the potential to increase to 10 % (Zerbe 2006). Currently, wood is used in the form of chips, pellets, and briquettes to produce heat or combined heat and generation of electricity (Saracoglu and Gunduz 2009). These basic energy carriers can be further transformed into liquid transportation fuels and gases (e.g., methane and hydrogen) (Demirbas 2007; Bessou et al. 2011). Conversion processes for these fuels require further development to improve efficiency and commercial viability. In addition, the potential exists to create high-value chemicals and other bioproducts from wood that would otherwise be made from fossil fuels, resulting in reduced emissions (Hajny 1981; USDOE 2009).

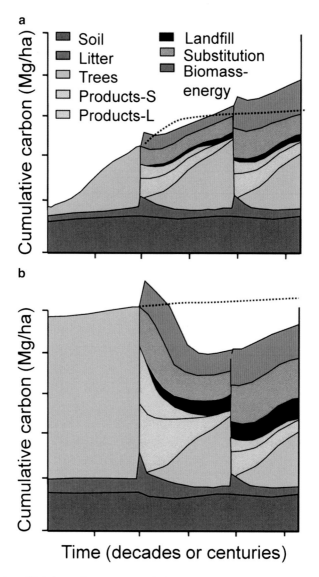

Fig. 7.5 Carbon (C) balance from two hypothetical management projects with different initial ecosystem C stocks and growth rates. Cumulative C stocks in forest, C removed from forest for use in wood projects (long [L]- and short-lived [S]), substitution, and biomass energy are shown on land that (**a**) has been replanted or afforested, or (**b**) has an established forest with high C stocks. The dotted line represents the trajectory of forest C stocks if no harvest occurred. Actual C pathways vary by project. Carbon stocks for trees, litter, and soils are net C stocks only. The scenario is harvested in x-year intervals, which in the United States could be as short as 15 years or longer than 100 years. This diagram assumes that all harvested biomass will be used and does not account for logging emissions. Carbon is sequestered by (1) increasing the average ecosystem C stock (tree biomass) by afforestation, or (2) accounting for C stored in wood products in use and in landfills, as well as preventing the release of fossil fuel C through product substitution or biomass energy. The product-substitution effect is assumed to be 2:1 on average. Biomass is assumed to be

Most biomass for energy is a byproduct of conventional forest product streams, such as milling residues (Gan and Smith 2006a), with some use of trees killed by insects, disease, fire, and wind (Peng et al. 2010; Tumuluru et al. 2010). Most of these residues, mainly sawdust and bark, are already used for direct heating in milling operations or used for other wood products (Ackom et al. 2010; Mälkki and Virtanen 2003; Nilsson et al. 2011); obtaining higher quantities of biomass feedstock would require using other residues. Residues that are generally not used are from logging, hazardous fuel reduction treatments, precommercial thinning, and urban areas (Mälkki and Virtanen 2003; Perlack et al. 2005; Gan and Smith 2006b; Gan 2007; Smeets and Faaij 2007; Ackom et al. 2010; Repo et al. 2011).

If forest harvesting for biomass energy is expanded, roundwood from standing trees will increasingly be used for energy, and short-rotation plantations (e.g., poplars) devoted to biomass feedstock production (Solomon et al. 2007) may become more common (Tuskan 1998; Fantozzi and Buratti 2010). Carbon emissions from increased use of roundwood for energy may be offset over time by a subsequent increase in forest C, which can be done through increased forest growth on land where roundwood is harvested. The amount and speed of the offset are influenced by the time period considered, forest growth rate, initial stand C density, and the efficiency with which wood offsets fossil fuel emissions (Schlamadinger et al. 1995). The offset can also be done through increased landowner investment in forestry, including converting non-forest land to forest, retaining land in forest that would otherwise be converted to non-forest, and planting land in faster growing pulpwood or short-rotation plantations. Forest inventory and C projections indicate that for scenarios with high wood energy use, more land will be retained in forest plantations for the southern United States (USDA FS 2012b). However, landowner investment in revenue for biomass is expected to be low for other parts of the United States.

Reductions in GHG emissions from wood-to-energy pathways depend, in part, on how efficiently wood substitutes for fossil fuels. The energy value of wood (energy content per unit mass) is lower than for fossil fuels (Demirbas 2005; Patzek and Pimentel 2005), and is most pronounced when wood substitutes for fossil fuels with high energy values (e.g., natural gas). The risk of not attaining various levels of offset from use of wood for energy differs, depending on whether biomass is from residues or from roundwood (Schlamadinger et al. 1995; Zanchi et al. 2010). Risks for using residues are small, especially if harvests and supply chains are well managed. Risks associated with using roundwood differ by forest conditions, treatments, and landowner investment in forest management. Large increases in

Fig. 7.5 (continued) a 1:1 substitute for fossil fuels in terms of C, but this is not likely for many wood-to-energy options. This represents a theoretical maximum C benefit for these forest products and management practices. Carbon "debt" is any period of time at which the composition of forest products and remaining forest C stocks after harvest is lower than estimated C stocks under a no-harvest scenario (Adapted from Solomon et al. 2007; Pachauri and Reisinger 2007; McKinley et al. 2011)

demand could cause loss of C if natural forest with high C density were converted to plantations with lower C density.

Several studies report that using biomass instead of fossil fuels can significantly reduce net C emissions (Boman and Turnbull 1997; Spath and Mann 2000; Mann and Spath 2001; Cherubini et al. 2009; Jones et al. 2010; Malmsheimer et al. 2011). However, other studies report that the postharvest regrowth period during which forest C is initially low negates the benefits of wood energy in the near term (Schlamadinger et al. 1995; Fargione et al. 2008; Pimentel et al. 2008; Mathews and Tan 2009; Melillo et al. 2009; Searchinger et al. 2009; Cardellichio and Walker 2010; Manomet Center for Conservation Sciences 2010; Bracmort 2011; McKechnie et al. 2011; Melamu and von Blottnitz 2011; Repo et al. 2011). Depending on assumptions about processes included in system boundaries and period of evaluation, studies that used LCAs with biomass pathways and forest C dynamics over time calculated limited or substantial reductions in CO_2 emissions. For some cases and time periods, LCAs with biomass pathways and forest C dynamics indicate biomass emissions can be higher than fossil fuel emissions (Pimentel et al. 2008; Searchinger et al. 2008; Johnson 2009; Manomet Center for Conservation Sciences 2010; McKechnie et al. 2011).

These conflicting conclusions are the result of different assumptions and methods used in the LCAs (Cherubini et al. 2009, 2012; Matthews and Tan 2009). Emerging C accounting methods are increasingly focused on the effect of emissions on the atmosphere and climate over an extended time period, rather than assuming C neutrality (Cherubini et al. 2012). Evaluation frameworks are needed to accurately quantify overall C and climate effects of specific combinations of forest management and wood energy use.

7.7 Mitigation Strategies: Markets, Regulations, Taxes, and Incentives

Forests comprise about a third of the land area in the United States, but fragmentation and conversion of forest to other land uses is increasing, especially in the East (Drummond and Loveland 2010). Various mechanisms exist at national, regional, and local scales that can enhance mitigation efforts while providing incentives to keep forests intact. Markets and incentive programs can potentially play a role in ecosystem-enhancing mitigation on private and non-federal lands, providing a means for landowners to be financially compensated for voluntary activities that improve ecosystem services. Some of these mechanisms, such as C markets, are designed to encourage mitigation, while other mechanisms help maintain or augment C stores as an ancillary benefit.

7.7.1 Markets, Registries, and Protocols for Forest-Based Carbon Projects

Carbon markets are an emissions trading mechanism and are typically designed to create a multi-sector approach that encourages reductions and (often but not always) enhances sequestration of GHG emissions (measured in Mg CO_2 equivalent, or CO_2e) in an economically efficient manner. Registries exist to track and account for C, and protocols outline the specific methodologies that are a prerequisite to creating legitimate C offsets. The United States does not have a national-level regulatory market, but several mandatory regional efforts and voluntary over-the-counter markets provide limited opportunities for mitigation through forest-based C projects. Offsets generated from these projects can compensate for emissions generated elsewhere. Forest C projects generally take the following form:

- Avoided emissions—Avoided deforestation (or avoided conversion): projects that avoid emissions by keeping forests in forest..
- Enhanced sequestration
  - Afforestation/reforestation: projects that reforest areas that are currently non-forest but may have been forested historically.
  - Improved forest management: projects that offer enhanced C mitigation through better or more sustainable management techniques. These projects are compatible with sustainable levels of timber harvest.
  - Urban forestry: projects that plant trees in urban areas. Only sequestered C is eligible (avoided C emissions that result from energy savings are not eligible for credit).

The Regional Greenhouse Gas Initiative (RGGI) is a mandatory multi-state effort in New England and the Mid-Atlantic that allows offset credits to be generated through afforestation projects within RGGI member states. The Climate Action Reserve is a mandatory initiative in California but accepts forest projects from throughout the country. In addition, protocols created by the American Carbon Registry, Verified Carbon Standard provide quality assurance for forest C projects that may be sold on the voluntary market (Kollmuss et al. 2010; Peters-Stanley et al. 2011). In 2009, 5.1 Mg of CO_2e, or 38 % of the global share of forest-based C offsets, was generated in North America (Hamilton et al. 2010). However, factors such as substantial startup and transaction costs and restrictions on the long-term use and stewardship of forest land enrolled in C projects are often barriers to participation by private forest landowners (Diaz et al. 2009).

7.7.2 Tax and Incentive Programs

Tax incentives may be designed to maintain a viable timber industry and achieve open space objectives, but also help maintain or enhance forest C stores. Many

Table 7.4 Programs that influence carbon mitigation

Program	Agency	Land area (10^6 ha)	Purpose
Conservation Reserve Program and Continuous Conservation Reserve Program	Farm Service Agency	~13	Reduce erosion, increase wildlife habitat, improve water quality, and increase forested acres
Environmental Quality Incentives Program	Natural Resources Conservation Service (NRCS)	~6.9	Encourages active forest management including timber stand improvement, site preparation for planting, culverts, stream crossings, water bars, planting, prescribed burns, hazard reduction, fire breaks, pasture, fence, grade stabilization, plan preparation
Conservation Stewardship Program	NRCS	Not applicable	Incentives for sustainable forest management and conservation activities
Wildlife Habitat Incentive Program	NRCS	0.26	Provides incentives to develop or improve fish and wildlife habitat, including prairie and savanna restoration, in-stream fish structures, livestock exclusion, and tree planting
Forest Legacy Program	U.S. Forest Service (USFS)	~0.8	Provides incentives to preserve privately owned working forest land through conservation easements and fee acquisitions
Stewardship Program	USFS	~14	Encourages private landowners to create and implement forest stewardship plans

states offer reduced taxes on forest land if it is maintained in forest and managed responsibly. For example, private forest landowners enrolled in the Managed Forest Law Program in Wisconsin receive an 80–95 % tax reduction on land that is at least 80 % forested and is managed for sustainable production of timber resources. In the Use Value Appraisal Program in Vermont, C benefits from these programs are evaluated for specific circumstances; younger, fast growing forests have higher rates of C uptake, whereas older stands may have lower C uptake but higher C storage (Harmon 2001; Malmsheimer et al. 2008). Therefore, a no-harvest unmanaged forest may produce more or less C benefit than an actively managed forest, but much depends on current C stocks, likelihood of disturbance, and how harvested timber is used (Ingerson 2007; Nunnery and Keeton 2010). The timeframe of expected C benefits depends on both forest management plans and forest product pathways, both short term and long term (McKinley et al. 2011).

Federal programs administered by the Natural Resources Conservation Service, U.S. Forest Service, and Farm Service Agency (Table 7.4) provide cost-share and rental payment incentives for farm, forest, watershed, and wildlife habitat stewardship. These programs may also enhance C storage, although it is not an explicit goal. The area enrolled in each program fluctuates annually and depends on commodity prices, program funding, and authorization levels. In 2010, 13 million ha of U.S. farmland were enrolled in the Conservation Reserve Program, down from 15 million ha in 2005 (Claassen et al. 2008; USDA Farm Service Agency 2010).

If new policies were to favor land management that reduces atmospheric CO_2, existing programs can be modified to explicitly provide incentives that encourage C mitigation. For example, the overall objective of a program could remain as is (to determine general eligibility), but the financial incentives for enrollment could be related to estimated average C benefit per land unit. Carbon benefit per hectare could be estimated at a county or regional scale based on a combination of factors, including geographic location, land use, species planted, and overall landscape connectivity. This would help to ensure that priority lands for C management receive the highest potential benefits. Alternatively, a specific forest C incentive program could complement current incentive programs by targeting small forest owners and providing financial incentives to retain forest land in forest. Best management practices could be made available (e.g., for artificial regeneration, thinning, and insect control) (Table 7.5), and financial incentives could be based on estimated C benefits (Pinchot Institute for Conservation 2011). These estimated benefits would require only a credible verification of practices rather than annual site monitoring.

7.8 The Role of Public Lands in C Mitigation

Public lands contain about 37 % of the land area of the United States, with federally managed lands occupying 76 % of the total area managed by all public entities. Managing these lands for C benefits would involve multiple jurisdictions, social objectives, and political factors, and would be governed by laws mandating multiple uses of land in the public domain. The Council on Environmental Quality, which is responsible for overseeing environmental policy across the federal government, developed draft guidelines on how federal agencies can improve how they consider the effects of GHG emissions and climate change when evaluating proposals for federal actions under the National Environmental Policy Act (Sutley 2010). Executive Order 13514 (2009) requires agencies to set targets that focus on sustainability, energy efficiency, reduced fossil fuel use, and increased water efficiency. In addition, the order requires agencies to measure, report, and reduce GHG emissions from direct and indirect activities, including federal land management practices. Recent guidance and orders are being considered by land management agencies, but it is unclear how effective they will be in reducing GHGs, given the many other uses of federal lands. Large areas of forest land protected by conservation organizations (e.g., The Nature Conservancy) across the United States are managed for public

Table 7.5 Tools and processes to inform forest management

Organization	Relevant content	Internet site
U.S. Forest Service Forest Inventory and Analysis	Forest statistics by state, including carbon (C) estimates Sample plot and tree data Forest inventory methods and basic definitions	http://fia.fs.fed.us
U.S. Forest Service Forest Health Monitoring	Forest health status Regional data on soils, dead wood stocks Forest health monitoring methods	http://www.fhm.fs.fed.us
U.S. Department of Agriculture Greenhouse Gas (GHG) Inventory	State-by-state forest C estimates	http://www.usda.gov/oce/global_change/gg_inventory.htm
United Nations Framework Convention on Climate Change and Intergovernmental Panel on Climate Change	International guidance on C accounting and estimation	http://unfccc.int http://www.ipcc.ch
Natural Resources Conservation Service	Soil Data Mart—access to a variety of soil data	http://soildatamart.nrcs.usda.gov
U.S. Forest Service, Northern Research Station	Accounting, reporting procedures, and software tools for C estimation	http://www.nrs.fs.fed.us/carbon/tools
U.S. Energy Information Administration, Voluntary GHG Reporting	Methods and information for calculating sequestration and emissions from forestry	http://www.eia.gov/oiaf/1605/gdlinshtml
U.S. Environmental Protection Agency	Methods and estimates for GHG emissions and sequestration	http://www.epa.gov/climatechange/emissions/usinventoryreport.html

benefits, and because they are often not subject to the regulatory issues above, they may be able to contribute to C mitigation more quickly than is possible on other public lands.

References

Ackom, E. K., Mabee, W. E., & Saddler, J. N. (2010). Industrial sustainability of competing wood energy options in Canada. *Applied Biochemistry and Biotechnology, 162*, 2259–2272.

Ager, A. A., Finney, M. A., McMahan, A., & Cathcart, J. (2010). Measuring the effect of fuel treatments on forest carbon using landscape risk analysis. *Natural Hazards and Earth System Sciences, 10*, 2515–2526.

Akbari, H. (2002). Shade trees reduce building energy use and CO_2 emissions from power plants. *Environmental Pollution, 116*, S119–S1260.

Albaugh, T. J., Allen, H. L., Dougherty, P. M., et al. (1998). Leaf area and above- and belowground growth responses of loblolly pine to nutrient and water additions. *Forest Science, 44*, 317–328.

Albaugh, T. J., Allen, H. L., Zutter, B. R., & Quicke, H. E. (2003). Vegetation control and fertilization in midrotation *Pinus taeda* stands in the southeastern United States. *Annals of Forest Science, 60*, 619–624.

Albaugh, T. J., Allen, H. L., Dougherty, P. M., & Johnsen, K. H. (2004). Long term growth responses of loblolly pine to optimal nutrient and water resource availability. *Forest Ecology and Management, 192*, 3–19.

Albaugh, T. J., Allen, H. L., & Fox, T. R. (2007). Historical patterns of forest fertilization in the southeastern United States from 1969 to 2004. *Southern Journal of Applied Forestry, 31*, 129–137.

Alig, R. J., Plantinga, A. J., Ahn, S., & Kline, J. D. (2003). *Land use changes involving forestry in the United States: 1952 to 1997, with projections to 2050* (General Technical Report PNW-GTR-587, 92 p). Portland: U.S. Department of Agriculture, Forest Service, Pacific Northwest Research Station.

Allen, H. L. (2008). Silvicultural treatments to enhance productivity. In J. Evans (Ed.), *The forests handbook: Vol. 2, Applying forest science for sustainable management* (pp. 129–139). Oxford/Malden: Blackwell Science Ltd.

Amishev, D. Y., & Fox, T. R. (2006). The effect of weed control and fertilization on survival and growth of four pine species in the Virginia Piedmont. *Forest Ecology and Management, 236*, 93–101.

Aukland, L., Costa, P. M., & Brown, S. (2003). A conceptual framework and its application for addressing leakage: The case of avoided deforestation. *Climate Policy, 3*, 123–136.

Balboa-Murias, M. A., Rodriguez-Soalleíro, R., Merino, A., & Álvarez-González, J. G. (2006). Temporal variations and distribution of carbon stocks in aboveground biomass of radiata pine and maritime pine pure stands under different silvicultural alternatives. *Forest Ecology and Management, 237*, 29–38.

Balshi, M. S., McGuire, A. D., Duffy, P., et al. (2009). Vulnerability of carbon storage in North American boreal forests to wildfires during the 21st century. *Global Change Biology, 15*, 1491–1510.

Baral, A., & Guha, G. S. (2004). Trees for carbon sequestration or fossil fuel substitution: The issue of cost vs. carbon benefit. *Biomass and Bioenergy, 27*, 41–55.

Barlaz, M. A. (1998). Carbon storage during biodegradation of municipal solid waste components in laboratory-scale landfills. *Global Biogeochemical Cycles, 12*, 373–380.

Battaglia, M. A., Rocca, M. E., Rhoades, C. C., & Ryan, M. G. (2010). Surface fuel loadings within mulching treatments in Colorado coniferous forests. *Forest Ecology and Management, 260*, 1557–1566.

Bessou, C., Ferchaud, F., Gabrielle, B., & Mary, B. (2011). Biofuels, greenhouse gases and climate change. A review. *Agronomy for Sustainable Development, 31*, 1–79.

Birdsey, R. A., Plantinga, A. J., & Health, L. S. (1993). Past and prospective carbon storage in United States forests. *Forest Ecology and Management, 58*, 33–40.

Birdsey, R., Pregitzer, K., & Lucier, A. (2006). Forest carbon management in the United States: 1600–2100. *Journal of Environmental Quality, 35*, 1461–1469.

Boman, U. R., & Turnbull, J. H. (1997). Integrated biomass energy systems and emissions of carbon dioxide. *Biomass and Bioenergy, 13*, 333–343.

Borders, B. E., Will, R. E., Markewitz, D., et al. (2004). Effect of complete competition control and annual fertilization on stem growth and canopy relations for a chronosequence of loblolly pine plantations in the lower coastal plain of Georgia. *Forest Ecology and Management, 192*, 21–37.

Bracmort, K. (2011). *Is biopower carbon neutral?* (7-5700, 14 p). Washington, DC: Congressional Research Service.

Cardellichio, P., & Walker, T. (2010). Why the Manomet study got the biomass carbon accounting right. *The Forestry Source, 15*, 4.

Carpenter, S. R., Caraco, N. F., Correll, D. L., et al. (1998). Nonpoint pollution of surface waters with phosphorus and nitrogen. *Ecological Applications, 8*, 559–568.

Carter, M. C., & Foster, C. D. (2006). Milestones and millstones: A retrospective on 50 years of research to improve productivity in loblolly pine plantations. *Forest Ecology and Management, 227*, 137–144.

Chen, J., Colombo, S. J., Ter-Mikaelian, M. T., & Heath, L. S. (2008). Future carbon storage in harvested wood products from Ontario's Crown forests. *Canadian Journal of Forest Research, 38*, 1947–1958.

Cherubini, F., Bird, N. D., Cowie, A., et al. (2009). Energy- and green-house gas-based LCA of biofuel and bioenergy systems: Key issues, ranges and recommendations. *Resource Conservation and Recycling, 53*, 434–447.

Cherubini, F., Guest, G., & Stømman, A. H. (2012). Application of probability distributions to the modeling of biogenic CO_2 fluxes in life cycle assessment. *GCB Bioenergy*. doi:10.1111/j.1757-1707.2011.01156.x.

Churkina, G., Brown, D. G., & Keoleian, G. (2010). Carbon stored in human settlements: the conterminous United States. *Global Change Biology, 16*, 135–143.

Claassen, R., Cattaneo, A., & Johansson, R. (2008). Cost-effective design of agri-environmental payment program: U.S. experience in theory and practice. *Ecological Economics, 65*, 737–752.

Coleman, M., Page-Dumroese, D., Archuleta, J., et al. (2010). Can portable pyrolysis units make biomass utilization affordable while using bio-char to enhance soil productivity and sequester carbon. In T. B Jain, R. T Graham, & J. Sandquist, Tech. (Eds.), *Integrated management of carbon sequestration and biomass utilization opportunities in a changing climate. Proceedings RMRS-P-61* (pp. 159–168). Fort Collins: U.S. Department of Agriculture, Forest Service, Rocky Mountain Research Station.

Demirbas, A. (2005). Potential applications of renewable energy sources, biomass combustion problems in boiler power systems and combustion related environmental issues. *Progress in Energy and Combustion Science, 31*, 171–192.

Demirbas, A. (2007). Progress and recent trends in biofuels. *Progress in Energy and Combustion Science, 33*, 1–18.

Depro, B. M., Murray, B. C., Align, R. J., & Shanks, A. (2008). Public land, timber harvests, and climate mitigation: quantifying carbon sequestration potential on U.S. public timberlands. *Forest Ecology and Management, 255*, 1122–1134.

Diaz, S., Hector, A., & Wardle, D. A. (2009). Biodiversity in forest carbon sequestration initiatives: Not just a side benefit. *Current Opinion in Environmental Sustainability, 1*, 55–60.

Donato, D. C., Fontaine, J. B., Campbell, J. L., et al. (2009). Conifer regeneration in stand-replacement portions of a large mixed-severity wildfire in the Klamath-Siskiyou Mountains. *Canadian Journal of Forest Research, 39*, 823–838.

Dore, S., Kolb, T. E., Montes-Helu, M., et al. (2010). Carbon and water fluxes from ponderosa pine forests disturbed by wildfire and thinning. *Ecological Applications, 20*, 663–683.

Drummond, M. A., & Loveland, T. R. (2010). Land-use pressure and a transition to forest-cover loss in the eastern United States. *Bioscience, 60*, 286–298.

Exec. Order No. 13,514, 74 Fed. Reg. 194, 52117 (2009, October 8). *Federal leadership in environmental, energy, and economic performance.* http://www.whitehouse.gov/administration/eop/ceq/sustainability. 28 Dec 2011.

Fantozzi, F., & Buratti, C. (2010). Life cycle assessment of biomass chains: Wood pellet from short rotation coppice using data measured on a real plant. *Biomass and Bioenergy, 34*, 1796–1804.

Fargione, J., Hill, J., Tilman, D., et al. (2008). Land clearing and the biofuel carbon debt. *Science, 319*, 1235–1238.

Farley, K. A., Jobbágy, E. G., & Jackson, R. B. (2005). Effects of afforestation on water yield: A global synthesis with implications for policy. *Global Change Biology, 11*, 1565–1576.

Farley, K. A., Piñeiro, G., Palmer, S. M., et al. (2008). Stream acidification and base cation losses with grassland afforestation. *Water Resources Research, 44*, W00A03.

Finkral, A. J., & Evans, A. M. (2008). Effects of a thinning treatment on carbon stocks in a northern Arizona ponderosa pine forest. *Forest Ecology and Management, 255*, 2743–2750.

Food and Agriculture Organization of the United Nations [FAO]. (2007). *State of the world's forests 2007* (144 p). Rome: Food and Agriculture Organization of the United Nations, Communications Division, Electronic Publishing Policy and Support Branch. http://www.fao.org/docrep/009/a0773e/a0773e00.htm. November 2011.

Fox, T. R., Allen, H. L., Albaugh, T. J., et al. (2007). Tree nutrition and forest fertilization of pine plantations in the southern United States. *Southern Journal of Applied Forestry, 31*, 5–11.

Fox, T. R., Jokela, E. J., & Allen, H. L. (2007). The development of pine plantation silviculture in the southern United States. *Journal of Forestry, 105*, 337–347.

Gan, J. (2007). Supply of biomass, bioenergy, and carbon mitigation: Method and application. *Energy Policy, 35*, 6003–6009.

Gan, J., & McCarl, B. A. (2007). Measuring transnational leakage of forest conservation. *Ecological Economics, 64*, 423–432.

Gan, J. B., & Smith, C. T. (2006a). A comparative analysis of woody biomass and coal for electricity generation under various CO_2 emission reductions and taxes. *Biomass and Bioenergy, 30*, 296–303.

Gan, J., & Smith, C. T. (2006b). Availability of logging residues and potential for electricity production and carbon displacement in the USA. *Biomass and Bioenergy, 30*, 1011–1020.

Hajny, G. J. (1981). *Biological utilization of wood production for production of chemicals and foodstuffs* (Research Paper FPL-RP-385, 65 p). Madison: U.S. Department of Agriculture, Forest Service, Forest Products Laboratory.

Hamilton, K., Sjardin, M., Peters-Stanley, M., & Marcello, T. (2010). *Building bridges: State of the voluntary carbon markets 2010: A report by Ecosystem Marketplace and Bloomberg New Energy Finance* (108 p). New York/Washington, DC: Bloomberg New Energy Finance/Ecosystem Marketplace.

Harden, J. W., Trumbore, S. E., Stocks, B. J., et al. (2000). The role of fire in the boreal carbon budget. *Global Change Biology, 6*, 174–184.

Harmon, M. (2001). Carbon sequestration in forests—Addressing the scale question. *Journal of Forestry, 99*, 24–29.

Harmon, M. E., & Marks, B. (2002). Effects of silvicultural practices on carbon stores in Douglas-fir–Western hemlock forests in the Pacific Northwest, U.S.A.: Results from a simulation model. *Canadian Journal of Forest Research, 877*, 863–877.

Harmon, M. E., Harmon, J. M., Ferrell, W. K., & Brooks, D. (1996). Modeling carbon stores in Oregon and Washington forest products: 1900–1992. *Climatic Change, 33*, 521–550.

Harmon, M. E., Moreno, A., & Domingo, J. B. (2009). Effects of partial harvest on the carbon stores in Douglas-fir/western hemlock forests: A simulation study. *Ecosystems, 12*, 777–791.

Houghton, R. A. (2005). Aboveground forest biomass and the global carbon balance. *Global Change Biology, 11*, 945–958.

Hurteau, M. D., Koch, G. W., & Hungate, B. A. (2008). Carbon protection and fire risk reduction: Toward a full accounting of forest carbon offsets. *Frontiers in Ecology and the Environment, 6*, 493–498.

Ingerson, A. (2007). *U.S. forest carbon and climate change: Controversies and win-win policy approaches* (18 p). Washington, DC: The Wilderness Society.

Intergovernmental Panel on Climate Change [IPCC]. (2001). *Climate change 2001: Mitigation. Third assessment report of the Intergovernmental Panel on Climate Change, Working Group III* (753 p). Cambridge: Cambridge University Press.

Jackson, R. B., & Schlesinger, W. H. (2004). Curbing the U.S. carbon deficit. *Proceedings of the National Academy of Sciences, USA, 101*, 15827–15829.

Jackson, R. B., Jobbágy, E. G., Avissar, R., et al. (2005). Trading water for carbon with biological carbon sequestration. *Science, 310*, 1944–1947.

Jackson, R. B., Randerson, J. T., Canadell, J. G., et al. (2008). Protecting climate with forests. *Environmental Research Letters, 3*, 044006.

Jiang, H., Apps, M. J., Peng, C., et al. (2002). Modeling the influence of harvesting on Chinese boreal forest carbon dynamics. *Forest Ecology and Management, 169*, 65–82.

Jobbágy, E. G., & Jackson, R. B. (2004). Groundwater use and salinization with grassland afforestation. *Global Change Biology, 10*, 1299–1312.

Johnson, E. (2009). Good-bye to carbon neutral: Getting biomass footprints right. *Environmental Impact Assessment Review, 29*, 165–168.

Johnson, D. W., & Curtis, P. S. (2001). Effects of forest management on soil C and N storage: Meta analysis. *Forest Ecology and Management, 140*, 227–238.

Jones, G., Loeffler, D., Butler, E., et al. (2010). Emissions, energy return and economics from utilizing forest residues for thermal energy compared to onsite pile burning. In T. B. Jain, R. T. Graham, & J. Sandquist, Tech. (Eds.), *Integrated management of carbon sequestration and biomass utilization opportunities in a changing climate, Proceedings RMRS-P-61* (pp. 145–158). Fort Collins: U.S. Department of Agriculture, Forest Service, Rocky Mountain Research Station.

Kaipainen, T., Liski, J., Pussinen, A., & Karjalainen, T. (2004). Managing carbon sinks by changing rotation length in European forests. *Environmental Science and Policy, 7*, 205–219.

Keyser, T. L., Lentile, L. B., Smith, F. W., & Shepperd, W. D. (2008). Changes in forest structure after a large, mixed-severity wildfire in ponderosa pine forests of the Black Hills, South Dakota, USA. *Forest Science, 54*, 328–338.

Kollmuss, A., Lee, C., & Lazaru, M. (2010). How offset programs assess and approve projects and credits. *Carbon Management, 1*, 119–134.

Lippke, B., & Edmonds, L. (2006). Environmental performance improvement in residential construction: The impact of products, biofuels, and processes. *Forest Products Journal, 56*, 58–63.

Lippke, B., Oneil, E., Harrison, R., et al. (2011). Life cycle impacts of forest management and wood utilization on carbon mitigation: Knowns and unknowns. *Carbon Management, 2*, 303–333.

Liski, J., Pussinen, A., Pingoud, K., et al. (2001). Which rotation length is favourable to carbon sequestration? *Canadian Journal of Forest Research, 31*, 2004–2013.

Magnani, F., Dewar, R. C., & Borghetti, M. (2009). Leakage and spillover effects of forest management on carbon storage: Theoretical insights from a simple model. *Tellus Series B: Chemical and Physical Meteorology, 61*, 385–393.

Mälkki, H., & Virtanen, Y. (2003). Selected emissions and efficiencies of energy systems based on logging and sawmill residues. *Biomass and Bioenergy, 24*, 321–327.

Malmsheimer, R. W., Heffernan, P., Brink, S., et al. (2008). Forest management solutions for mitigating climate change in the United States. *Journal of Forestry, 106*, 115–173.

Malmsheimer, R. W., Bower, J. L., Fried, J. S., et al. (2011). Managing forest because carbon matters: Integrating energy, products, and land management policy. *Journal of Forestry, 109*, S7–S50.

Mann, M. K., & Spath, P. L. (2001). A life-cycle assessment of biomass cofiring in a coal-fired power plant. *Clean Products and Processes, 3*, 81–91.

Manomet Center for Conservation Sciences. (2010). *Massachusetts biomass sustainability and carbon policy study: Report to the Commonwealth of Massachusetts Department of Energy Resources* (Report NCI-201003, 182 p). Brunswick: Manomet Center for Conservation Sciences.

Marland, G., & Marland, S. (1992). Should we store carbon in trees? *Water, Air, and Soil Pollution, 64*, 181–195.

Marland, G., Schlamadinger, B., & Leiby, P. (1997). Forest/biomass based mitigation strategies: Does the timing of carbon reductions matter? *Critical Reviews in Environmental Science and Technology, 27*, S213–S226.

Mathews, J. A., & Tan, H. (2009). Biofuels and indirect land use change effects: The debate continues. *Biofuels, Bioproducts and Biorefining, 3*, 305–317.

McKeand, S. E., Jokela, E. J., Huber, D. A., et al. (2006). Performance of improved genotypes of loblolly pine across different soils, climates, and silvicultural inputs. *Forest Ecology and Management, 227*, 178–184.

McKechnie, J., Colombo, S., Chen, J., et al. (2011). Forest bioenergy or forest carbon? Assessing trade-offs in greenhouse gas mitigation with wood-based fuels. *Environmental Science and Technology, 45*, 789–795.

McKeever, D. B., Adair, C., & O'Connor, J. (2006). *Wood products used in the construction of low-rise nonresidential buildings in the United States, 2003* (68 p). Tacoma: Wood Products Council.

McKenzie, D., Peterson, D. L., & Littell, J. (2009). Global warming and stress complexes in forests of western North America. In A. Bytnerowicz, M. J. Arbaugh, A. R. Riebau, & C. Andersen (Eds.), *Wildland fires and air pollution* (pp. 317–337). The Hague: Elsevier Publishers.

McKinley, D. C., & Blair, J. M. (2008). Woody plant encroachment by *Juniperus virginiana* in a mesic native grassland promotes rapid carbon and nitrogen accrual. *Ecosystems, 11*, 454–468.

McKinley, D. C., Ryan, M. G., Birdsey, R. A., et al. (2011). A synthesis of current knowledge on forests and carbon storage in the United States. *Ecological Applications, 21*, 1902–1924.

Melamu, R., & von Blottnitz, H. (2011). 2nd generation biofuels a sure bet? A life cycle assessment of how things could go wrong. *Journal of Cleaner Production, 19*, 138–144.

Melillo, J. M., Reilly, J. M., Kicklighter, D. W., et al. (2009). Indirect emissions from biofuels: How important? *Science, 326*, 1397–1399.

Meyfroidt, P., Rudel, T. K., & Lambin, E. F. (2010). Forest transitions, trade, and the global displacement of land use. *Proceedings of the National Academy of Sciences, USA, 107*, 20917–20922.

Mitchell, S. R., Harmon, M. E., & O'Connell, K. E. B. (2009). Forest fuel reduction alters fire severity and long-term carbon storage in three Pacific Northwest ecosystems. *Ecological Applications, 19*, 643–655.

Murray, B. C., McCarl, B. A., & Lee, H.-C. (2004). Estimating leakage from forest carbon sequestration programs. *Land Economics, 80*, 109–124.

Nave, L. E., Vance, E. D., Swanston, C. W., & Curtis, P. S. (2010). Harvest impacts on soil carbon storage in temperate forests. *Forest Ecology and Management, 259*, 857–866.

Nilsson, U., & Allen, H. L. (2003). Short- and long-term effects of site preparation, fertilization and vegetation control on growth and stand development of planted loblolly pine. *Forest Ecology and Management, 175*, 367–377.

Nilsson, D., Bernesson, S., & Hansson, P. A. (2011). Pellet production from agricultural raw materials—A systems study. *Biomass and Bioenergy, 35*, 679–689.

Nowak, D. J., & Crane, D. E. (2002). Carbon storage and sequestration by urban trees in the USA. *Environmental Pollution, 116*, 381–389.

Nowak, D. J., Kuroda, M., & Crane, D. E. (2004). Tree mortality rates and tree population projections in Baltimore, Maryland, USA. *Urban Forestry and Urban Greening, 2*, 139–147.

Nunnery, J. S., & Keeton, W. S. (2010). Forest carbon storage in the northeastern United States: net effects of harvesting frequency, post-harvest retention, and wood products. *Ecological Applications, 259*, 1363–1375.

Pacala, S. W., Hurtt, G. C., Baker, D., et al. (2001). Consistent land- and atmosphere-based U.S. carbon sink estimates. *Science, 292*, 2316–2320.

Pachauri, R. D., & Reisinger, A. (Eds.). (2007). *Climate change 2007: Synthesis report. Contribution of Working Groups I, II and III to the fourth assessment report of the Intergovernmental Panel on Climate Change* (104 p). Geneva: Intergovernmental Panel on Climate Change.

Pataki, D. E., Alig, R. J., Fung, A. S., et al. (2006). Urban ecosystems and the North American carbon cycle. *Global Change Biology, 12*, 2092–2102.

Patzek, T. W., & Pimentel, D. (2005). Thermodynamics of energy production from biomass. *Critical Reviews in Plant Sciences, 24*, 327–364.

Peng, J. H., Bi, H. T., Sokhansanj, S., et al. (2010). An economical and market analysis of Canadian wood pellets. *International Journal of Green Energy, 7*, 128–142.

Perlack, R. D., Wright, L. L., Turhollow, A. F., et al. (2005). *Biomass as feedstock for a bioenergy and bioproducts industry: The technical feasibility of a billion-ton annual supply* (60 p). Oak Ridge: U.S. Department of Energy, Oak Ridge National Laboratory.

Peterson, D. L., Johnson, M. C., Agee, J. K., et al. (2005). *Forest structure and fire hazard in dry forests of the western United States* (General Technical Report PNW-GTR-628, 30 p). Portland: U.S. Department of Agriculture, Forest Service, Pacific Northwest Research Station.

Peters-Stanley, M., Hamilton, K., Marcello, T., & Sjardin, M. (2011). *Back to the future: State of the voluntary carbon markets 2011* (78 p). Washington, DC: Ecosystem Marketplace.

Pimentel, D., Marklein, A., Toth, M. A., et al. (2008). Biofuel impacts on world food supply: Use of fossil fuel, land and water resources. *Energies, 1*, 41–78.

Pinchot Institute for Conservation. (2011). *The Pinchot letter, 16*(1). Available at http://www.pinchot.org/pubs/c34

Reich, P. B., Grigal, D. F., Aber, J. D., & Gower, S. T. (1997). Nitrogen mineralization and productivity in 50 hardwood and conifer stands on diverse soils. *Ecology, 78*, 335–347.

Reinhardt, E. D., Keane, R. E., Calkin, D. E., & Cohen, J. D. (2008). Objectives and considerations for wildland fuel treatment in forested ecosystems of the interior western United States. *Forest Ecology and Management, 256*, 1997–2006.

Reinhardt, E. D., Holsinger, L., & Keane, R. (2010). Effects of biomass removal treatments on stand-level fire characteristics in major forest types of the Northern Rocky Mountains. *Western Journal of Applied Forestry, 25*, 34–41.

Repo, A., Tuomi, M., & Liski, J. (2011). Indirect carbon dioxide emissions from producing bioenergy from forest harvest residues. *GCB Bioenergy, 3*, 107–115.

Restaino, J. C., & Peterson, D. L. (2013). Wildfire and fuel treatment effects on forest carbon dynamic in the western United States. *Forest Ecology and Management, 303*, 46–60.

Saracoglu, N., & Gunduz, G. (2009). Wood pellets—Tomorrow's fuel for Europe. *Energy Sources: Part A: Recovery Utilization and Environmental Effects, 31*, 1708–1718.

Sathre, R., & O'Connor, J. (2008). *A synthesis of research on wood products and greenhouse gas impacts* (Technical Report TR-19). Vancouver: FPInnovations, Forintek Division.

Schlamadinger, B., & Marland, G. (1996). The role of forest and bioenergy strategies in the global carbon cycle. *Biomass and Bioenergy, 10*, 275–300.

Schlamadinger, B., Spitzer, J., Kohlmaier, G. H., & Lüdeke, M. (1995). Carbon balance of bioenergy from logging residues. *Biomass and Bioenergy, 8*, 221–234.

Schönau, A. P. G., & Coetzee, J. (1989). Initial spacing, stand density and thinning in eucalypt plantations. *Forest Ecology and Management, 29*, 245–266.

Schwarze, R., Niles, J. O., & Olander, J. (2002). Understanding and managing leakage in forest-based greenhouse-gas-mitigation projects. *Philosophical Transactions of the Royal Society of London, Series A: Mathematical, Physical and Engineering Sciences, 360*, 1685–1703.

Scott, J. H., & Reinhardt, E. D. (2001). *Assessing crown fire potential by linking models of surface and crown fire behavior* (Research Paper RMRS-RP-29, 59 p). Fort Collins: U.S. Department of Agriculture, Forest Service, Rocky Mountain Research Station.

Searchinger, T., Heimlich, R., Houghton, R. A., et al. (2008). Use of U.S. croplands for biofuels increases greenhouse gases through emissions from land-use change. *Science, 319*, 1238–1240.

Searchinger, T. D., Hamburg, S. P., Melillo, J., et al. (2009). Fixing a critical climate accounting error. *Science, 326*, 527–528.

Seely, B., Welham, C., & Kimmins, H. (2002). Carbon sequestration in a boreal forest ecosystem: Results from the ecosystem simulation model, FORECAST. *Forest Ecology and Management, 169*, 123–135.

Skog, K. E. (2008). Sequestration of carbon in harvested wood products for the United States. *Forest Products Journal, 58*, 56–72.

Smeets, E. M. W., & Faaij, A. P. C. (2007). Bioenergy potentials from forestry in 2050: An assessment of the drivers that determine the potentials. *Climatic Change, 81*, 353–390.

Smithwick, E. A. H., Harmon, M. E., Remillard, S. M., et al. (2002). Potential upper bounds of carbon stores in forests of the Pacific Northwest. *Ecological Applications, 12*, 1303–1317.

Sohngen, B., & Brown, S. (2008). Extending timber rotations: Carbon and cost implications. *Climate Policy, 8*, 435–451.

Sohngen, B., Mendelsohn, R., & Sedjo, R. (1999). Forest management, conservation, and global timber markets. *American Journal of Agricultural Economics, 81*, 1–13.

Sohngen, B., Beach, R. H., & Andrasko, K. (2008). Avoided deforestation as a greenhouse gas mitigation tool: Economic issues. *Journal of Environmental Quality, 37*, 1368–1375.

Solomon, B. D., Barnes, J. R., & Halvorsen, K. E. (2007). Grain and cellulosic ethanol: History, economics, and energy policy. *Biomass and Bioenergy, 31*, 416–425.

Spath, P. L., & Mann, M. K. (2000). *Life cycle assessment of a natural gas combined-cycle power generation system* (NREL/TP-570-27715, 55 p). Golden: National Renewable Energy Laboratory.

Stephens, S. L., Moghaddas, J. J., Hartsough, B. R., et al. (2009). Fuel treatment effects on stand-level carbon pools, treatment-related emissions, and fire risk in a Sierra Nevada mixed-conifer forest. *Canadian Journal of Forest Research, 39*, 1538–1547.

Sutley, N. H. (2010). *Draft NEPA guidance on consideration of the effects of climate change and greenhouse gas emissions* (12 p). Washington, DC: Council on Environmental Quality. http://ceq.hss.doe.gov/nepa/regs/Consideration_of_Effects_of_GHG_Draft_NEPA_Guidance_FINAL_02182010.pdf

Thornley, J. H. M., & Cannell, M. G. R. (2000). Managing forests for wood yield and carbon storage: A theoretical study. *Tree Physiology, 20*, 477–484.

Tumuluru, J. S., Sokhansanj, S., Lim, C. J., et al. (2010). Quality of wood pellets produced in British Columbia for export. *Applied Engineering in Agriculture, 26*, 1013–1020.

Tuskan, G. A. (1998). Short-rotation woody crop supply systems in the United States: What do we know and what do we need to know? *Biomass and Bioenergy, 14*, 307–315.

U.S. Climate Change Science Program. (2007). *The first state of the carbon cycle report (SOCCR): The North American carbon budget and implications for the global carbon cycle: A report by the U.S. Climate Change Science Program and the Subcommittee on Global Change Research* (242 p). Asheville: National Oceanic and Atmospheric Administration, National Climatic Data Center.

U.S. Department of Agriculture, Farm Service Agency. (2010). *Conservation reserve program: Annual summary and enrollment statistics*. https://www.fsa.usda.gov/Internet/FSA_File/annual2010summary.pdf

U.S. Department of Agriculture, Forest Service [USDA FS]. (2012a). *Future of America's forest and rangelands: Forest Service 2010 Resources Planning Act assessment* (General Technical Report WO-87, 198 p). Washington, DC: USDA FS.

U.S. Department of Agriculture, Forest Service. [USDA FS]. (2012b). *Southern forest futures project*. http://www.srs.fs.usda.gov/futures. 20 June 2012.

U.S. Department of Energy [USDOE]. (2009). *Annual energy review 2008* (407 p). Washington, DC: Energy Information Administration.

U.S. Environmental Protection Agency [USEPA]. (2005). *Greenhouse gas mitigation potential in U.S. forestry and agriculture* (EPA 430-R-05-006, 157 p). Washington, DC: U.S. Environmental Protection Agency, Office of Atmospheric Programs.

U.S. Environmental Protection Agency [USEPA]. (2008). *Inventory of U.S. greenhouse gas emissions and sinks: 1990–2006* (430-R-08-005). Washington, DC: U.S. Environmental Protection Agency, Office of Atmospheric Programs.

U.S. Environmental Protection Agency [USEPA]. (2009a). *Land use, land use change and forestry. Inventory of U.S. greenhouse gas emissions and sinks: 1990–2007*. Washington, DC: U.S. Environmental Protection Agency, Office of Atmospheric Programs.

U.S. Environmental Protection Agency [USEPA]. (2009b). *Waste. Inventory of U.S. greenhouse gas emissions and sinks: 1990–2007*. Washington, DC: U.S. Environmental Protection Agency, Office of Atmospheric Programs.

U.S. Environmental Protection Agency [USEPA]. 2010. *Inventory of U.S. greenhouse gas emissions and sinks: 1990–2008* (EPA 430-R-09-006). Washington, DC: U.S. Environmental Protection Agency, Office of Atmospheric Programs.

U.S. Environmental Protection Agency [USEPA]. (2011). *Draft inventory of U.S. greenhouse gas emissions and sinks: 1990–2009* (EPA 430-R-11-005). Washington, DC: U.S. Environmental Protection Agency, Office of Atmospheric Programs. http://epa.gov/climatechange/emissions/usinventoryreport.html. 12 Mar 2011.

United Nations. (1992). *United Nations framework convention on climate change.* FCCC/INFORMAL/84, GE.05-62220 (E) 200705. http://unfccc.int/resource/docs/convkp/conveng.pdf. 20 Sept 2012.

Upton, B., Miner, R., Spinney, M., & Heath, L. S. (2008). The greenhouse gas and energy impacts of using wood instead of alternatives in residential construction in the United States. *Biomass and Bioenergy, 32*, 1–10.

Van Auken, O. W. (2000). Shrub invasions of North American semiarid grasslands. *Annual Review of Ecology and Systematics, 31*, 197–215.

Woodall, C. W., D'Amato, A. W., Bradford, J. B., & Finley, A. O. (2011). Effects of stand and inter-specific stocking on maximizing standing tree carbon stocks in the eastern United States. *Forest Science, 57*, 365–378.

Zanchi, G., Pena, N., & Bird, N. (2010). The upfront carbon debt of bioenergy. Graz: Joanneum Research. http://www.birdlife.org/eu/pdfs/Bioenergy_Joanneum_Research.pdf. 23 Nov 2011.

Zerbe, J. I. (2006). Thermal energy, electricity, and transportation fuels from wood. *Forest Products Journal, 56*, 6–14.

Chapter 8
Adapting to Climate Change

Constance I. Millar, Christopher W. Swanston, and David L. Peterson

8.1 Principles for Forest Climate Adaptation

Forest ecosystems respond to natural climatic variability and human-caused climate change in ways that are adverse as well as beneficial to the biophysical environment and to society. Adaptation can be defined as responses or adjustments made—passive, reactive, or anticipatory—to climatic variability and change (Carter et al. 1994). Many adjustments occur whether humans intervene or not; for example, plants and animals shift to favorable habitats, and gene frequencies may change to favor traits that enable persistence in a warmer climate.

Here we assess (general) strategies and (specific) tactics that resource managers can use to reduce forest vulnerability and increase adaptation to changing climate (Peterson et al. 2011). Plans and activities range from short-term, stop-gap measures, such as removing conifers that are progressively invading mountain meadows, to long-term, proactive commitments, such as vegetation management to reduce the likelihood of severe wildfire or of beetle-mediated forest mortality.

C.I. Millar
Pacific Southwest Research Station, U.S. Forest Service, Albany, CA, USA
e-mail: cmillar@fs.fed.us

C.W. Swanston
Northern Research Station, U.S. Forest Service, Houghton, MI, USA
e-mail: cswanston@fs.fed.us

D.L. Peterson (✉)
Pacific Northwest Research Station, U.S. Forest Service, Seattle, WA, USA
e-mail: peterson@fs.fed.us

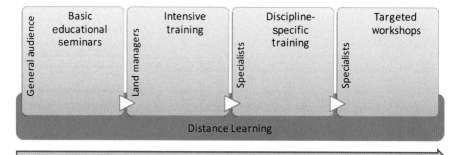

Fig. 8.1 Conceptual diagram of educational and training efforts leading to increased complexity of adaptation planning and activities. These elements are integrated but need not be taken consecutively. Distance learning can be incorporated into all activities (From Peterson et al. 2011)

8.1.1 Adaptation Planning and Implementation

Adaptation strategies, plans, and management actions are generally tied to broad goals of ecosystem sustainability. Restoration, maintenance, and promotion of natural ecological processes and ecosystem services define the mission of most public land-management agencies and many private lands where production forestry is not dominant. Successful implementation of climate adaptation plans occurs when projects are developed and deployed for specific places with concrete treatments and prescriptions, explicit objectives, and for definitive time periods. Successful implementation also implies that monitoring and adaptive management (in a general sense) will continue for the duration of the adaptation effort.

8.1.2 Education and Training

Training for land management professionals in the fundamental concepts of climatology and physical and ecological sciences related to climate change is essential. Such knowledge will increase the institutional capacity to understand potential effects of climate change and associated uncertainty, and to construct appropriate strategies and actions. A multi-level approach facilitates climate change education and dialogue. Recently developed education programs (Peterson et al. 2011; Swanston and Janowiak 2012) have incorporated several elements including basic education, intensive training, and discipline-specific and targeted workshops (Fig. 8.1). Short (1- to 2-day) basic educational seminars convey fundamental principles of climate change and the effects of climate change on ecosystems and generate discussion of how different resources under management consideration can adapt to projected changes. Intensive training includes week-long courses providing detailed information on fundamental climate processes and interactions, as well as mechanisms of forest response to climate stressors. Participants have the opportunity to evaluate issues or resources by using available (e.g., online) tools. Discipline-specific trainings allow focused presentation and discussion of climate change implications for specific resource issues (e.g., silviculture, wildlife).

8.1.3 Science-Management Partnerships

Partnerships between scientists and resource managers are a critical foundation for understanding climate science and developing adaptation strategies. These collaborations can develop in different forms. For example, science information might reside within an agency, but in different program areas than those traditionally involved with forest management. University extension specialists have a long history of spanning boundaries between science and applications (e.g., providing genetic expertise in developing seed-transfer rules), and can be included in partnerships. Research scientists with universities and agencies increasingly participate in resource management collaborations. Interactive dialogue is a key element in these collaborations, with managers and scientists reciprocally learning from and informing each other about relevance. A short-term commitment, typically two years or more, will be needed to develop adaptation strategies and tactics, and a longer commitment is advisable to ensure that new science is considered and adaptation effectiveness is evaluated over time.

8.1.4 Risk and Uncertainty

Given the environmental complexities of forest ecosystems and the diverse and often conflicting societal issues associated with forests, resource managers and decision makers are accustomed to the challenges of risk and uncertainty. Climate change adds new dimensions of uncertainty, increasing the complexity of risk analyses. Trends in climate and ecosystem response can be bounded with probabilistic envelopes that describe what is likely to occur in the future, but unexpected conditions and surprises are likely, especially at local scales. Effective forest adaptation strategies need to (1) be aware of risks, (2) assess vulnerabilities, (3) develop adaptation responses that are realistic yet minimize uncertainties, and (4) incorporate new knowledge and over time to modify decisions as appropriate (essentially the adaptive management process) (Moser and Luers 2008). Adaptation responses to risk include (1) no action—continue conventional practices, (2) contingency planning—develop a response strategy (e.g., to anticipated major disturbance), and (3) anticipatory and proactive strategies—curtail or diminish potential impacts (e.g., of a major disturbance) while optimizing attainment of goals (Joyce et al. 2008).

8.1.5 Toolkit Approach

Novelty and surprise in climate change effects, combined with multiple management objectives at different spatial and temporal scales, mean that no single approach will fit all situations. A toolkit approach to adaptation strategies recognizes that the best strategy will require selecting appropriate methods for the specific situation. Tools include resource management practices, educational and reference modules, decision support aids, and qualitative and quantitative models that address adaptation of natural and cultural resources to climate change (Peterson et al. 2011). Tools include existing management practices, perhaps used in new ways, as well as novel approaches developed to meet climate challenges.

8.1.6 No-Regrets Decision Making

"No-regrets" decision making refers to actions that result in a variety of benefits under multiple scenarios and have little or no risk of undesirable outcomes. This can include (1) implementing fuel treatments in dry forests to reduce fire hazard and facilitate ecological restoration, while creating resilience to increased fire occurrence in a warmer climate, and (2) installing new, larger culverts in locations where peak flows during flooding are expected to be higher in a warmer climate, thus protecting roads and reducing maintenance costs. These types of actions benefit resources and values regardless of climate change effects and can be implemented in the near term (Swanston and Janowiak 2012).

8.1.7 Flexibility and Adaptive Learning

Because future climates and ecosystem responses are uncertain, our experience in developing forest adaptation strategies is limited, flexibility, experimentation, and adaptive learning should be incorporated in adaptation strategies. Although a formal adaptive management program should normally be developed in conjunction with implemented projects, other approaches to monitoring that facilitate modified management practices are also appropriate.

8.1.8 Mixed-Models Approach

Climate- and ecosystem-response models are proliferating, and downscaled climate change scenarios may seem useful for conducting vulnerability analyses and developing adaptation responses at local to regional scales. However, given uncertainty in both climate models and response models, output from projections should be used cautiously. Models are often useful for examining forest response to recent historical events and for attributing causality (e.g., identifying climatic factors that influence large wildfires or insect outbreaks); however, they are often less useful for forecasting at small spatial scales or over long time periods. Output from models is useful as background information for envisioning a range of potential futures rather than to project a single outcome. The use of different types of models—with different assumptions, process interactions, and input data—to address the same issue is recommended. Both quantitative (algorithm based) and qualitative (e.g., flow charts, indices, and verbal tools) models should are useful, and projected futures can be compared. In recent years, it has been suggested that, if a model (or several models) hindcasts observed historical conditions well, it will also accurately predict future conditions. This is not necessarily true, because models can produce a correct historical reconstruction for the wrong reasons (Crook and Forster 2011), which means that forecasts could also be wrong. Given the limitations of models, resource professionals should not hesitate to use their experience and judgment to evaluate model projections of future climate and ecosystem responses. Daniels et al. (2012) provide a straightforward guide for effective use of models.

8.1.9 Integration with Other Priorities and Forest Management Objectives

Adaptation strategies need to be integrated with mitigation activities (actions to reduce human influence on the climate system) (Metz et al. 2001). Adaptation and mitigation goals are preferably considered concurrently, although in some situations strategies may conflict, and compromise choices may be required. Climate change is only one of many challenges confronting forest management, and other priorities must be evaluated at different temporal scales. For example, managing under the Endangered Species Act of 1973 (ESA) can invoke actions that are legally required in the short term but are illogical, given long-term projections of the effects of climate change. For forest lands where ecological sustainability is the central goal, ecosystem-based management as practiced in land management since the late 1980s (e.g., Lackey 1995; Kohm and Franklin 1997) provides a foundation for addressing most climate change effects. Ecosystem-based management acknowledges that natural systems change continuously and that such dynamics bring high levels of uncertainty. Ecosystem-based management concepts are therefore an appropriate foundation for forest adaptation.

8.2 The Context for Adaptation

Adaptation strategies will differ for different forest ecosystems as a function of the diversity of biophysical characteristics and biosocial issues associated with each forest. Climate change affects forest ecosystems at many temporal and spatial

Table 8.1 Factors that affect the relevance of information for assessing vulnerability to climate change of large, intermediate, and small spatial scales

	Relevance by spatial scale		
Factors	Large[a]	Intermediate[b]	Small[c]
Availability of information on climate and climate change effects	High for future climate and general effects on vegetation and water	Moderate for river systems, vegetation, and animals	High for resource data, low for climate change
Accuracy of predictions of climate change effects	High	Moderate to high	High for temperature and water, low to moderate for other resources
Usefulness for specific projects	Generally not relevant	Relevant for forest density management, fuel treatment, wildlife, and fisheries	Can be useful if confident that information can be downscaled accurately
Usefulness for planning	High if collaboration across management units is effective	High for a wide range of applications	Low to moderate

Modified from Peterson et al. (2011)
[a]More than 10 000 km² (e.g., basin, multiple national forests)
[b]100–10 000 km² (e.g., subbasin, national forest, ranger district)
[c]Less than 100 km² (e.g., watershed)

scales, for example, from its influence on timing of bud burst to the evolution of leaf morphology, and from trophic interactions on a rotting log to shifts in biome distribution across continents. The longevity of forest trees, their influence on the physical landscape (e.g., soil development, watershed quality), and role as habitat add complexity to scale issues. Analysis at the correct spatial scales is especially important for assessing trends of climate change and ecological response, given that averages and trends on broad scales (e.g., continental) can mask variability at fine scales (e.g., watershed) (Wiens and Bachelet 2010).

An adaptation framework based on appropriate temporal and spatial scales (e.g., Peterson and Parker 1998) ensures that plans and activities address climate effects and responses effectively. Because scales are nested, the best strategies focus on the scale of the relevant project and include evaluation of conditions and effects at scales broader than the project level, as well as analysis of effects at finer scales (Tables 8.1, 8.2). Broad-scale analysis establishes context, including recognition of processes and effects observed only at large scales (e.g., species decline, cumulative watershed effects) and possible adverse consequences that could be alleviated by early action.

Most public forest lands are managed for long-term ecological sustainability, although emphasis differs by designation for protection level (parks, wilderness, and reserves) and ecosystem services (national and state forests, Bureau of Land Management [BLM] forest and woodlands, and tribal forest lands). Conservation on

Table 8.2 Factors that affect the relevance of information for assessing vulnerability to climate change of large, intermediate, and small time scales

Factor	Relevance by time scale		
	Large[a]	Intermediate[b]	Small[c]
Availability of information on climate and climate change effects	High for climate, moderate for effects	High for climate and effects	Not relevant for climate change and effects predictions
Accuracy of predictions of climate change effects	High for climate and water, low to moderate for other resources	High for climate and water, moderate for other resources	Low
Usefulness for specific projects	High for temperature and water, low to moderate for other resources	High for water, moderate for other resources	Low owing to inaccuracy of information at this scale
Usefulness for planning	High	High for water, moderate for other resources	Low

Modified from Peterson et al. (2011)
[a]More than 50 years
[b]5 to 50 years
[c]Less than 5 years

U.S. public lands is subject to legal and regulatory direction, such as the National Environmental Policy Act [NEPA] of 1969, Clean Air Act of 1970, Clean Water Act of 1977, and ESA. Goals and time horizons of adaptation strategies for public lands differ from those for private lands. Adaptation on industrial forest land focuses on sustaining productive output over a given period of economic analysis (Sedjo 2010), whereas adaptation on nonindustrial private forest lands differs according to the goals and capacities of individual landowners.

8.3 The Adaptation Process

8.3.1 Overview of Forest Adaptation Strategies

The literature on forest adaptation strategies (Baron et al. 2008; Joyce et al. 2008; Peterson et al. 2011; Swanston and Janowiak 2012) (Table 8.3) includes broad conceptual frameworks, approaches to specific types of analyses (e.g., vulnerability assessments, scenario planning, adaptive management), and tools and guidance for site-specific and issue-specific problems. Adaptation at the highest conceptual level in forest ecosystems focuses on resistance, resilience, response, and realignment strategies (Millar et al. 2007) (Box 8.1). These general principles help to identify the scope and scale of appropriate options at the broadest levels (Spittlehouse 2005),

Table 8.3 Climate adaptation guides relevant to the forest sector

Category	Emphasis	Reference
Adaptation framework	General options for wildlands	Millar et al. (2007)
	Options for protected lands	Baron et al. (2008, 2009)
	Adaptation guidebooks	Snover et al. (2007), Peterson et al. (2011), Swanston and Janowiak (2012)
Vulnerability analysis	Climate change scenarios	Cayan et al. (2008)
	Scenario exercises	Weeks et al. (2011)
	Forest ecosystems	Aubry et al. (2011), Littell et al. (2010)
	Watershed analysis	Furniss et al. (2010)
Genetic management	Seed transfer guidelines	McKenney et al. (2009)
	Risk assessment	Potter and Crane (2010)
Assisted migration	Framework for translocation	McLachlan et al. (2007), Ricciardi and Simberloff (2009)
Decision making	Silvicultural practices	Janowiak et al. (2011b)
	Climate adaptation workbook	Janowiak et al. (2011a)
Priority setting	Climate project screening tool	Morelli et al. (2011b)

Box 8.1: A General Framework for Adaptation Options Suitable for Forested Ecosystems

Options range from short-term, conservative strategic approaches to strategies for long-term, proactive plans (from Millar et al. 2007):

Promote Resistance

Actions that enhance the ability of species, ecosystems, or environments to resist forces of climate change and that maintain values and ecosystem services in their present or desired states and conditions

Increase Resilience

Actions that enhance the capacity of ecosystems to withstand or absorb increasing impact without irreversible changes in important processes and functionality

Enable Ecosystems to Respond

Actions that assist climatically driven transitions to future states by mitigating and minimizing undesired and disruptive outcomes

Realign Highly Altered Ecosystems

Actions that use restoration techniques to enable ecosystem processes and functions (including conditions that may or may not have existed in the past) to persist through altered climates and in alignment with changing conditions but they do not provide guidance for developing site-specific plans. In some cases, it may be necessary to consider overarching issues that affect the scientific context for adaptation, such as historical variability, ecological change over time, and use of historic targets in management and restoration (Harris et al. 2006; Milly et al. 2008; Jackson 2012).

Special concerns for adaptation in parks and protected areas (Baron et al. 2008, 2009; Stephenson and Millar 2012) emphasize that future ecosystems will differ from the past, and that fundamental changes in species and their environments will be inevitable. Effective adaptation will need to identify resources and processes at risk, define thresholds and reference conditions, and establish monitoring and assessment programs (adaptive management). Preparing for and adapting to climate change is as much a cultural and intellectual challenge as an ecological issue. Diverse regulations and values dictate desired future ecosystem conditions, which in turn drive decisions about goals, strategies, and actions (Baron et al. 2009).

The reality of change and novelty in future forest ecosystems underscores the importance of vulnerability assessments in developing adaptation strategies (Littell and Peterson 2005; Spittlehouse 2005; Johnstone and Williamson 2007; Nitschke and Innes 2008; Lindner et al. 2010; Littell et al. 2010; Aubry et al. 2011). Vulnerability assessments can differ in terms of subject matter, geographic focus, level of detail, and quantitative rigor. Regional-scale assessments can be cautiously downscaled to smaller management units, recognizing there will be tradeoffs in accuracy. Watersheds have been shown to be a particularly good geographic focus for vulnerability assessment (Furniss et al. 2010). Scenario planning as a tool for vulnerability assessment has been well developed for forested ecosystems in U.S. national parks (Weeks et al. 2011). Tools developed for setting priorities in forest planning and for assessing risks are especially applicable for near-term decision making (Janowiak et al. 2011a; Morelli et al. 2011b).

Recent comprehensive approaches that incorporate both conceptual strategies and specific tools in guidebooks for developing adaptation strategies (Peterson et al. 2011; Swanston and Janowiak 2012) have proven to be useful for both resource managers and scientists. These guidebooks encourage education and training in the basic climate sciences and describe how to proceed from assessment to on-the-ground practices.

Box 8.2: Setting Management Goals and Strategies is Necessary to Develop Site-Specific Forest Adaptation Projects (From Swanston and Janowiak 2012)

Management Goals Management goals are broad, general statements that express a desired state or process to be achieved. They are often not attainable in the short term and provide the context for more specific objectives. Examples of management goals include:

- Maintain and improve forest health and vigor.
- Maintain wildlife habitat for a variety of species.

(continued)

(continued)

Management Objectives Management objectives are concise, time-specific statements of measurable planned results that correspond to pre-established goals in achieving a desired outcome. These objectives include information on resources to be used for planning that defines precise steps to achieve identified goals. Examples of management objectives include:

- Regenerate a portion of the oldest aspen forest type through clearcut harvest in the next year to improve forest vigor in young aspen stands.
- Identify and implement silvicultural treatments within five years to increase the oak component of selected stands and enhance wildlife habitat.

8.3.2 Strategic Steps for Adaptation

The following steps represent a broad consensus on how to develop forest climate adaptation strategies (Swanston and Janowiak 2012):

Step 1: Define location (spatial extent), management goals and objectives, and timeframes—Determine spatial and temporal scales and site-specific locations for appropriate strategies. Management goals and objectives (Box 8.2) for climate adaptation should be explicit and integrated with mitigation and other management goals. Goals are not necessarily stated in narrowly specific quantitative terms; rather, many forest adaptation goals and objectives can be defined broadly (e.g., sustaining ecosystem services).

Step 2: Analyze vulnerabilities—Vulnerability to climate change is "the degree to which geophysical, biological, and socio-economic systems are susceptible to, and unable to cope with, adverse impacts of climate change" (Solomon et al. 2007). Vulnerability is a function of the degree to which a system is exposed to a change in climatic conditions, its sensitivity to that change, and its adaptive capacity (IPCC 2001; Gallopín 2006; Solomon et al. 2007). Vulnerability assessments, which can take different forms (Glick et al. 2011; USGCRP 2011), determine how climatic variability and change might affect natural resources, and inform the development of appropriate priorities, strategies, and timeframes for action.

Step 3: Determine priorities—Priority actions for climate adaptation may differ from those for traditional forest management, and if conditions are changing rapidly, priorities need to be re-assessed regularly. When conditions are urgent and resources limited (e.g., a species in rapid decline), triage methods can be useful (Joyce et al. 2008). In longer term planning, no-regrets assessments (National Research Council 2002; Overpeck and Udall 2010) minimize risk.

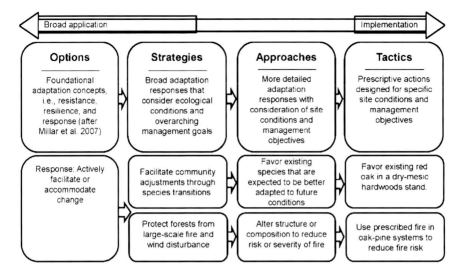

Fig. 8.2 A continuum of adaptation options to address needs at appropriate scales, and examples of each (*shaded boxes*) (From Janowiak et al. 2011a)

Step 4: Develop options, strategies, and tactics—This process begins at a broad conceptual level and steps down to regional and local, site-specific project planning (Swanston and Janowiak 2012), as reflected by the increasing specificity of the following terms (Fig. 8.2). *Adaptation options* are the broadest and most widely applicable level in a continuum of management responses to climate change. Options include resistance, resilience, response, and realignment; they can be short or long term depending on how they are applied (Millar et al. 2007) (Box 8.1), and they can be general or specific and focused on a local situation. *Adaptation strategies* illustrate ways that options can be used. Similar to options, strategies are broad and can be applied in many ways across different forest landscapes (Table 8.4). *Approaches* provide greater detail on how forest managers can respond, with differences in application among specific forest types and management goals becoming evident. *Tactics* are the most specific adaptation response, providing prescriptive direction in how actions are applied on the ground. The culmination of this process is development of a plan, such as a NEPA document or other project plan, prescription, or treatment description.

Step 5: Implement plans and projects—Where possible, project implementation should include replication, randomization, and other experimental design elements, which increases the value of the final step.

Step 6: Monitor, review, adjust—Adaptive management, a key element in climate-adaptation planning (Baron et al. 2008, 2009; Joyce et al. 2008), involves a comprehensive set of steps developed in an experimental framework. Monitoring is tied to predefined thresholds and other target goals developed to test

Table 8.4 Climate change adaptation strategies under broad adaptation options

Strategy	Resistance	Resilience	Response
Sustain fundamental ecological conditions	X	X	X
Reduce the impact of existing ecological stressors	X	X	X
Protect forests from large-scale fire and wind disturbance	X		
Maintain or create refugia	X		
Maintain or enhance species and structural diversity	X	X	
Increase ecosystem redundancy across the landscape		X	X
Promote landscape connectivity		X	X
Enhance genetic diversity		X	X
Facilitate community adjustments through species transitions			X
Plan for and respond to disturbance			X

From Butler et al. (2012)

hypotheses about project effectiveness and appropriateness; if thresholds are exceeded, plans need to be reviewed and adjusted (Walters 1986; Margoluis and Salafsky 1998; Joyce et al. 2008, 2009). Many constraints exist to effective implementation of adaptive management, but at least some informal monitoring keyed to assessing treatment effectiveness is essential for addressing dynamic conditions driven by climate change.

8.4 Tools and Resources for Adaptation and Implementation

Until recently, few guides to implementing climate adaptation plans were available, but many active projects now exist, including in the forest sector. The examples in Table 8.5 are not exhaustive, but represent the type of tools available and the meta-level databases and Web resources that assist in finding relevant tools for specific locations and needs.

8.5 Institutional Responses

Executive Order 13514 (2009), "Federal Leadership in Environmental, Energy, and Economic Performance," directs each federal agency to evaluate climate change risks and vulnerabilities to manage the short- and long-term effects of climate change on the agency's mission and operations. An interagency climate change adaptation task force includes 20 federal agencies and develops recommendations for agency actions in support of a national climate change adaptation strategy. Some of the more successful adaptation efforts to date have involved collaboration among different institutions. Collaboration can take many forms, and effective collaborations will differ by landscape and local institutional relationships.

Table 8.5 Resources that can assist climate change adaptation in forest ecosystems

	Description	Web site (reference)
Web sites: Climate Change Resource Center	U.S. Forest Service portal containing comprehensive information and resources relevant to forest resource managers	http://www.fs.fed.us/ccrc (USDA FS 2011a)
Climate Adaptation Knowledge Exchange	Knowledge base with an interactive online platform, adaptation case studies, directory of practitioners, and summaries of tools and information from other sites	http://www.cakex.org (CAKE 2011)
NaturePeopleFuture.org	The Nature Conservancy (TNC) knowledge base for climate adaptation summarizing adaptation projects, related conservation projects, and adaptation tools	http://conserveonline.org/workspaces/climateadaptation (TNC 2011a)
Tribes and Climate Change	Information to help Native people understand climate change and its effects, including information on climate science, tribal engagement in climate change, and resources to assist adaptation	http://www4.nau.edu/tribalclimatechange (NAU 2011)
Tools: Climate Wizard	Web-based tool that uses climate projections relevant to the time and space resolution of inquiries, enabling users to visualize modeled changes at several time and spatial scales	http://www.climatewizard.org (TNC 2011b)
Vegetation Dynamics Development Tool	User-friendly state-and-transition landscape model for examining the role of various disturbance agents and management actions in vegetation change, allowing users to test sensitivity of vegetation dynamics to climate	http://essa.com/tools/vddt (ESSA 2011)
Template for Assessing Climate Change Impacts and Management Options	Web-based tool that connects forest planning to climate change science, providing access to relevant projections and links to scientific literature on climate effects and management options	http://www.forestthreats.org/research/projects/project-summaries/taccimo (North Carolina State University 2011)
Climate Project Screening Tool	Verbal interview and priority-setting tool for exploring options that ameliorate the effects of climate in resource projects, allowing managers to assess relative vulnerabilities and anticipate effects of different actions	(Morelli et al. 2011b)

(continued)

Table 8.5 (continued)

	Description	Web site (reference)
Climate Change Adaptation Workbook	Using a 5-step process, the workbook incorporates regionally specific climate change information in resource management at different spatial scales and levels of decision making	(Janowiak et al. 2011a)
System for Assessing Vulnerability of Species	Verbal index tool that identifies relative vulnerability or resilience of vertebrate species to climate change, based on a questionnaire with 22 predictive criteria to create vulnerability scores	(Bagne et al. 2011)

8.5.1 U.S. Forest Service

The U.S. Forest Service has the best developed national strategy and on-the-ground implementation of adaptation of all federal agencies (USDA FS 2008). They are led by the climate change advisor's office, which develops guidance and evaluates progress toward climate adaptation. Forest Service research and development also has a climate change strategic plan (Solomon et al. 2009). The National Roadmap for Responding to Climate Change (USDA FS 2011b) summarizes tactical approaches and implementation, including 10 steps along four dimensions: agency and organizational capacity, partnerships and conservation education, adaptation, and mitigation (Fig. 8.3). The process includes (1) science-based assessments of risk and vulnerability; (2) evaluation of knowledge gaps and management outcomes; (3) engagement of staff, collaborators, and partners through education, science-based partnerships, and alliances; and (4) management of resources via adaptation and mitigation.

The Climate Change Resource Center (USDA FS 2011a) (Table 8.5) serves as a reference Web site with information and tools to address climate change in planning and project implementation. Climate change coordinators are designated for each Forest Service region and national forest. Current initiatives from research and management branches of the agency provide climate science, develop vulnerability assessments, prepare adaptive monitoring plans, and align planning, policy, and regulations with climate challenges (Box 8.3). The Performance Scorecard (USDA FS 2011c) (Table 8.6) is used to document progress of national forests, regions, and research stations on adaptation plans and "climate smart" actions.

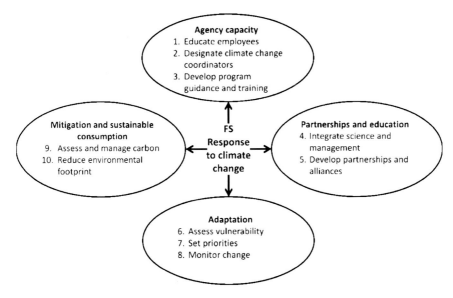

Fig. 8.3 Four dimensions of action outlined by the U.S. Forest Service roadmap for responding to climate change (From USDA FS 2011b)

> **Box 8.3: U.S. Forest Service Initiatives to Promote Progress Toward Achieving Goals of the National Roadmap for Responding to Climate Change (From USDA FS 2011b)**
>
> **Furnish predictive information on climate change and variability**, both immediate and longer term, building on current research capacity and partnerships with the National Oceanic and Atmospheric Administration, National Aeronautics and Space Administration, U.S. Geological Survey, and other scientific agencies
>
> - Develop, interpret, and deliver spatially explicit scientific information on recent shifts in temperature and moisture regimes, including incidence and frequency of extreme events
> - Provide readily interpretable forecasts at regional and subregional scales
>
> **Develop vulnerability assessments**, working through research and management partnerships and collaboratively with partners
>
> - Assess the vulnerability of species, ecosystems, communities, and infrastructure and identify potential adaptation strategies
> - Assess the impacts of climate change and associated policies on tribes, rural communities, and other resource-dependent communities
>
> (continued)

(continued)
- Collaborate with the U.S. Fish and Wildlife Service and National Marine Fisheries Service to assess the vulnerability of threatened and endangered species and to develop potential adaptation measures

Tailor monitoring to facilitate adaptive responses

- Expand observation networks, intensify sampling in some cases, and integrate monitoring systems across jurisdictions (see, for example, the national climate tower network on the experimental forests and ranges)
- Monitor the status and trends of key ecosystem characteristics, focusing on threats and stressors that may affect the diversity of plant and animal communities and ecological sustainability. Link the results to adaptation and genetic conservation efforts

Align Forest Service policy and direction with the Forest Service strategic response to climate change

- Revise National Forest System land management plans using guidance established in the Planning Rule, which requires consideration of climate change and the need to maintain and restore ecosystem and watershed health and resilience
- Review Forest Service manuals and other policy documents to assess their support for the agency's strategic climate change direction. Evaluate current policy direction for its ability to provide the flexibility and integration needed to deal with climate change
- Develop proposals for addressing critical policy gaps

8.5.2 U.S. Department of the Interior (DOI)

A U.S. DOI secretarial order (2009) provides a framework to coordinate climate change activities among DOI bureaus and to integrate science and management expertise with DOI partners. Climate Science Centers and Landscape Conservation Cooperatives form the cornerstones of the framework (DOI FWS 2011). Each has a distinct role, but they share complementary capabilities in support of DOI resource managers and of integrated climate solutions with federal, state, local, tribal, and other stakeholders.

The National Park Service (NPS) climate change response strategy (NPS 2010) provides direction for addressing effects of climate change in NPS units. The broad goals of the strategy include developing effective natural resource adaptation

Table 8.6 Performance scorecard used by the U.S. Forest Service for annual review of progress and compliance, and to identify deficit areas in implementation of the national roadmap for responding to climate change

Scorecard element	Questions addressed
Organizational capacity:	
Employee education	Are all employees provided with training on the basics of climate change, impacts on forests and grasslands, and the Forest Service response?
	Are resource specialists made aware of the potential contribution of their own work to climate change response?
Designated climate change coordinators	Is at least one employee assigned to coordinate climate change activities and be a resource for climate change questions and issues?
	Is this employee provided with the time, training, and resources to make his/her assignment successful?
Program guidance	Does the unit have written guidance for progressively integrating climate change considerations and activities into unit-level operations?
Engagement:	
Science and management partnerships	Does the unit actively engage with scientific organizations to improve its ability to respond to climate change?
Other partnerships	Have climate change-related considerations and activities been incorporated into existing or new partnerships (other than science partnerships)?
Adaptation:	
Assessing vulnerability	Has the unit engaged in developing relevant information about the vulnerability of key resources, such as human communities and ecosystem elements, to the impacts of climate change?
Adaptation actions	Does the unit conduct management actions that reduce the vulnerability of resources and places to climate change?
Monitoring	Is monitoring being conducted to track climate change impacts and the effectiveness of adaptation activities?
Mitigation and sustainable consumption:	
Carbon (C) assessment and stewardship	Does the unit have a baseline assessment of C stocks and an assessment of the influence of disturbance and management activities on these stocks?
	Is the unit integrating C stewardship with the management of other benefits being provided by the unit?
Sustainable operations	Is progress being made toward achieving sustainable operations requirements to reduce the environmental footprint of the agency?

Adapted from USDA FS (2011b, c)

plans and promoting ecosystem resilience, requiring that units (1) develop adaptive capacity for managing natural and cultural resources, (2) inventory resources at risk and conduct vulnerability assessments, (3) prioritize and implement actions and monitor the results, (4) explore scenarios, associated risks, and possible management options, and (5) integrate climate change effects in facilities management. Ecosystem dynamics associated with climate change have forced rethinking of the NPS preservation legacy, and new paradigms are emerging to incorporate ecological change in adaptation philosophies (Cole and Yung 2010; Stephenson and Millar 2012).

The Bureau of Land Management (BLM) focuses on a landscape approach to climate change adaptation, working within ecosystems at large scales and across agency boundaries to assess natural resource conditions and trends, natural and human influences, and opportunities for resource conservation and development. The BLM uses (1) rapid ecoregional assessments (REA), which synthesize information about resource conditions and trends, emphasizing areas of high ecological value (e.g., important wildlife habitats); (2) ecoregional direction, which uses the REAs to identify management priorities for public lands and guide adaptation actions; (3) monitoring for adaptive management, which relies on monitoring and mapping programs to meet understand resource conditions and trends, and evaluate and refine implementation actions; and (4) science integration, which relies on Climate Science Centers to provide management-relevant science. To date, no operational adaptation plans have been produced.

8.5.3 Regional Integrated Sciences and Assessment (RISA)

Funded by the National Oceanographic and Atmospheric Administration's Climate Program Office, the RISA program supports research and stakeholder interaction to improve understanding of climate effects in various regions of the United States, and facilitates the use of climate information in decision making. RISA teams analyze climate data; apply, provide, and interpret climatic information for resource managers and policymakers; and provide information on climate change and regional effects of climate change.

8.5.4 State and Local Institutions

Climate-adaptation responses of state and local institutions are diverse, ranging from minimal action to fully developed and formal programs. State responses that focus on forest-sector issues include the following.

8.5.4.1 Western Governors' Association (WGA)

A nonpartisan organization of governors from 19 Western states, 2 Pacific territories, and 1 commonwealth, the WGA addresses the effects of climate on forest health, wildfire, water and watersheds, recreation, and forest products. The WGA supports integration of climate adaptation science in Western states (WGA 2009) and published a report on priorities for climate response in the West (WGA 2010), including sharing climate-smart practices for adaptation, developing science to be used in decision making, and coordinating with federal entities and other climate adaptation initiatives. The WGA is focusing on developing training to help states incorporate new protocols and strategies relative to climate change, and improving coordination of state and federal climate adaptation initiatives.

8.5.4.2 Washington State Climate Response Strategy

Building on the Washington State Climate Change Impacts Assessment (McGuire et al. 2009; Washington State Department of Ecology 2012), the response strategy is a collaborative effort involving both public and private stakeholders. Recommendations for climate adaptation efforts in major forest ecological systems have been developed (Helbrecht et al. 2011) (Box 8.4), including for fire management and genetic preservation (Jamison et al. 2011). Strategies consistent with adaptation on forest lands include (1) preserve and protect existing working forest, (2) assess how land management decisions help or hinder adaptation, (3) foster interagency collaboration, (4) promote sociocultural and economic relations between eastern and western Washington to improve collaboration, (5) develop options that address major disturbance events, and (6) incorporate state decisions with global and local factors when adapting to climate change (Washington State Department of Ecology 2012).

Box 8.4: Interim Recommendations for the Washington State Climate Change Response Strategy on Species, Habitats, and Ecosystems (From Helbrecht et al. 2011)

Facilitate the resistance, resilience, and response of natural systems

- Provide for habitat connectivity across a range of environmental gradients.
- For each habitat type, protect and restore areas most likely to be resistant to climate change.
- Increase ecosystem resilience to large-scale disturbances, including pathogens, invasive species, wildfire, flooding, and drought.
- Address stressors contributing to increased vulnerability to climate change.

(continued)

(continued)
- Incorporate climate change projections in plans for protecting sensitive species.

  Build scientific and institutional readiness to support effective adaptation
- Fill critical information gaps and focus monitoring on climate change.
- Build climate change into land use planning.
- Develop applied tools to assist land managers.
- Strengthen collaboration and partnerships.
- Conduct outreach on the values provided by natural systems at risk from climate change.

8.5.4.3 Minnesota State Climate Response

The Minnesota Department of Natural Resources is building intellectual and funding capacity to implement policies that address climate change and renewable energy issues, including vulnerability assessments that identify risks and adaptation strategies for forest ecosystems. The Minnesota Forest Resources Council is developing recommendations to the governor and federal, state, county, and local governments on policies and practices that result in the sustainable management of forest resources. Regional landscape committees establish landscape plans that identify local issues, desired future forest conditions, and strategies to attain these goals (MFRC 2011).

8.5.4.4 North Carolina State Climate Response

The North Carolina Department of Environment and Natural Resources (DENR) is developing an adaptation strategy to identify and address potential effects on natural resources, with emphasis on climate-sensitive ecosystems and land use planning and development. The North Carolina Natural Heritage Program is evaluating likely effects of climate change on state natural resources, including 14 forest ecosystems that are likely to respond to climate change in similar ways. The DENR is coordinating with other agencies on an integrated climate response and climate change response plan.

8.5.4.5 State University and Academic Responses

The University of Washington Climate Impacts Group (CIG) has a strong focus on climate science in the public interest. Besides conducting research and assessing climate effects on water, forests, salmon, and coasts, the CIG applies scientific information in regional decisions (e.g., Snover et al. 2007). The CIG works closely with stakeholders and has been a key coordinator for forest climate adaptation projects (e.g., Halofsky et al. 2011; Littell et al. 2011). The Alaska Coastal Rainforest Center, based at the University of Alaska-Southeast, in partnership with the University of Alaska-Fairbanks and other stakeholders, provides educational opportunities, facilitates research, and promotes learning about temperate rain forests. The center facilitates dialogue on interactions among forest ecosystems, communities, and social and economic. The Center for Island Climate Adaptation and Policy, based at the University of Hawai'i at Mānoa, promotes interdisciplinary research and solutions to public and private sectors, with a focus on science, planning, indigenous knowledge, and policy relative to climate adaptation. Recent projects focus on education, coordinating with state natural resource departments on adapting to climate change (CICAP 2009), and policy barriers and opportunities for adaptation. Forest-related climate issues include effects of invasive species, forest growth and decline, migration and loss of forest species, and threats to sustainability of water resources.

8.5.5 Industrial Forestry

The response from forest industries in the United States to climate change has to date focused mostly on carbon sequestration, energy conservation, the role of biomass, and other climate-mitigation issues. Detailed assessments and efforts to develop adaptation strategies for forest industry have mostly been at the global to national scale (Sedjo 2010; Seppälä et al. 2009a, b). Many forestry corporations promote stewardship forestry focused on adaptability of forest ecosystems to environmental challenges, but most ongoing adaptation projects are small scale and nascent. For example, Sierra Pacific Industries (SPI) in California is evaluating the potential for giant sequoia (*Sequoiadendron giganteum* [Lindl.] J. Buchholz) plantations to serve as a safeguard against a changing climate. Giant sequoia currently grows in small groves scattered in the Sierra Nevada. Germplasm would be collected by SPI from the native groves and planted in riparian corridors on productive industry land, then managed as reserves that would benefit from the resilience of giant sequoia to climatic variability and its ability to regenerate after disturbance.

8.5.6 Native American Tribes and Nations

Many Native American tribes and nations have been actively developing detailed forest adaptation plans in response to climate change. Overall goals commonly relate to promoting ecosystem sustainability and resilience, restoration of forest ecosystems, and maintenance of biodiversity, especially of elements having historical and legacy significance to tribes. Maintenance of cultural tradition within the framework of changing times is also inherent in many projects.

An exceptional example of a tribal response is the climate change initiative of the Swinomish Tribe in Washington (SITC 2010). The Swinomish Reservation (3,900 ha) is located in northwestern Washington and includes 3,000 ha of upland forest. The initiative focuses on building understanding among the tribal community about climate change effects, including support from tribal elders and external partners. A recent scientific assessment summarizes vulnerabilities of forest resources to climate change, and outlines potential adaptation options (Rose 2010). Tribes have been active partners in collaborative forest adaptation plans. An example is the Confederated Tribes and Bands of the Yakama Nation, whose reservation occupies 490,000 ha in south-central Washington. Tribal lands comprise forest, grazing, and farm lands in watersheds of the Cascade Range. The Yakama Nation has extensive experience in managing dry forest ecosystems and implementing forest action plans, and belongs to the Tapash Sustainable Forest Collaborative, in partnership with the U.S. Forest Service, Washington State Departments of Fish and Wildlife and of Natural Resources, and The Nature Conservancy. The collaborative encourages coordination among landowners to respond to common challenges to natural resources (Tapash Collaborative 2010). Climate change was ranked as a significant threat to forest productivity, leading to a proposal to incorporate specific adaptation strategies and tactics across the Tapash landscape.

8.5.7 Nongovernmental Organizations

Nongovernmental organizations and professional organizations serve a wide range of special interests, and thus respond to climate adaptation challenges in diverse ways.

8.5.7.1 Pacific Forest Trust (PFT)

A nonprofit organization dedicated to conserving and sustaining America's productive forest landscapes, PFT provides support, knowledge, and coordination on private forest lands in the United States. Through its Working Forests, Winning Climate program, PFT has created policy and market frameworks to expand conservation stewardship of U.S. forests to help sustain ecosystem services (PFT 2011). The PFT also supports climate adaptation by working with private forest owners to promote stewardship forestry, whereby forests are managed to provide goods and services that society has come to expect.

8.5.7.2 The Nature Conservancy (TNC)

A science-based conservation organization, TNC has a mission to preserve plants, animals, and natural communities by protecting the lands and waters they need to survive. The TNC climate change adaptation program seeks to enhance the resilience of people and nature to climate change effects by protecting and maintaining ecosystems that support biodiversity and deliver ecosystem services. The program promotes ecosystem-based approaches for adaptation through partnerships, policy strategies for climate adaptation, tools to assist resource managers, and research. The Canyonlands Research Center (Monticello, Utah), a TNC initiative in the Colorado Plateau region, focuses on forest-climate concerns such as woodland ecosystem restoration, invasive species, and effects of drought.

8.5.7.3 Trust for Public Land (TPL)

A conservation organization that helps agencies and communities conserve land for public use and benefit, TPL uses vulnerability assessments, resilience and connectivity data, and other tools to realign its conservation planning at different spatial scales. The TPL is also designing and implementing restoration to enhance the climate resilience of protected tracts. As a member of the Northern Institute of Applied Climate Science, TPL provides guidance to federal and nonfederal partners on strategic planning and on-the-ground management.

8.5.7.4 The Wilderness Society (TWS)

The Wilderness Society leads efforts to fund natural resource adaptation and manage lands so they are more resilient under stresses of climate change, and is a leader in the Natural Resources Adaptation Coalition, which focuses on maintaining and restoring wildlands that include forest wilderness. Specific TWS goals relative to adaptation in forests include (1) restoring native landscapes to increase ecosystem resiliency, (2) protecting rural communities and providing flexibility in wildland fire management, (3) removing invasive species from ecosystems, and (4) repairing damaged watersheds.

8.5.8 Ski Industry

Although not a direct member of the forest sector, the ski industry relies on mountainous terrain, usually forested land leased from federal landowners, and is concerned about reduced snow, rising temperatures, and extreme weather events that may affect the profitability of ski areas. Adaptation options used by the ski industry (Scott and McBoyle 2007) include (1) snowmaking to increase the duration of the ski season (Scott et al. 2006), (2) optimizing snow retention (slope development and operational practices such as slope contouring, vegetation management, and glacier protection), and (3) cloud seeding. Forest vigor and stand conditions within and adjacent to ski area boundaries are important, because forests burned by wildfire or killed by insect outbreaks affect snow retention, wind patterns, and aesthetic value.

8.6 Regional Responses

Although general guidance and strategic plans about climate adaptation exist for many land management agencies, strategies for specific places and resource issues are in the early stages. Here we summarize recent efforts to develop forest adaptation strategies for specific locations.

8.6.1 Western United States

8.6.1.1 Olympic National Forest/Olympic National Park (ONFP), Washington

This case study covers a large landscape within a geographic mosaic of lands managed by federal and state agencies, tribes, and private landowners (Littell et al. 2011). The ONFP supports a diverse set of ecosystem services, including recreation, timber, water supply to municipal watersheds, pristine air quality, and abundant fish and wildlife. Management of Olympic National Forest focuses on "restoration forestry," which emphasizes facilitation of late-successional characteristics, biodiversity, and watershed values in second-growth forest. Collaboration with adjoining Olympic National Park, which has a forest protection and preservation mission, is strong. Development of the ONFP adaptation approach employed a science-management partnership, including scientific expertise from the CIG, to implement education, analysis, and recommendations for action. Analysis focused on hydrology and roads, vegetation, wildlife, and fish—a vulnerability assessment workshop for each resource area was paired with a workshop to develop adaptation options based on the assessment. Emphasis in adaptation was on conserving biodiversity while working to restore late-successional forest structure through active management. The process used in the case study has been adopted by local resource managers to incorporate climate change issues in forest plans and projects (Halofsky et al. 2011) and is currently being used to catalyze climate-change education, vulnerability assessment, and adaptation planning across 2.5 million ha in Washington state (North Cascadia Adaptation Partnership 2011; Raymond et al. 2013).

8.6.1.2 Inyo National Forest and Devils Postpile National Monument, California

Inyo National Forest (INF) in eastern California contains Mediterranean and dry forest ecosystems, grading from alpine through forest to shrub-steppe vegetation. Much of the national forest is wilderness with a high degree of biodiversity. Water is scarce, fire and insects are important issues, and recreation is the dominant use of public lands. Devils Postpile National Monument (DEPO) is a small national park unit surrounded by INF lands, and collaboration with INF is strong. Ongoing projects focus on vulnerability of INF resources to climate effects that might affect DEPO, and climate adaptation is a high priority in the DEPO general management plan. A science-management partnership facilitated sharing of knowledge about climate change and effects through targeted workshops (Peterson et al. 2011), and assessment reports developed by scientists (Morelli et al. 2011a) assisted managers in considering climate effects relevant to specific resource responsibilities. For INF, the Climate Project Screening Tool (Morelli et al. 2011b) was developed, providing a screening process to rapidly assess if climate change would affect resources in the queue for current-year management implementation. For DEPO, where ecosystem protection is prioritized, managing the monument as a climate refugium (Joyce et al. 2008; Peterson et al. 2011) is being evaluated. Because DEPO is at the bottom of a large canyon with cold-air drainage, it contains high biodiversity, and the potential for cold-air drainage to increase in the future may ameliorate the effects of a warmer climate (Daly et al. 2009).

8.6.1.3 Shoshone National Forest, Wyoming

Resource managers in Shoshone National Forest worked with Forest Service scientists to write a synthesis on climate change effects and a vulnerability assessment of key water and vegetation resources. The synthesis (Rice et al. 2012) describes what is currently understood about local climate and the surrounding Greater Yellowstone Ecosystem and how future climate change may affect local ecosystems. The assessment highlights components of local ecosystems considered most vulnerable to projected changes in climate and will be integrated in resource-related decision making processes of forest management through collaborative workshops to train managers.

8.6.1.4 The Strategic Framework for Science in Support of Management in the Southern Sierra Nevada, California (SFS)

The SFS addresses collaborative climate adaptation for the southern Sierra Nevada bioregion of California (Nydick and Sydoriak 2011), including the southern and western slopes of the Sierra Nevada, three national parks, a national monument, three national forests, tribal lands, state and local public lands, forest industry, and other private lands. This landscape spans ecosystems from alpine through diverse conifer and hardwood forests to woodland and chaparral. The effort is coordinated by a coalition of federal resource managers and academic and agency scientists, and was launched with a public symposium to review the state of science on climate issues and adaptation options. Interactions among climate change and habitat fragmentation, encroaching urbanization, shifting fire regimes, invasive species, and increasing air pollution are important issues in this region. The SFS collaborative has generated a list of ideas to provide knowledge and tools regarding agents of change and potential responses, and a framework document (Exline et al. 2009) is being used to guide adaptation.

8.6.2 Southern United States

Uwharrie National Forest (North Carolina) (UNF) represents a typical national forest context in the southeastern United States, containing 61 parcels mixed with private land and near metropolitan areas (Joyce et al. 2008). Providing a wide range of ecosystem services, the region is undergoing a rapid increase in recreational demand. The UNF identified forest mortality, wildfire, insect outbreaks, soil erosion, stream sedimentation, and water shortages as key issues relative to climate effects. Revision of the forest plan explicitly considers climate change effects. Opportunities for adaptation in UNF focus on reestablishing longleaf pine (*Pinus palustris* Mill.) through selective forest management (Joyce et al. 2008). Replanting of drought-tolerant species could provide increased resistance to potential future drought and intense wildlife. Selective harvest and prescribed burns also could target restoration of longleaf pine savannas, mitigating water stress, fuel loads, and wildfire risk anticipated under warming conditions.

8.6.3 Northern United States

The U.S. Forest Service in the Northeast and upper Midwest is pursuing a comprehensive program of adaptation to climate change (Fig. 8.4), including education and training, partnership building, vulnerability assessment and synthesis, planning and decision support, and implementation of demonstration projects. The Forest Service Northern Research Station, Northeastern Area State and Private Forestry, and Northern Institute of Applied Climate Science work collectively to respond to climate change needs. The Climate Change Response Framework (CCRF) developed by these entities augments the institutional capacity of national forests to adapt to climate change by providing a model for collaborative management and climate change response that can accommodate multiple locations, landscapes, and organizations (Fig. 8.5).

8 Adapting to Climate Change

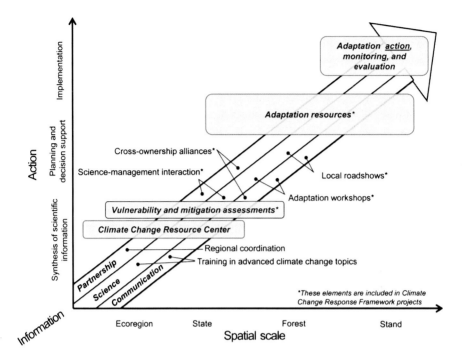

Fig. 8.4 The U.S. Forest Service Eastern Region approach to climate change response works from ecoregional scales down to the stand scale by moving information to action through partnerships, science, and communication

Individual projects focus on building science-management partnerships, developing vulnerability assessments and synthesis of existing information, and establishing a standardized process for considering management plans and activities in the context of the assessment. First, an ecosystem vulnerability assessment and synthesis evaluates ecosystem vulnerabilities and management implications under a range of plausible future climates. Second, a shared landscape initiative promotes dialogue among stakeholders and managers about climate change, ecosystem response, and management. Third, a science team encourages rapid dissemination of information. Fourth, an adaptation resources document includes relevant strategies and a process for managers to devise appropriate tactics. Fifth, demonstration projects incorporate project information and tools in adaptation activities. The CCRF emphasizes an all-lands approach, including national forests, other agencies, and other landowners and stakeholders.

The Northwoods CCRF Project covers 26 million ha of forest in Michigan, Minnesota, and Wisconsin, including six national forests, the Forest Service Northern Research Station, state resource agencies, universities, and other stakeholders. Products to date focus on northern Wisconsin, including a vulnerability assessment (Swanston et al. 2011), a forest adaptation resources document (Swanston and Janowiak 2012), and initiation of demonstration projects in Chequamegon-Nicolet

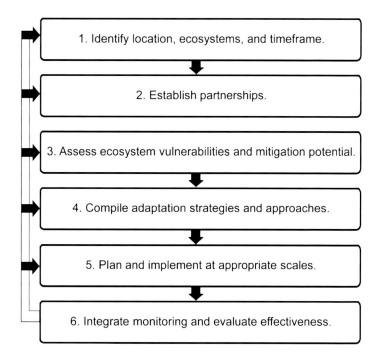

Fig. 8.5 The Climate Change Response Framework uses an adaptive management approach to help land managers understand the potential effects of climate change on forest ecosystems and integrate climate change considerations into management (From Swanston et al. 2012)

National Forest, where each district was asked to integrate climate change considerations into forest activities. The Central Hardwoods CCRF, which covers 17 million ha of hardwood forest in Missouri, Illinois, and Indiana, has formed a regional coordinating team with partners from three national forests, the Northern Research Station, and other stakeholders. The Central Appalachians CCRF, which covers 11 million ha of central Appalachian forest in West Virginia and Ohio, includes partners from two national forests and state forestry agencies.

8.6.4 National Example—Watershed Vulnerability Assessment

In 2010, a watershed vulnerability assessment process was tested in 11 national forests (Furniss et al. 2010, 2013), with the goal of quantifying current and projected future condition of watersheds as affected by climate change. These forests developed a general process that can be tailored to local data availability and resource investment (Box 8.5). Design of useful strategies for reducing the effects of climate change on ecosystem services requires the ability to (1) identify watersheds of highest priority for protecting amenity values, (2) identify watersheds in which climate-related risk to those values is greatest and least, (3) detect evidence of the magnitudes of change as early as possible, and (4) select actions appropriate for reducing effects in particular watersheds (Peterson et al. 2011).

Hydrologic specialists from participating forests developed an approach for quantifying watershed vulnerability within a relatively short period, and four national forests completed the process within 8 months. Acquiring suitable climate exposure data (the magnitude of deviation in climate that a system experiences), which had not been previously used by the participants, was challenging. Threshold values for species and water use differed across the forests. For example, brook trout (*Salvelinus fontinalis* Mitchill) was viewed as a stressor in one forest and a valued resource in another. These differences suggest that, whereas information on processes and resource conditions can be shared among forests, local (forest- and watershed-scale) assessments have the greatest value.

Box 8.5: Steps Defining the Watershed Vulnerability Assessment Process and the Types of Questions to be Addressed (From Furniss et al. 2010)

Step 1—Set up the analysis and establish the scope and water resource values that will drive the assessment
Step 2—Assess exposure
Step 3—Assess sensitivity
Step 4—Evaluate and categorize vulnerability
Step 5—Recommend responses
Step 6—Critique the vulnerability assessment

Typical questions to be addressed in a watershed vulnerability assessment:

- Which places are vulnerable?
- Which places are resilient?
- Where are the potential refugia?
- Where will conflicts arise first, and worst?
- Which factors can exacerbate or ameliorate local vulnerability to climate change?
- What are the priorities for adaptive efforts?
- How can context-sensitive adaptations be designed?
- What needs tracking and monitoring?

8.7 Assessment: Challenges and Opportunities

8.7.1 Assessing Adaptation Response

In recent years, several organizations have produced climate change response strategies that define adaptation goals and describe a framework for action in field units. These strategies, intended to inform and guide consistent agency-wide responses, emphasize (1) staff training and education in climate sciences, (2) science-management partnerships, (3) assessment of vulnerabilities and risks, (4) maintenance of ecosystem sustainability and biodiversity conservation, (5) integration of climate challenges with other forest disturbance agents and stressors, (6) integration of adaptation with greenhouse gas mitigation, (7) all-lands and collaborative approaches (working with whole ecosystems and across jurisdictional borders), (8) recognition of short- and long-term planning perspectives, (9) setting priorities, and (10) monitoring and adaptive management.

Adaptation strategies have been advanced unevenly by federal agencies at regional and local levels (e.g., national forests and national parks). Successful implementation in individual management units has been facilitated by motivated leaders, support from local leadership, and the involvement of constituencies. Some units have worked with local scientists to analyze regional climate projections, develop ecosystem vulnerability assessments, and develop intellectual capacity through staff and constituency education. Collaborative partnerships that extend across ownerships and jurisdictions have been developed as a foundation for some adaptation projects, promoting communication across ownerships. A few progressive units have implemented climate adaptation projects on the ground, but only a few site-specific adaptation projects, as described above, have been accomplished and tiered to local and regional strategies. Responses of state governments have also been variable, with forest-sector states in the western and northern United States leading the way with adaptation strategies. As with federal agencies, concepts and frameworks for adaptation are sometimes available, but site-specific projects are rare. Adaptation responses by tribes and nongovernmental organizations have focused on education, vulnerability assessments, collaborative partnerships, and biodiversity protection, but again with limited on-the-ground activity.

Some organizations have made progress on adaptation by considering climate response strategies as equal or subordinate to more established objectives of ecosystem sustainability, forest and watershed restoration, and biodiversity conservation. Therefore, climate change is not perceived as a primary driver, or even a "lightning rod" issue, and adaptation goals can be accomplished through projects that address high-priority management goals, such as management of fuels, invasive species, insects, and watershed condition.

Implementation of site-specific adaptation plans has been uneven and often superficial across the forest sector, often failing to corroborate the output of climate models and ecosystem response models with local ecosystems (Millar et al. 2007). A subtle danger in using complex, downscaled models is that users may accept model output as a single future, rather than one of several possible outcomes. It is preferable to use models to understand processes and cautiously project climate and ecosystem responses for specific landscapes and time scales, then develop adaptation options for those outcomes.

8.7.2 Adaptation Challenges

Numerous barriers have made it difficult for forest management organizations to develop and implement plans that would promote widespread preparation of U.S. forests for a warmer climate. We view these barriers as challenges that need to be addressed as quickly as possible.

8.7.2.1 Education, Awareness, and Empowerment

University curricula now include courses on climate science, ecosystem responses to climate change, and implications for resource management. Connecting these relatively new educational curricula with historical climatology would improve understanding of concepts like "100-year floods" or "restoration to historic conditions," that assume stationary long-term conditions. Development of appropriate management responses to climate change will need to incorporate a more dynamic perspective on climate and ecosystems (Milly et al. 2008). In general, if resource managers acquire a better understanding of climate science, they will have greater confidence in taking management action. Even if resource managers are knowledgeable about climate science, they may lack support from leadership to implement adaptation, so organizations will benefit if climate education propagates through the highest levels.

Despite widespread public engagement in land management over the past 30 years, pressure to act on climate change has not been as prominent as for other resource issues. Minimal support exists for implementing adaptation projects, and opposition often exists to projects that address indirect effects of climate, such as forest thinning, postfire logging, and road improvements for watershed protection. Reaching out to the public with educational programs on climate change may improve local support for adaptation planning and management.

8.7.2.2 Policy, Planning, and Regulations

Both public and private lands are subject to policy, planning, and regulatory direction. Federal agencies are constrained by hierarchies of laws and internal policy and direction, whereas private forest landowners have greater flexibility to determine actions on their land but are still bound by local, state, and federal laws. In federal agencies, site-specific projects are tiered to levels of planning at higher levels in the organization.

In national forests, site-specific projects tier to each forest's land management plan, which guide management activities to ensure that sustainable management considers the broader landscape and resource values. The U.S. Forest Service has developed procedures through a new national planning rule (Federal Register vol. 76, no. 30; 36 CFR Part 219) to amend, revise, and develop land management plans. The planning rule gives the Forest Service the ability to complete plan revisions more quickly and reduce costs, while using current science, collaboration, and an all-lands approach to produce better outcomes for federal lands and local communities. The planning rule enables management in the context of climate change and other stressors, requiring plans to address maintenance and restoration of ecosystem health and resilience, protection of key resources (e.g., water, air, and soil), and protection and restoration of water quality and riparian areas.

Facing the challenge of working at spatial and temporal scales compatible with climate change requires integration of goals and projects from small to large scales, which may be challenging across a mix of ownerships, making collaboration among multiple organizations essential. As noted above, progress has been made by collaborative efforts that overcame perceived barriers in the regulatory and policy environment. Even at small scales, such as a single national forest or national park, traditional planning approaches dissect lands into discrete units, subject to standards and guidelines for each type of management unit (e.g., watershed protection, timber harvest, wilderness). A more flexible approach that works across the current land classification and regulatory environment will be more compatible with the dynamism of climate and ecosystem responses.

Most environmental laws developed over the past 40 years assume climatic stationarity and thus lack capacity (or legal authority) to accommodate dynamic climate-related changes. For example, endangered species laws often reference native species ranges prior to Euro-American settlement. Climate change will likely catalyze range shifts that will define new native ranges, and enforced maintenance of species in the prior range could be counteradaptive. The National Forest Management Act (NFMA 1976) implies maintenance of the status quo based on historical conditions, usually defined as pre-settlement (nineteenth century) ranges. Because regeneration is the most effective period for changing forest trajectories, planting nursery stock from outside the current seed zone, non-traditional mixes of species, or new species might be a defensible adaptation response (Joyce et al. 2008).

8.7.2.3 Monitoring and Adaptive Management

Future climates and environmental conditions will likely be different than the past, and the imprint of human land use has fragmented and altered forest ecosystems for over a century, making it difficult to determine which forest conditions might be "natural" or "normal." Forest adaptation can meet this challenging set of conditions with innovative approaches informed by monitoring and adaptive management ("learn as you go"). Unfortunately adaptive management in public agencies has been implemented slowly, owing to lack of funding commitment. Modifying objectives and issues to include climate change will be required for monitoring and adaptive management to be a successful partner with adaptation.

8.7.2.4 Financial Barriers

Significant additional funding will be needed for a full national response to forest climate adaptation. Education and training, development of science-management partnerships, vulnerability assessments, and development of adaptation strategies can in many cases be integrated with other aspects of management, although effective consideration of climate requires additional time and effort. Collaboration across organizations and leveraging of institutional capacities can improve efficiency and stretch budgets, allowing at least some progress to be made in landscapes that may be regarded as particularly sensitive to climate change.

8.7.3 A Vision for Climate Smart Forest Management

Facilitating long-term sustainability of ecosystem function is the foundation of climate change adaptation. Effective climate change adaptation will differ by ecosystem, management goals, human community, and regional climate. If adaptation is addressed in a piecemeal fashion (ecological, geographic, and social), some components of the forest sector may suffer the consequences of slow response and inefficiencies. We offer a vision of successful adaptation across U.S. forests within the next 20 years, guided by the statement "A proactive forest sector makes the necessary investments to work across institutional and ownership boundaries to sustain ecosystem services by developing, sharing, and implementing effective adaptation approaches." The following actions are needed to accomplish this vision:

- **Investment**—Invest in (1) basic and applied research; (2) adequate staffing to accommodate increased planning, monitoring complexity, and interaction with partners; and (3) internal and external communication on the dynamic nature of climate and forests. Share monitoring data across multiple agencies and ownerships. Support resource centers, instructional courses, and professional meetings that encourage rapid communication of adaptation management and science. Ensure that planning and other functions facilitate the implementation of on-the-ground activities.
- **Development**—Continue research on forest ecosystem sciences to provide insights into forest responses to climate change, including the effectiveness of climate-adaptation strategies and policies. Update relevant information that allows resource managers to (1) assess vulnerability of ecosystem components,

(2) incorporate a range of climate projections, (3) use multiple modeling approaches to project ecosystem response, and (4) incorporate skills and experience of scientists and land managers. Institutionalize active learning through (1) adaptive management trials that evaluate adaptation techniques, (2) working forests, especially national forests, that serve as "living laboratories" for testing adaptation techniques, and (3) documentation of broad landscape conditions and trends.
- **Sharing**—Clearly state management goals in forest planning documents, including options for sustaining ecosystem function under a range of plausible future climates. Identify vulnerable ecosystems and ecosystem components in vulnerability assessments and management plans, clearly state adaptation options, and identify in potential risks to ecosystem services. Increase investment in local programs that assist small landowners. Share climate information across boundaries of public and private lands, and encourage collaborative management across administrative and ownership boundaries. Institutionalize science-management partnerships to ensure long-term dialogue and collaboration around climate change science and practice.
- **Implementation**—Incorporate climate change in planning activities, and adjust on-the-ground prescriptions to include adaptation where necessary. Provide feedback to the scientific community with feedback on the relevance and clarity of tools and information. Integrate monitoring across multiple scales and institutions, and identify indicators that are sensitive to changes in key ecosystem components, and provide a link from monitoring to decision making. Include the increased potential of extreme events and novel climates in management plans, and ensure that decision making can accommodate multiple potential futures. Use active management to promote resistance and resilience where appropriate, managing some forests to "soften the landing" as they transition to new species assemblages and forest structures. Quickly restore forests affected by extreme events, considering the potential effects of climate on species composition and ecological processes.

We are confident that the U.S. forest sector can make significant progress toward a vision of sustained forest ecosystem function in the face of climate change. This can be accomplished by embracing education and communication about the central role of climatic dynamics in ecosystem processes for resource professionals, stakeholders, and the general public. Accountability for infusing climate into all organizational efforts will ensure that management plans, projects, and decisions are "climate smart." Knowledge about climate is not an independent staff area, but a context through which resource issues can be evaluated. An all-lands approach to climate change and forest management will make collaboration is the norm, ensuring diverse organizational and social perspectives. "Early adapter" collaborations show how regulations, traditions, cultures, and organizational legacies can be navigated successfully. Organizations need to be nimble and flexible to develop effective adaptive responses to climatic challenges. A more streamlined planning process will ensure that projects are implemented in a timely way; planning that prioritizes project implementation including uncertainty, risk, and provisions for experimentation will have the most success.

The challenge of climate change adaptation will require creativity by future generations of forest resource managers. No one agency or organization can fully meet the challenge, but this task is within reach if willing partners work collaboratively toward sustainable management grounded in knowledge of climate science and dynamic ecosystems.

References

Aubry, C., Devine, W., Shoal, R., et al. (2011). *Climate change and forest biodiversity: A vulnerability assessment and action plan for national forests in western Washington* (310 p). Portland: U.S. Department of Agriculture, Forest Service, Pacific Northwest Region. http://ecoshare.info/wp-content/uploads/2011/05/CCFB.pdf

Bagne, K. E., Friggens, M. M., Finch, D. M. (2011). *A system for assessing vulnerability of species (SAVS) to climate change* (General Technical Report RMRS-GTR-257, 28 p). Fort Collins: U.S. Department of Agriculture, Forest Service, Rocky Mountain Research Station. http://www.treesearch.fs.fed.us/pubs/37850

Baron, J. S., Allen, C. D., Fleishman, E., et al. (2008). National parks. In S. H. Julius & J. M. West (Eds.), *Preliminary review of adaptation options for climate-sensitive ecosystems and resources: A report by the U.S. Climate Change Science Program and the subcommittee on Climate Change Research* (pp. 4-1–4-68). Washington, DC: U.S. Environmental Protection Agency.

Baron, J. S., Gunderson, L., Allen, C. D., et al. (2009). Options for national parks and reserves for adapting to climate change. *Environmental Management, 44*, 1033–1042.

Butler, P. R., Swanston, C. W., Janowiak, M. K., et al. (2012). Adaptation strategies and approaches. In C. W. Swanston & M. K. Janowiak (Eds.), *Forest adaptation resources: Climate change tools and approaches for land managers* (General Technical Report NRS-87, pp. 15–34). Newtown Square: U.S. Department of Agriculture, Forest Service, Northern Research Station.

Carter, T. R., Parry, M. L., Harasawa, H., & Nishioka, S. (1994). *IPCC technical guidelines for assessing climate change impacts and adaptation. Working group 2 of the Intergovernmental Panel on Climate Change* (pp. 1–72). Cambridge: Cambridge University Press.

Cayan, D. R., Maurer, E. P., Dettinger, M. D., et al. (2008). Climate change scenarios for the California region. *Climatic Change, 87*, S21–S42.

Center for Island Climate Adaptation and Policy [CICAP]. (2009). *A framework for climate change adaptation in Hawaii* (30 p). Honolulu: State of Hawaii, Ocean Resources Management Plan Working Group; University of Hawaii, Center for Island Climate Adaptation and Policy.

Clean Air Act of 1970; 42 U.S.C. s/s 7401 et seq.

Clean Water Act of 1977; 33 U.S.C. s/s 1251 et seq.

Climate Adaptation Knowledge Exchange [CAKE]. (2011). CAKE: Climate adaptation knowledge exchange. http://www.cakex.org. 28 Dec 2011.

Cole, D. N., & Yung, L. (2010). *Beyond naturalness: Rethinking park and wilderness stewardship in an era of rapid change* (304 p). Washington, DC: Island Press.

Crook, J. A., & Forster, P. M. (2011). A balance between radiative forcing and climate feedback in the modeled 20th century temperature response. *Journal of Geophysical Research, 116*, D17108.

Daly, C., Conklin, D. R., & Unsworth, M. H. (2009). Local atmospheric decoupling in complex topography alters climate change impacts. *International Journal of Climatology, 30*, 1857–1864.

Daniels, A. E., Morrison, J. F., Joyce, L. A., et al. (2012). *Climate projections FAQ* (General Technical Report RMRS-GTR-277WWW, 32 p). Fort Collins: U.S. Department of Agriculture, Forest Service, Rocky Mountain Research Station.

Endangered Species Act of 1973 [ESA], 16 U.S.C. 1531–1536, 1538–1540.

ESSA. (2011). Vegetation dynamics development tool (VDDT). http://essa.com/tools/vddt. 28 Dec 2011.

Exec. Order No. 13,514, 74 Fed. Reg. 194, 52117 (2009, October 8). Federal leadership in environmental, energy, and economic performance. http://www.whitehouse.gov/administration/eop/ceq/sustainability. 28 Dec 2011.

Exline, J., Graber, D., Stephenson, N., et al. (2009). *A strategic framework for science in support of management in the Southern Sierra Nevada ecoregion: A collaboratively developed approach* (24 p). Three Rivers: Southern Sierra Nevada Ecoregion. http://www.nps.gov/seki/naturescience/upload/strategic-framework-2.pdf. 27 Dec 2011.

Furniss, M. J., Roby, K. B., Cenderelli, D., et al. (2013). *Assessing the vulnerability of watersheds to climate change* (General Technical Report PNW-GTR-884, 308 p). Portland: U.S. Department of Agriculture, Forest Service, Pacific Northwest Research Station.

Furniss, M. J., Staab, B. P., Hazelhurst, S., et al. (2010). *Water, climate change, and forests: Watershed stewardship for a changing climate* (General Technical Report GTR-PNW-812, 75 p). Portland: U.S. Department of Agriculture, Forest Service, Pacific Northwest Research Station.

Gallopín, G. C. (2006). Linkages between vulnerability, resilience, and adaptive capacity. *Global Environmental Change, 16*, 293–303.

Glick, P., Stein, B. A., Edelson, N. A., (Eds.). (2011). *Scanning the conservation horizon: A guide to climate change vulnerability assessment* (176 p). Washington, DC: National Wildlife Federation. http://www.nwf.org/vulnerabilityguide. 27 Dec 2011.

Halofsky, J. E., Peterson, D. L., O'Halloran, K. A., & Hawkins Hoffman, C. (2011). *Adapting to climate change at Olympic National Forest and Olympic National Park* (General Technical Report PNW-GTR-844, 130 p). Portland: U.S. Department of Agriculture, Forest Service, Pacific Northwest Research Station.

Harris, J. A., Hobbs, R. J., Higgs, E., & Aronson, J. (2006). Ecological restoration and global climate change. *Restoration Ecology, 14*, 170–176.

Helbrecht, L., Jackson, A., Speaks, P., co-chairs. (2011). *Washington State climate change response strategy: Interim recommendations from Topic Advisory Group 3 on species, habitats and ecosystems.* (Topic Advisory Group 3 interim report, February 2011, 95 p). Olympia: State of Washington, Department of Ecology.

Intergovernmental Panel on Climate Change [IPCC]. (2001). *Climate change 2001: Mitigation. Third assessment report of the Intergovernmental Panel on Climate Change, Working Group III* (753 p). Cambridge: Cambridge University Press.

Jackson, S. T. (2012). Conservation and resource management in a changing world: Extending historical range of variation beyond the baseline. In J. Wiens, C. Regan, G. Hayward, & H. Safford (Eds.), *Historical environmental variation in conservation and natural resource management* (pp. 92–110). New York: Springer.

Jamison, R., Cook, K., Sandison, D., co-chairs. (2011). *Washington state integrated climate change response strategy: Interim recommendations of the natural resources: Working lands and water topic advisory group.* http://www.ecy.wa.gov/climatechange/2011TAGdocs/R2011_interimreport.pdf

Janowiak, M. K., Butler, P. R., Swanston, C. W., et al. (2011a). Adaptation workbook. In C. W. Swanston & M. K. Janowiak (Eds.), *Forest adaptation resources: Climate change tools and approaches for land managers* (General Technical Report NRS-87, pp. 35–56). Newtown Square: U.S. Department of Agriculture, Forest Service, Northern Research Station. Chapter 3.

Janowiak, M. K., Swanston, C. W., Nagel, L. M., et al. (2011b). *Silvicultural decision-making in an uncertain climate future: A workshop-based exploration of considerations, strategies, and approaches* (General Technical Report NRS-81, 14 p). Newtown Square: U.S. Department of Agriculture, Forest Service, Northern Research Station.

Johnstone, M., & Williamson, T. (2007). A framework for assessing climate change vulnerability of the Canadian forest sector. *The Forestry Chronicle, 83*, 358–361.

Joyce, L. A., Blate, G. M., Littell, J. S., et al. (2008). National forests. In S. H. Julius & J. M. West (Eds.), *Preliminary review of adaptation options for climate-sensitive ecosystems and resources: A report by the U.S. Climate Change Science Program and the Subcommittee on Climate Change Research* (pp. 3-1–3-127). Washington, DC: U.S. Environmental Protection Agency. Chapter 3.

Joyce, L. A., Blate, G. M., McNulty, S. G., et al. (2009). Managing for multiple resources under climate change. *Environmental Management, 44*, 1022–1032.

Kohm, K. A., Franklin, J. F., (Eds.). (1997). *Creating a forestry for the 21st century: The science of ecosystem management* (475 p). Washington, DC: Island Press.

Lackey, R. T. (1995). Seven pillars of ecosystem management. *Landscape and Urban Planning, 40*, 21–30.

Lindner, M., Maroschek, M., Netherer, S., et al. (2010). Climate change impacts, adaptive capacity, and vulnerability of European forest ecosystems. *Forest Ecology and Management, 259*, 698–709.

Littell, J. S., & Peterson, D. L. (2005). A method for estimating vulnerability of Douglas-fir growth to climate change in the northwestern U.S. *The Forestry Chronicle, 81*, 369–374.

Littell, J. S., Oneil, E. E., McKenzie, D., et al. (2010). Forest ecosystems, disturbance, and climatic change in Washington State, USA. *Climatic Change, 102*, 129–158.

Littell, J. S., Peterson, D. L., Millar, C. I., & O'Halloran, K. A. (2011). U.S. National forests adapt to climate change through science-management partnerships. *Climatic Change, 110*, 269–296.

Margoluis, R., Salafsky, N. (1998). *Measures of success: Designing, managing, and monitoring conservation and development projects* (384 p). Washington, DC: Island Press.

McGuire, E. M., Littell, J., Whitely Binder, L., (Eds.). (2009). *The Washington climate change impacts assessment* (414 p). Seattle: University of Washington, Center for Science in the Earth System, Joint Institute for the Study of the Atmosphere and Oceans, Climate Impacts Group.

McKenney, D., Pedlar, J., & O'Neill, G. (2009). Climate change and forest seed zones: Past trends, future prospects and challenges to ponder. *The Forestry Chronicle, 85*, 258–266.

McLachlan, J. S., Hellmann, J. J., & Schwart, M. W. (2007). A framework for debate of assisted migration in an era of climate change. *Conservation Biology, 21*, 297–302.

Metz, B., Davidson, O., Swart, R., Pan, J., (Eds.). (2001). *Climate change 2001: Mitigation. Contribution of Working Group III to the third assessment report of the Intergovernmental Panel on Climate Change* (700 p). Cambridge: Cambridge University Press.

Millar, C. I., Stephenson, N. L., & Stephens, S. L. (2007). Climate change and forests of the future: Managing in the face of uncertainty. *Ecological Applications, 17*, 2145–2151.

Milly, P. C. D., Betancourt, J., Falkenmark, M., et al. (2008). Stationarity is dead: Whither water management? *Science, 319*, 573–574.

Minnesota Forest Resources Council [MFRC]. (2011). *Landscape management documents.* http://www.frc.state.mn.us/resources_documents_landscape.html. 28 Dec 2011.

Morelli, T. L., McGlinchy, M. C., & Neilson, R. P. (2011a). *A climate change primer for land managers: An example from the Sierra Nevada* (Research Paper PSW-RP-262, 44 p). Albany: U.S. Department of Agriculture, Forest Service, Pacific Southwest Research Station.

Morelli, T. L., Yeh, S., Smith, N., et al. (2011b). *Climate project screening tool: An aid for climate change adaptation* (Research Paper PSW-RP-263, 29 p). Albany: U.S. Department of Agriculture, Forest Service, Pacific Southwest Research Station.

Moser, S. C., & Luers, A. L. (2008). Managing climate risks in California: The need to engage resource managers for successful adaptation to change. *Climatic Change, 87*, S309–S322.

National Environmental Policy Act of 1969 [NEPA]; 42 U.S.C. 4321 et seq.

National Forest Management Act of 1976 [NFMA]. (1976). Act of October 22, 1976; 16 U.S.C. 1600.

National Park Service [NPS]. (2010). *Climate change response strategy: Science, adaptation, mitigation, communication* (26 p). Fort Collins/Washington, DC: U.S. Department of the Interior, National Park Service, Climate Change Response Program/U.S. Department of the Interior, National Park Service.

National Research Council, Committee on Abrupt Climate Change. (2002). *Abrupt climate change: Inevitable surprises* (230 p). Washington, DC: National Academy Press.

Nitschke, C. R., & Innes, J. L. (2008). Integrating climate change into forest management in south-central British Columbia: An assessment of landscape vulnerability and development of a climate-smart framework. *Forest Ecology and Management, 256*, 313–327.

North Carolina State University [NCSU]. (2011). *TACCIMO: Template for assessing climate change impacts and management options.* http://www.sgcp.ncsu.edu:8090

North Cascadia Adaptation Partnership. (2011). *North Cascadia Adaptation Partnership: Preparing for climate change through science-management collaboration.* http://northcascadia.org

Northern Arizona University [NAU]. (2011). *Tribes and climate change.* http://www4.nau.edu/tribalclimatechange

Nydick, K., & Sydoriak, C. (2011). The strategic framework for science in support of management in the Southern Sierra Nevada, California. *Park Science, 28*, 41–43.

Overpeck, J., & Udall, B. (2010). Dry times ahead. *Science, 328*, 1642–1643.

Pacific Forest Trust [PFT]. (2011). *Working forests, winning climates.* http://www.pacificforest.org/Working-Forests-Winning-Climate.html. 27 Dec 2011.

Peterson, D. L., Parker, V. T., (Eds.). (1998). *Ecological scale: Theory and applications* (608 p). New York: Columbia University Press.

Peterson, D. L., Millar, C. I., Joyce, L. A., et al. (2011). *Responding to climate change on national forests: A guidebook for developing adaptation options* (General Technical Report PNW-GTR-855, 109 p). Portland: U.S. Department of Agriculture, Forest Service, Pacific Northwest Research Station.

Potter, K. M., & Crane, B. S. (2010). *Assessing forest tree genetic risk across the southern Appalachians: A tool for conservation decision-making in changing times.* http://www.forestthreats.org/current-projects/project-summaries/genetic-risk-assessment-system.

Raymond, C. L., Peterson, D. L., & Rochefort, R. M. (2013). The North Cascadia adaptation partnership: A science-management collaboration for responding to climate change. *Sustainability, 5*, 136–159.

Ricciardi, A., & Simberloff, D. (2009). Assisted colonization is not a viable conservation strategy. *Trends in Ecology and Evolution, 24*, 248–253.

Rice, J., Tredennick, A., & Joyce, L. A. (2012). *Climate change on the Shoshone National Forest, Wyoming: A synthesis of past climate, climate projections, and ecosystem implications* (General Technical Report RMRS-GTR-264, 60 p). Fort Collins: U.S. Department of Agriculture, Forest Service, Rocky Mountain Research Station.

Rose, K. A. (2010). *Tribal climate change adaptation options: A review of the scientific literature* (86 p). Seattle: U.S. Environmental Protection Agency, Region 10, Office of Air, Waste and Toxics.

Scott, D., & McBoyle, G. (2007). Climate change adaptation in the ski industry. *Mitigation and Adaptation Strategies for Global Change, 12*, 1411–1431.

Scott, D., McBoyle, G., Minogue, A., & Mills, B. (2006). Climate change and the sustainability of ski-based tourism in Eastern North America: A reassessment. *Journal of Sustainable Tourism, 14*, 376–398.

Sedjo, R. (2010). *Adaptation of forests to climate change: Some estimates* (Discussion Paper RFF DP 10-06, 51 p). Washington, DC: Resources for the Future.

Seppälä, R., Buck, A., & Katila, P. (Eds.). (2009a). *Adaptation of forests and people to climate change: A global assessment report* (IUFRO world series, 224 p, Vol. 22). Vienna: International Union of Forest Research Organizations.

Seppälä, R., Buck, A., & Katila, P. (Eds.). (2009b). *Making forest lands fit for climate change: A global view of climate-change impacts on forests and people and options for adaptation*

(Policy brief, 39 p). Helsinki: Ministry of Foreign Affairs of Finland: International Union of Forest Research Organizations.

Snover, A., Whitley Binder, L., Lopez, J., Willmott, E., Kay, J., Howell, D., Simmonds, J. (2007). *Preparing for climate change: A guidebook for local, regional, and state governments* (172 p). Oakland: In association with and published by ICLEI–Local Governments for Sustainability.

Solomon, S., Quin, D., Manning, M. et al., (Eds.). (2007). *Climate change 2007: The physical science basis. Contribution of Working Group I to the fourth assessment report of the Intergovernmental Panel on Climate Change* (996 p) Cambridge: Cambridge University Press.

Solomon, A., Birdsey, R., Joyce, R., Hayes, J. (2009). *Forest Service global change research strategy overview, 2009–2019. FS-917a* (18 p). Washington, DC: U.S. Department of Agriculture, Forest Service, Research and Development. http://www.fs.fed.us/climatechange/documents/global-change-strategy.pdf

Spittlehouse, D. L. (2005). Integrating climate change adaptation into forest management. *The Forestry Chronicle, 81*, 691–695.

Stephenson, N. L., & Millar, C. I. (2012). Climate change: Wilderness' greatest challenge. *Park Science, 28*, 34–38.

Swanston, C. W., & Janowiak, M. K. (2012). *Forest adaptation resources: Climate change tools and approaches for land managers* (General Technical Report NRS-87, 121 p). Newtown Square: U.S. Department of Agriculture, Forest Service, Northern Research Station.

Swanston, C. W., Janowiak, M., Iverson, L., et al. (2011). *Ecosystem vulnerability assessment and synthesis: A report from the climate change response framework project in northern Wisconsin* (General Technical Report NRS-82, 142 p). Newtown Square: U.S. Department of Agriculture, Forest Service, Northern Research Station.

Swanston, C. W., Janowiak, M. K., & Butler, P. R. (2012). Climate change response framework overview. In C. W. Swanston & M. K. Janowiak (Eds.), *Forest adaptation resources: Climate change tools and approaches for land managers* (General Technical Report NRS-87, pp. 8–14). Newtown Square: U.S. Department of Agriculture, Forest Service, Northern Research Station.

Swinomish Indian Tribal Community [SITC]. (2010). *Swinomish climate change initiative: Climate adaptation action plan* (144 p). La Conner: Swinomish Tribal Community, Office of Planning and Community Development.

Tapash Sustainable Forest Collaborative. (2010). *The Tapash sustainable forest collaborative: Collaborative forest landscape restoration program.* http://www.tapash.org/cflra.htm. 27 Dec 2011.

The Nature Conservancy [TNC]. (2011a). *Climate Wizard.* http://www.climatewizard.org. 28 Dec 2011.

The Nature Conservancy [TNC]. (2011b). NaturePeopleFuture.org: *TNC's knowledge base for climate change adaptation.* http://www.naturepeoplefuture.org. 28 Dec 2011.

U.S. Department of Agriculture, Forest Service [USDA FS]. (2008). *Forest Service strategic framework for responding to climate change (Version 1).* Washington, DC: U.S. Department of Agriculture, Forest Service. http://www.fs.fed.us/climatechange/documents/strategic-framework-climate-change-1-0.pdf

U.S. Department of Agriculture, Forest Service [USDA FS]. (2011a). *Climate Change Resource Center (CCRC).* http://www.fs.fed.us/ccrc

U.S. Department of Agriculture, Forest Service [USDA FS]. (2011b). *National roadmap for responding to climate change. FS-957B* (32 p). Washington, DC: U.S. Department of Agriculture, Forest Service. http://www.fs.fed.us/climatechange/pdf/roadmapfinal.pdf

U.S. Department of Agriculture, Forest Service [USDA FS]. (2011c, August). *Navigating the performance scorecard: A guide for national forests and grasslands (Version 2)* (104 p). Washington, DC: U.S. Department of Agriculture, Forest Service. http://www.fs.fed.us/climatechange/advisor/scorecard/scorecard-guidance-08-2011.pdf. 27 Dec 2011.

U.S. Department of the Interior [DOI]. (2009). Secretarial order no. 3289. Addressing the impacts of climate change on America's water, land, and other natural and cultural resources.

U.S. Department of the Interior, Fish and Wildlife Service [DOI FWS]. (2011, March). Landscape conservation cooperatives: Shared science for a sustainable future. *Fact sheet.* http://www.fws.gov/science/shc/pdf/LCC_Fact_Sheet.pdf

U.S. Global Change Research Program [USGCRP]. (2011). *The United States national climate assessment: Uses of vulnerability assessments for the national climate assessment* (NCA Report Series, vol 9). http://downloads.globalchange.gov/nca/workshop-reports/vulnerability-assessments-workshop-report.pdf

Walters, C. (1986). *Adaptive management of renewable resources* (374 p). New York: Macmillan.

Washington State Department of Ecology. (2012). *Preparing for a changing climate: Washington state's integrated climate response strategy.* Olympia: Washington State Department of Ecology, Climate Policy Group. http://www.ecy.wa.gov/climatechange/ipa_responsestrategy.htm

Weeks, D., Malone, P., & Welling, L. (2011). Climate change scenario planning: A tool for managing parks into uncertain futures. *Park Science, 28,* 26–33.

Western Governors Association [WGA]. (2009). *Supporting the integration of climate change adaptation science in the West* (Policy Resolution 09-2, 3 p). Denver: Western Governors' Association.

Western Governors Association [WGA]. (2010). *Climate adaptation priorities for the western states: Scoping report* (20 p). Denver: Western Governors Association.

Wiens, J. A., & Bachelet, D. E. (2010). Matching the multiple scales of conservation with the multiple scales of climate change. *Conservation Biology, 24,* 51–62.

Chapter 9
Risk Assessment

Dennis S. Ojima, Louis R. Iverson, Brent L. Sohngen, James M. Vose,
Christopher W. Woodall, Grant M. Domke, David L. Peterson,
Jeremy S. Littell, Stephen N. Matthews, Anantha M. Prasad,
Matthew P. Peters, Gary W. Yohe, and Megan M. Friggens

9.1 A Risk-Based Framework

D.S. Ojima (✉)
Natural Resource Ecology Laboratory, Colorado State University, Fort Collins,
CO, USA
e-mail: dojima@nrel.colostate.edu

L.R. Iverson
Northern Research Station, U.S. Forest Service, Delaware, OH, USA
e-mail: liverson@fs.fed.us

B.L. Sohngen
Department of Agricultural, Environmental, and Development Economics,
Ohio State University, Columbus, OH, USA
e-mail: sohngen.1@osu.edu

What is "risk" in the context of climate change? How can a "risk-based framework" help assess the effects of climate change and develop adaptation priorities? *Risk*

D.S. Ojima
Natural Resource Ecology Laboratory, Colorado State University, Fort Collins, CO, USA
e-mail: dojima@nrel.colostate.edu

L.R. Iverson • S.N. Matthews • A.M. Prasad • M.P. Peters
Northern Research Station, U.S. Forest Service, Delaware, OH, USA
e-mail: liverson@fs.fed.us; snmatthews@fs.fed.us; aprasad@fs.fed.us; matthewpeters@fs.fed.us

B.L. Sohngen
Department of Agricultural, Environmental, and Development Economics, Ohio State University, Columbus, OH, USA
e-mail: sohngen.1@osu.edu

can be described by the likelihood of an impact occurring and the magnitude of the consequences of the impact (Yohe 2010) (Fig. 9.1). High-magnitude impacts are always risky, even if their probability of occurring is low; low-magnitude impacts are not very risky, even if their probability of occurring is high. Applying this approach to forest management is challenging because both the likelihood of occurrence and the magnitude of the effects may be difficult to estimate (especially at local scales) and often depend on past and current land use, and the timing, frequency, duration, and intensity of multiple chronic and acute climate-related disturbances.

Despite these challenges, there is much that we do know and it is possible to begin thinking about how to develop a risk-based framework for evaluating the effects of climate change on forests. A risk management framework simply means that risks are identified and estimates are made for their probability of occurrence and their impact. Where we have sufficient knowledge, this framework provides a means to quantify what is known, identify where uncertainties exist, and help managers and decision makers develop strategies with better knowledge of risks.

Climate change will affect forest ecosystems, and the risk of negative consequences to forests and associated biosocial systems will probably increase (Ryan and Archer 2008). However, predicting these risks is difficult because of uncertainty in almost all aspects of the problem. How can we incorporate uncertainty into an analysis of risks and subsequent management decisions? Regional and local projections of climate change are uncertain (Baron et al. 2008; Joyce et al. 2008; Fagre et al. 2009). Despite these uncertainties, climate science has advanced to provide a set of robust climate change projections: the climate is warming, the probability of large precipitation events is increasing, seasonal patterns will be altered, and extreme events are more likely (Solomon et al. 2007). These tendencies

---

J.M. Vose
Southern Research Station, U.S. Forest Service, Raleigh, NC, USA
e-mail: jvose@fs.fed.us

C.W. Woodall • G.M. Domke
Northern Research Station, U.S. Forest Service, St. Paul, MN, USA
e-mail: cwoodall@fs.fed.us; gmdomke@fs.fed.us

D.L. Peterson (✉)
Pacific Northwest Research Station, U.S. Forest Service, Seattle, WA, USA
e-mail: peterson@fs.fed.us

J.S. Littell
Alaska Climate Science Center, U.S. Geological Survey, Anchorage, AK, USA
e-mail: jlittell@usgs.gov

G.W. Yohe
Economics Department, Wesleyan University, Middletown, CT, USA
e-mail: gyohe@wesleyan.edu

M.M. Friggens
Rocky Mountain Research Station, U.S. Forest Service, Albuquerque, NM, USA
e-mail: meganfriggens@fs.fed.us

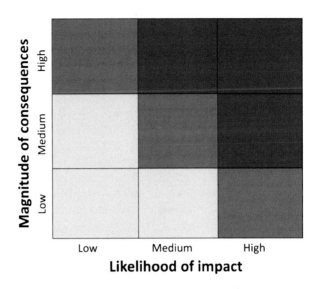

Fig. 9.1 A conceptual risk framework used to help identify risks associated with climate change and prioritize management decisions (Yohe and Leichenko 2010). Colors represent varying degrees of risk (*red* = highest, *yellow* = lowest). In a qualitative definition of consequence, low = climate change is unlikely to have a measurable effect on structure, function, or processes within a specified timeframe (e.g., 2030s, 2050s 2090s); medium = climate change will cause at least one measurable effect on structure, function, or processes within a specified timeframe; and high = climate change will cause multiple or irreversible effects on structure, function, or processes within a specified timeframe. In a qualitative definition of likelihood, low = climate change impacts are unlikely to be measurable within the specified timeframe, medium = climate change impacts are likely to be measurable within the specified timeframe, and high = climate change impacts are very likely (or have already been observed) within or before the specified timeframe

are becoming more apparent in observations across the United States and will affect forest resources nationwide (Karl et al. 2009).

A key challenge is to determine how climate change will alter local biosocial systems, trigger threshold-dependent events, and create nonlinear interactions across interconnected stressors on forest resources (Fagre et al. 2009; Allen et al. 2010), and further, how climate change effects can be addressed by local management actions. Forest managers have experience adapting forest management practices to climatic variability and disturbance regimes. For example, conifer plantations are often managed in short rotations, which limits exposure to risks from insects, wildfires, and windstorms. In mixed-age hardwood forests where management is often less intensive (e.g., where partial harvests are the norm), managers simultaneously choose trees to remove and trees in the understory to release for the next generation of growth. Hence, by using silvicultural techniques to select the species, density, and age class distribution of the next generation of forest, susceptibility to a range of future threats can be modified.

Given what we know about climate change, a robust decision-making approach is needed that acknowledges uncertainty, incorporates system vulnerabilities, and evaluates assets critical for making management decisions (Australian Government 2005; Baron et al. 2008; Joyce et al. 2008; Fagre et al. 2009; Ranger and Garbett-Shiels 2011). A risk management approach provides a framework for identifying management options for climate change, where uncertainties are recognized and management objectives and priorities are explicitly addressed (McInerney and Keller 2008; Yohe and Leichenko 2010; Dessai and Wilby 2011; Ranger and Garbett-Shiels 2011; Iverson et al. 2012). This approach incorporates vulnerability assessment, identifies priority actions relative to management goals, identifies critical information needs, and provides a vision of short- and long-term strategies to enhance the flexibility of management decisions and reduce the probability of poor decisions (Australian Government 2005; Peterson et al. 2011a). This approach also promotes a shift from reactive adaptation to proactive adaptation and coping management (Ranger and Garbett-Shiels 2011) (see Chap. 8), including the following general strategy:

- Identify actions to avoid, that is, avoid choices that lead to less flexibility to adjust to changing conditions.
- Implement "no regrets" management to cope with current stresses and increase resilience to anticipated climate-related stresses.
- Make decisions that integrate across landscapes and governance and that include all concerned and affected stakeholders.
- Develop activities that have strong links among observations, research, and management to understand how ecosystems and social systems are changing, help make decisions, understand thresholds, and help adjust future management and research.

The risk framework must consider the biosocial context of the system being evaluated, reflecting the contribution of forest ecosystem services to different communities and the capability of forest systems to withstand different climate stresses. Providing a more thorough consideration of sources of uncertainty allows for improved development of management strategies, which include key socioeconomic properties. This integrated and multi-sectoral approach will incorporate an improved assessment of risk and current management capacity, and will identify critical uncertainties that may exist under future scenarios if novel consequences emerge.

Case studies using a risk-based framework and concepts are discussed in the following sections on water, carbon, fire, forests, and birds. They are intended as examples, using different approaches to convey risk assessment, and will hopefully create interest by scientists and land managers in developing risk assessments for the effects of climate change on a wide range of forest resources.

9.2 Risk Case Studies

9.2.1 *Water Resources*

J.M. Vose (✉)
Southern Research Station, U.S. Forest Service, Raleigh, NC, USA
e-mail: jvose@fs.fed.us

The importance of forest watersheds for producing and maintaining high quality water flows is well accepted in the scientific literature (Barten et al. 2008). High quality flows are a function both high water quality (e.g., low nutrients and suspended sediment) and regulated flows (e.g., dampened extremes, stable base flows). Climate change will interact with and alter watershed processes in ways that may affect the ability of forests to maintain high quality water flow (Milly et al. 2008; Vörösmarty et al. 2010; see Chap. 3). Some of these interactions will be direct, for example, changes in total precipitation and extreme precipitation events that alter rainfall-runoff relationships. Others will be indirect, such as climate driven disturbances and changing forest species that can alter evapotranspiration and hydrologic flow paths. Altered precipitation and disturbance regimes will interact, for example, through the effects of a combination of extreme wildfires and more intense storms on water quality.

The strong dependency of humans and aquatic organisms on forest watersheds for drinking water (ecosystem services) and habitat (ecological flows), respectively, adds an inherently high level of risk to any climate-based changes in hydrologic processes (Milly et al. 2008; McDonald et al. 2011). For example, an increased frequency of low (or zero) flows could have severe impacts on aquatic species and municipal water supplies. Climatic, biophysical, socioeconomic, and demographic conditions differ greatly across the United States, so the vulnerability of forest-derived water resources (i.e., where water flows are insufficient to meet human needs or sustain aquatic ecosystems) to climate change is not uniform. Integrated water balance models have been used to identify vulnerable regions across the globe and in the United States and to evaluate how changes in climate and human demography will affect future vulnerabilities (Vörösmarty et al. 2010; USDA FS 2012).

In this case study, we develop a risk-based assessment approach based on the assumption that the ratio of precipitation (P) to potential forest evapotranspiration (PET) provides a simple index of water supply (Fig. 9.2). PET sets a theoretical upper limit on plant water use (transpiration) and evaporative losses and is driven by climatic factors (e.g., air temperature, solar radiation, wind speed, etc.). Because plant physiognomy affects water use, PET is usually referenced to specific vegetation types (e.g., forest vs. grass) (Allen et al. 1994). When P is greater than PET,

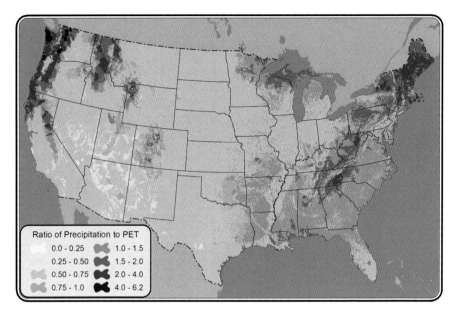

Fig. 9.2 Ratio of precipitation (P) to potential evapotranspiration (PET) for forests in the continental United States. PET was calculated using the Hamon (1961) model

excess water is available for streamflow and groundwater recharge. This excess water provides surface water and groundwater recharge for potential human use, and supports aquatic ecosystems.

Using a ratio of P and PET as an index of vulnerability, risk for both ecosystem services and ecological flows in areas where P/PET is less than 1 would increase if P decreases (Fig. 9.3). In contrast, areas where P/PET is considerably higher than 1 may be less vulnerable to lower P and higher PET, but could in some cases be more vulnerable if higher P is associated with more extreme rainfall and flooding. Vulnerability is also a function of the socioeconomic and ecological ability to rapidly mitigate or adapt to impacts. For municipal water supply, examples of socioeconomic responses include reduced demand through conservation, increased available water supply through more storage capacity, and redistribution via intra- and inter-basin transfers. For ecological flows, aquatic species will be especially vulnerable to changes in both annual flow and intra-annual flow because of limited capacity for mitigation and adaptation. Therefore, the negative consequences of reduced ecological flows on aquatic species would potentially be severe.

Using a risk-based framework in combination with the P/PET map for forests in the continental United States, we project that the Southwest has the highest risk for detrimental effects of lower precipitation and extended droughts. Some areas

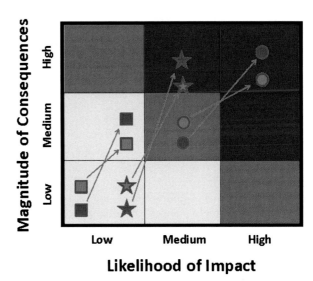

Fig. 9.3 Changes in risk (*arrows* indicate transition from current risk to future risk) as a consequence of increased drought frequency and severity. P/PET < 1 = ○; P/PET > 1 = □; P/PET = 1 = ☆; *green* = ecological flows, *blue* = ecosystem service flows. Risks are higher for ecological flows than for ecosystem service flows because some of the risks to the latter can be offset by engineering and conservation

in the Southwest already employ conservation, storage, and inter-basin transfers to meet current needs and offset current risks, although these measures are unlikely to be sufficient to offset the effects of climate change. In contrast, we project that areas in the upper Lake States, Northwest, and Northeast, where P/PPT is considerably higher than 1, have much lower risk from higher temperature and lower precipitation, because P is already in excess. However, these areas may have higher risk from extreme rainfall events that increase flood frequency and severity (e.g., Halofsky et al. 2011). In this case, responses may require re-examining current flood zones and riparian buffer widths, changing road and culvert designs to accommodate higher flows, enhancing storm water management, and changing designs for roads and infrastructure.

In areas where P/PET is near 1 (e.g., eastern portions of the southern United States), direct and indirect climatic changes can tip the P/PET balance in either direction (Jackson et al. 2009). If large deviations from a P/PET ratio near 1 have been historically infrequent, then neither aquatic organisms nor socioeconomic systems may have the capacity to withstand extreme events (droughts, heavy rainfall), and they may be at even greater risk to climate change than areas that have developed under frequently dry (P < PET) or frequently wet (P > PET) conditions.

9.2.2 A Framework for Assessing Climate Change Risks to Forest Carbon Stocks

C.W. Woodall (✉) • G.M. Domke
Northern Research Station, U.S. Forest Service, St. Paul, MN, USA
e-mail: cwoodall@fs.fed.us; gmdomke@fs.fed.us

Forest ecosystems can reduce the effects of climate change through sequestration of carbon (C) (Pan et al. 2011) as well as contribute to net emissions through tree mortality, wildfires, and other disturbances (Kurz et al. 2008). A conceptual framework for assessing climate change risks to forest ecosystem C stocks facilitates efficient allocation of efforts to monitor and mitigate climate change effects. For example, the U.S. National Greenhouse Gas Inventory (NGHGI) of forest C stocks (Heath et al. 2011) can be used as a basis for developing a climate change risk framework for forest C stocks (Woodall et al. n.d.).

A risk framework for forest C stock incorporates consequence and likelihood as components of risk (Fig. 9.4; compare to Fig. 3.1). One of the most critical future consequences of climate change on forest C stocks is the shift from C sink (net annual sequestration) to C source (net annual emission). Although global forests currently sequester more C than they emit on an annual basis (Pan et al. 2011), it is unclear if or for how long this trend will continue in the future (Birdsey et al. 2006; Reich 2011). If the strength of the C sink decreases and forests became net emitters of C and other greenhouse gasses (GHG) (e.g., methane) a positive feedback loop may be created in which climate change effects may further exacerbate forest C emissions. Likelihood can be phrased as the probability of a C stock becoming a net emitter of C. Likelihoods would be minimal for individual C stocks that are least affected over short timespans (e.g., 50–100 years). Taken together, the C risk framework hinges on the concepts of a "status change" in which forest C stocks transition between C source or sink and a "tipping point" at which forest systems might collapse with concomitant emission of C and potential positive feedbacks that may exacerbate climate change.

We assert that the consequences of a C stock becoming a net emitter of C is directly related to its population estimate over a region of interest. In this case study, it is the C stocks of individual forest pools for the entire United States as reported to the Intergovernmental Panel on Climate Change to meet United Nations Framework Convention on Climate Change requirements (USEPA 2011a, b). If a pool is largest in the United States, then that pool has the largest consequence on global climate change if it is entirely emitted. All current U.S. forest C stocks represent nearly 25 years of U.S. GHG emissions at current emission rates (Woodall et al. 2011). The pools and estimates (Tg C) of C stocks in 2008 (Heath et al. 2011) are ordered as: soil organic C (17,136 Tg C), aboveground live biomass (16,854 Tg C), forest floor (4,925 Tg C), belowground biomass (3,348 Tg C), and dead wood (3,073 Tg C).

The likelihood of any individual C stock becoming a net emitter of C is an emerging area of research. For the purposes of this risk framework (Fig. 9.4), it

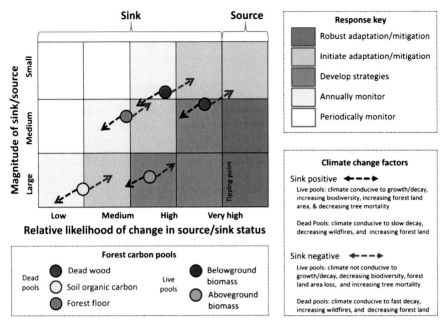

Fig. 9.4 Climate change risk matrix for forest ecosystem carbon (C) pools in the United States, in which climate change may cause C pools to move in a positive (sink = net annual sequestration) or negative (source = net annual emission) direction. Likelihood of change in C stocks is based on the coefficient of variation across the national Forest Inventory and Assessment plot network (x-axis). Size of C stocks is based on the U.S. National Greenhouse Gas Inventory (y-axis). Societal response (e.g., immediate adaptive response or periodic monitoring) to climate change events depends on the size and relative likelihood of change in stocks. The dead wood pool, a relatively small stock, exhibits increasingly high variability across the landscape and therefore may be affected by climate change and disturbance events such as wildfire. In contrast, the forest floor is a relatively small C stock, and has low variability. Potential climate change effects are not incorporated in the matrix, because they represent many complex feedbacks both between C stocks (e.g., live aboveground biomass transitioning to the dead wood pool) and the atmosphere (e.g., forest floor decay)

is proposed that the likelihood of a C stock becoming a net emitter is related to the empirical variation in the stock across the diverse ecosystems and climates of the United States. If climate change occurs such that a mesic boreal forest ecosystem becomes a xeric mixed-hardwood shrubland, then the contemporary range in variation in C stocks between those systems indicates likelihood of C emission. For example, if forest floor C stocks change minimally regardless of climate, then in turn climate change would least affect these stocks. As an initial appraisal of empirical variation in C stocks across the United States, the coefficients of variation (percentage) of individual plot-scale measurements of C stocks (Forest Inventory and Analysis; Heath et al. 2011) across the United States are ordered as dead wood (126.9 Tg C), belowground biomass (107.8 Tg C), aboveground live biomass (104.5 Tg C), forest floor (73.7 Tg C), and soil organic C (67.6 Tg C).

Although climate change events can alter natural variation in C stocks, when compared to contemporary levels, these estimates of variation provide a starting point for a risk framework.

When the consequences and likelihoods of forest C stocks becoming net emitters of C are viewed together, a cohesive approach to monitoring and managing risk emerges. Given the magnitude of potential emissions coupled with the natural variability in these stocks at the continental scale, annual monitoring of dead wood and aboveground live biomass C stocks are needed. In addition, strategies to mitigate negative climate change events (e.g., droughts) can be undertaken. The major research gap in such an approach is how far a pool would move within the risk framework after a climate-related event (the length and direction of the negative/positive arrows in Fig. 9.4). For example, if forest lands convert to grasslands as a result of reduced precipitation and lack of tree regeneration, how would the aboveground biomass pool align itself within the risk framework? Despite the qualitative nature and research gaps within the forest C stock risk framework, this approach provides a conceptual means of identifying priority research needs and a decision system for mitigating climate change.

9.2.3 Risk Assessment for Wildfire in the Western United States

D.L. Peterson (✉)
Pacific Northwest Research Station, U.S. Forest Service, Seattle, WA, USA
e-mail: peterson@fs.fed.us

J.S. Littell
Alaska Climate Science Center, U.S. Geological Survey, Anchorage, AK, USA
e-mail: jlittell@usgs.gov

Wildfire is one of the two most significant disturbance agents (the other being insects) in forest ecosystems in the western United States, and in a warmer climate, will drive changes in forest composition, structure, and function (Dale et al. 2001; McKenzie et al. 2004). Although wildfire is highly stochastic in space and time, sufficient data exist to establish clear relationships between some fire characteristics and some climatic parameters. An assessment of wildfire risk in response to climate change requires brief definitions of the terms "fire hazard" and "fire risk," which are often confused in the scientific literature and other applications (Hardy 2005). *Fire hazard* is the potential for the structure, condition, and arrangement of a fuelbed to affect its flammability and energy release. *Fire risk* is the probability that a fire will ignite, spread, and potentially affect one or more resources valued by people. The most common means of expressing wildfire risk are (1) frequency, (2) a combination of intensity (energy release) and severity (effects on forests, structures, and other values), and (3) area burned.

Fire frequency, the number of fires for a particular location and period of time, differs by region as a function of both lightning and human ignitions, with the requirement that fuels are sufficiently dry and abundant to burn. Lightning ignitions dominate mountainous regions that have convective weather patterns (e.g., most of the Rocky Mountains), whereas human ignitions dominate regions with little lightning and high human populations (e.g., southern California). Modeling studies (+4.2 °C scenario) (Price and Rind 1994) and empirical studies (+1.0 °C scenario) (Reeve and Toumi 1999) suggest that lightning frequency will increase up to 40 % globally in a warmer climate. Although no evidence exists to suggest that recent climate change has caused an increase in lightning or fire frequency in the West, lightning may increase as the temperature continues to rise (Price and Rind 1994; Reeve and Toumi 1999). Assuming that human population will increase throughout the West, it is reasonable to infer that human ignitions will also increase in most regions. Even if the sources and numbers of potential ignitions do not change, a warmer climate may facilitate increased drying of fine surface fuels (less than 8 cm diameter) over a longer period (on a daily and seasonal basis) than currently exists (Littell and Gwozdz 2011), allowing more potential ignitions to become actual ignitions that will become wildfires.

Fire intensity, or energy released during active burning, is directly proportional to *fire severity* in most forests, and can be expressed as effects on vegetation, habitat, and in some cases, human infrastructure. Results of modeling based on a doubled carbon dioxide (CO_2) emission scenario suggest that fire intensity will increase significantly by 2070 in the northern Rocky Mountains, Great Basin, and Southwest (Brown et al. 2004). Fire severity and biomass consumption have increased in boreal forests of Alaska during the past 10 years (Turetsky et al. 2010), and large, intense fires have become more common in California (Miller et al. 2008) and the southwestern United States during the past 20 years. However, interannual and longer term variability in climate-fire relationships can affect trends, making it difficult to infer whether climate change is responsible. Longer time series of fire occurrence, when available, will allow better quantification of the influence of multidecadal modes of climatic variability (e.g., the Pacific Decadal Oscillation, Atlantic Multi-decadal Oscillation). Fire intensity and severity are a function of both climate and land use history, especially the effects of fire exclusion on elevated fuel loads, and forests with high fuel loading will continue to be susceptible to crown fire in the absence of active management (see below).

Fire area has a stronger relationship with climate in the western United States than does either fire frequency or severity/intensity. An empirical analysis of annual area burned (1916–2003) for federal lands in the West projected that, for a temperature increase of 1.6 °C, area burned will increase two to three times in most states (McKenzie et al. 2004). In contrast, a mechanistic model projected that, for the same temperature increase, area burned will increase by only 10 % in California (Lenihan et al. 2003). Using the 1977–2003 portion of the same data set used by McKenzie et al. (2004), Littell et al. (2009) stratified fire area data by Bailey's ecoprovinces (Bailey 1995) to account for fire-climate sensitivities. On average, the model explained 66 % of the variability in historical area burned by combinations of seasonal temperature, precipitation, and Palmer Drought Severity Index. In most forest ecosystems in the northern mountainous portions of the West, fire area was

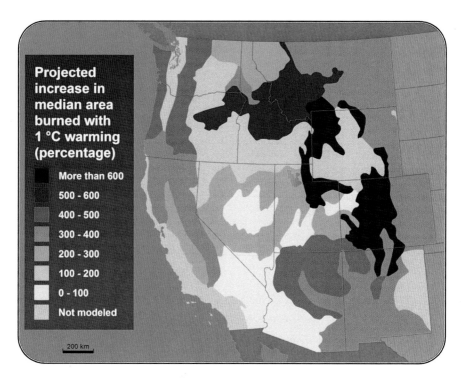

Fig. 9.5 Percentage of increase (relative to 1950–2003) in median area burned for western United States ecoprovinces for a 1 °C temperature increase. Color intensity is proportional to the magnitude of the projected increase in area burned (From Littell (n.d.))

primarily associated with drought conditions, specifically, increased temperature and decreased precipitation in the year of fire and seasons before the fire season. In contrast, in arid forests and woodlands in the Southwest, fire area was influenced primarily by the production of fuels in the year prior to fire and secondarily by drought in the year of the fire.

Littell (n.d.) projected the statistical models of Littell et al. (2009) forward for a 1 °C temperature increase, calculating median area burned and probabilities that annual fire area would exceed the maximum annual area burned in the historical record (1950–2003). Fire area is projected to increase significantly in most ecoprovinces (Fig. 9.5); probability of exceeding the historical maximum annual burn area varied greatly by ecoprovince (range 0–0.44). For the Northwest, the projected increases in area burned are consistent with those found by Rogers et al. (2011) using the MC1 simulation model. A weakness of the statistical models is that, if the projected increased area burned were sustained over several decades, then at some point the large areas burned and decreasing fuel loads would result in less area burned than projected by the models. Neither statistical nor process-based models can satisfactorily account for the effects of extreme fire years and biophysical thresholds that may be exceeded in a much warmer climate.

Based on information summarized above and on expert judgment of the authors, the effects of climate change on fire risk are summarized for fire regimes that occur in forests of the western United States (Table 9.1). We estimate risk for a 2 °C increase, which is more likely by mid-twenty-first century than the more conservative temperature scenarios used by McKenzie et al. (2004) and Littell et al. (n.d.). All fire regimes in forest ecosystems would experience some increase in fire risk. Low-severity and mixed-severity fire regimes dominate dry forest ecosystems of the West and would incur the greatest overall risk in terms of land area. High-severity regimes cover less land area, so they would have less influence on large-scale ecological changes; however, local effects could be significant, particularly where high-severity fire regimes occur close to large population centers, where socioeconomic exposure could be high even if probability of an event were low.

Management of fire risk is a standard component of fire management in the United States. Fire suppression has traditionally been used on both public and private lands to reduce fire area and fire severity. Increasing area burned will provide significant challenges for federal agencies and other organizations that fight fire because of the high cost of suppression and difficulty of deploying firefighters to multiple large fires that may burn concurrently and over a longer fire season. Fuel treatments in dry forest ecosystems of the West can greatly reduce the severity of wildfires (Johnson et al. 2011) (see Sect. 6.5), although funding is available to treat only a small percentage of the total area with elevated fuel loadings. Fuel treatments that include mechanical thinning and surface fuel removal are expensive, especially in the wildland-urban interface, and in a warmer climate, more fuel may need to be removed to attain the same level of reduction in fire severity as is achieved under current prescriptions (Peterson et al. 2011b). Allowing more wildfires to burn unsuppressed is one way to achieve resource benefits while reducing risk, although this approach is often politically unacceptable, especially when fire threatens human infrastructure and other values. Managing fire risk will be one of the biggest challenges for forest resource managers in the West during the next several decades.

9.2.4 Risk Assessment for Forest Habitats: Case Study in Northern Wisconsin

L.R. Iverson (✉) • S.N. Matthews • A.M. Prasad • M.P. Peters
Northern Research Station, U.S. Forest Service, Delaware, OH, USA
e-mail: liverson@fs.fed.us; snmatthews@fs.fed.us; aprasad@fs.fed.us; matthewpeters@fs.fed.us

G.W. Yohe
Economics Department, Wesleyan University, Middletown, CT, USA
e-mail: gyohe@wesleyan.edu

We used a risk matrix to assess risk from climate change for multiple forest species by discussing an example that depicts a range of risk for three tree species in

Table 9.1 Likelihood and magnitude of increased wildfire risk for fire regimes in forests of the western United States, based on a temperature increase of 2 °C[a,b]

Risk parameter	Fire regime			Rationale for risk ratings
	Low severity	Moderate severity	High severity	
Frequency:				More fires will occur in all forests because of longer fire seasons and higher human population. In low-severity systems with low fuel loads, more fires will maintain resilience to fire and climate change; in low-severity systems with high fuel loads, more fires will cause more crown fires. In moderate-severity systems, more fires could convert them to low-severity systems. In high-severity systems, even a small increase in fire frequency will have a large effect on forest structure, function, and carbon dynamics.
Likelihood	Moderate	Moderate	Moderate	
Magnitude	Low	Moderate	High	
Overall risk and potential action	Low; no action recommended	Moderate; encourage fire prevention in high population areas	Moderate; encourage fire prevention in high population areas	
Intensity/severity:				In low-severity and high-severity systems, fire intensity and severity will probably be higher because of more extreme fire weather and elevated fuel loads for the next few decades. In high-severity systems, fuel moisture, not quantity, is limiting, so intensity and severity will not change much; crown fires are always intense and kill much of the overstory.
Likelihood	Moderate[c]	Moderate	Low	
Magnitude	Moderate[c]	Moderate	Low	
Overall risk and potential action	Moderate; increase fuel treatment area and fuel removal	Moderate; increase fuel treatment area and fuel removal	Low; no action recommended	
Area burned:				All fire regimes will experience more area burned. This will be especially prominent in drier, low-severity and moderate-severity systems. In high-severity systems, more area will burn, and although the percentage increase will be less than in other systems, it will have significant local ecological effects.
Likelihood	High	High	Moderate	
Magnitude	High	Moderate	Moderate	
Overall risk and potential action	High; greatly increase fuel treatment area, allow some fires to burn	Moderate; increase fuel treatment area, allow some fires to burn	Moderate; no action recommended	

[a] Risk ratings are qualitative estimates based on information summarized above and on expert judgment of the authors
[b] Fire regimes are defined as (1) low severity: 5- to 30-year frequency, less than 20 % overstory mortality (dry mixed conifer forests and woodlands); (2) mixed severity: 30- to 100-year frequency, patchy and variable overstory mortality (mesic mixed conifer and drier high-elevation forests); and (3) high severity: more than 100-year frequency, more than 80 % overstory mortality (low-elevation conifer and wetter subalpine forests)
[c] Fire intensity/severity are expected to increase in the next few decades, but they may decrease if fuel loadings are sufficiently reduced over time

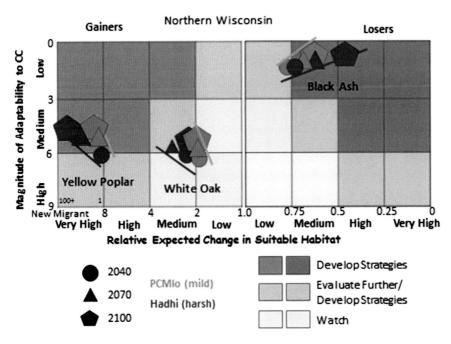

Fig. 9.6 Risk matrix of potential change in suitable habitat for three tree species in northern Wisconsin that are expected to either lose habitat (*black ash*), gain habitat (*white oak*), or become a potential new migrant because of newly appearing habitat (*yellow poplar*)

northern Wisconsin. We define risk as the product of the likelihood of an event occurring and the consequences or effects of that event. In the context of species habitats, likelihood is related to potential changes in suitable habitat at various times in the future. Consequences are related to the adaptability of a species to cope with the changes, especially the increasing intensity or frequency of future disturbance events. Data were generated from an atlas of climate change for tree species of the eastern United States (USDA FS 2011).

A risk matrix allows resource managers to determine which species need adaptation strategies, further evaluation, or monitoring programs. We adopted an established risk matrix structure (Yohe 2010; Yohe and Leichenko 2010; Iverson et al. 2012) to assess the likelihood of exposure and magnitude of vulnerability (or consequences) for three tree species in northern Wisconsin (Fig. 9.6). Much of the climate change literature focuses on potential decreases in forest species ("losers"), but increases may also pose management challenges, so the matrix was modified to include species or forest assemblages that are projected to increase in suitable habitat in the future ("gainers") (Fig. 9.6). The risk matrix is demonstrated for black ash (*Fraxinus nigra* Marsh.) (loser), white oak (*Quercus alba* L.) (gainer), and yellow poplar (*Lireodendron tulipifera* L.) (new migrant).

Black ash carries more risk because, among other disadvantageous traits, it has low resistance to emerald ash borer (*Agrilus planipennis* Fairmaire), which currently threatens all ash species in North America (Prasad et al. 2010). White oak is expected to gain habitat in northern Wisconsin, because it is well adapted to drier conditions and increased disturbance. Relative to other species, projected risk over time for this species is relatively low. Yellow poplar is not now recorded in northern Wisconsin, and as a potential new migrant into the region, it may provide new opportunities for habitat and wood products.

Using methods described in the DISTRIB system (Iverson et al. 2008, 2011; Prasad et al. 2009), data for the likelihood (x-axis) are based on a series of species distribution models to assess habitat suitability for 134 tree species in the eastern United States, for current and future (2040, 2070, and 2100) climatic conditions. "Likelihood" in this context is, for any point in time, the potential that a section of forest within a specified region will have suitable habitat for a given species relative to its current suitable habitat. In this example, we used 2 global change models and 2 emission scenarios (PCMlo and Hadhi) to elicit a range of possible risks, from low to high, associated with future climates. The matrix shows high variation between the modeled output, with Hadhi causing larger changes in suitable habitat for all species. For black ash, which loses habitat, the x-axis ranges from 0 (complete loss of habitat over time) to $+1$ (no change in habitat over time). For white oak, which gains habitat, the x-axis ranges from $+1$ to $+8$. For yellow poplar, a species entering new habitat, the range is confined to the leftmost column of the graph. These numbers themselves are not a direct scale of "likelihood," but rather are scales of future:current importance values.

Consequences in this context are related to the adaptability of a species or forest assemblage under climate change, based on a literature assessment of species biological traits and capacity to respond to disturbances that are likely to occur within the twenty-first century, including how those disturbances will be affected by climate change. Data for this axis come from a literature-based scoring system, called "modification factors," to capture species response to climate change (Matthews et al. 2011a). This approach was used to assess the capacity for each species to adapt to 12 disturbance types and to assess nine biological characteristics related to species adaptability. Each character was scored individually from -3 to $+3$ as an indication of the adaptability of the species to climate change. The mean, scaled values for biological and disturbance characteristics were each rescaled to 0–6 and combined as a hypotenuse of a right triangle; the resulting metric (ranging from 0 to 8.5) was used for the y-axis of the risk matrix (Fig. 9.6). Because several disturbances (e.g., floods, droughts, insect attacks) are expected to increase over time, we also used a formula based on modification factors to enhance relevance for certain factors from 2040 to 2100.

The risk matrix provides a visual tool for comparing species risks relative to changing habitats associated with climate change. Trajectories displayed in the matrix reveal insights about species response to climate change and can be considered in the development of potential adaptation strategies, although they cannot account for non-linear responses to extreme climate and altered disturbance regimes. The risk matrix can also help organize "climate change thinking" on a resource management team and communicate information to stakeholder groups and the general public. Finally, the risk matrix can be used to assess climate change risk for a variety of resource disciplines, and although the metrics may not be derived from the same methodologies, the capacity to rate one species against another, or one location against another, provides a consistent approach to managing climate change risk.

9.2.5 Risk Assessment for Bird Species: A Case Study in Northern Wisconsin

M.M. Friggens (✉)
Rocky Mountain Research Station, U.S. Forest Service, Albuquerque, NM, USA
e-mail: meganfriggens@fs.fed.us

S.N. Matthews
Northern Research Station, U.S. Forest Service, Delaware, OH, USA
e-mail: snmatthews@fs.fed.us

Species distribution models for 147 bird species have been derived using climate, elevation, and distribution of current tree species as potential predictors (Matthews et al. 2011b). In this case study, a risk matrix was developed for two bird species (Fig. 9.7), with projected change in bird habitat (the x-axis) based on models of altered suitable habitat resulting from changing climate and tree species habitat. Risk was evaluated for three time steps (2040, 2070, 2100) and based on two climate models and two emissions scenarios (Hadhi, PCMlo).

To assess the y-axis of the matrix (Fig. 9.7), we used the System for Assessing Vulnerability of Species (SAVS) (Bagne et al. 2011; Davison et al. 2011) to estimate species adaptability to future changes, including disturbances. The SAVS tool is based on 22 traits that represent potential areas of vulnerability or resilience with respect to future climate change. Each trait forms the basis of a question that is scored according to predicted effect (reduced, neutral or increased population). By selecting responses for each question, a user creates a score that represents relative vulnerability to climate change effects, with higher positive values indicating higher vulnerability. Scores were calculated considering all 22 traits and divided among 4 categories: habitat, physiology, phenology, and biotic interactions. To calculate a baseline that could be used to compare current versus future vulnerability, we zeroed out individual questions for traits relating to exposure to future conditions and calculated a score based on the intrinsic characteristics of a species that reflect its sensitivity to population declines as a result of stochastic or other events.

Northern Wisconsin is near the edge of the distribution of the northern cardinal (*Cardinalis cardinalis* L.) and offers relatively limited habitat opportunities because of current winter climatic conditions. However, with projected increases in temperatures for northern Wisconsin, the habitat for the northern cardinal is

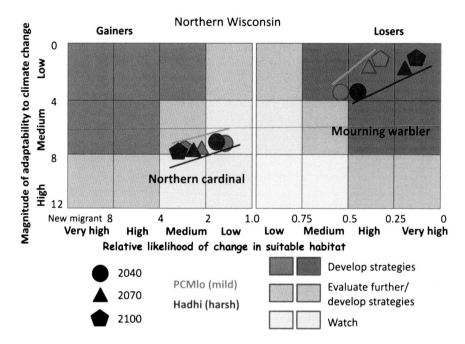

Fig. 9.7 Risk of the effects of climate change on the northern cardinal and mourning warbler, expressed as a combination of likelihood of habitat change (x-axis) and magnitude of adaptability (y-axis). Values are rescaled from calculations that used the approach in SAVS (Bagne et al. 2011; Davison et al. 2011)

projected to double by the end of the century (future:current habitat ratio of 2.2). The northern cardinal uses habitats ranging from shrublands to forests, has a broad diet, and has been shown to be positively associated within an urbanizing landscape (Rodewald and Shustack 2008). The SAVS baseline scores indicate less vulnerability (−0.91) and that the species does not show increased vulnerability risk under climate change (−1.82). Characteristics such as adaptability of nesting locations and flexibility in reproductive time contribute to the less vulnerable score.

In contrast, the mourning warbler (*Oporornis philadelphia* A. Wilson) shows higher risk based on its more specialist nature, specificity to breeding habitats, and Neotropical migration life history. These innate traits make the mourning warbler more susceptible under current conditions (SAVS +3.64) and is also considered at an increased risk of exposure to negative effects of climate change (+5.45). The mourning warbler is primarily a boreal species and despite its use of early successional habitats and a positive response to some human disturbances such as timber harvest (Hobson and Schieck 1999), its occurrence in northern Wisconsin declined over a recent 16-year interval (Howe and Roberts 2005). Moving beyond contemporary changes, its habitat is projected to decrease to one third of its current range by the end of the century (future:current ratio as low as 0.13 or 0.33, depending on climate model). These potential changes in habitat are attributed to higher temperatures and loss of boreal forest habitat (Iverson et al. 2008). In addition, the premontane and montane tropical life zones inhabited by the mourning warbler during winter are predicted to be highly sensitive to climatic affects (Enquist 2002). Therefore, when viewed together, the likelihood and magnitude of projected climate change suggest high risk for this species, and an increased opportunity for the northern cardinal, whose habitat will expand into northern Wisconsin.

The general approach used here can be applied to a wide range of species, using either quantitative information or qualitative logic. The empirical statistical models used here provide insights on the broad-scale determinants of species distributions, but with some limiting assumptions. Models derived from mechanistic relationships that explore processes regulating population dynamics also demonstrate the importance of local climatic conditions on avian populations (Rodenhouse 1992; Anders and Post 2006), but they are available for only a limited number of species. The detailed parameterizations of process models also have important assumptions and can be difficult to apply across a broad array of species. Thus, more refined inferences on how climate change may affect avian populations will require careful consideration of both empirical and mechanistic approaches to modeling species distributions, including the influence of ecological disturbances on habitat, and threshold values for minimum habitat quantity and quality.

References

Allen, R. D., Smith, M., Perrier, A., & Pereira, L. S. (1994). An update for the definition of reference evapotranspiration. *ICID Bulletin, 43*, 1–34.

Allen, C. D., Macalady, A. K., Chenchouni, H., et al. (2010). A global overview of drought and heat-induced tree mortality reveals emerging climate change risks for forests. *Forest Ecology and Management, 259*, 660–684.

Anders, A. D., & Post, E. (2006). Distribution-wide effects of climate on population densities of a declining migratory landbird. *Journal of Animal Ecology, 75*, 1365–2656.

Australian Government. (2005). *Climate change risk and vulnerability: Promoting an efficient adaptation response in Australia* (159pp). Canberra: Department of the Environment and Heritage, Australian Greenhouse Office.

Bagne, K. E., Friggens, M. M., & Finch, D. M. (2011). *A system for assessing vulnerability of species (SAVS) to climate change* (Gen. Tech. Rep. RMRS-GTR-257, 28pp). Fort Collins: U.S. Department of Agriculture, Forest Service, Rocky Mountain Research Station.

Bailey, R. G. (1995). *Description of the ecoregions of the United States [1:7,500,000]* (2nd ed., Misc. Pub. 1391). Washington, DC: U.S. Department of Agriculture, Forest Service.

Baron, J. S., Allen, C. D., Fleishman, E., et al. (2008). National parks. In S. H. Julius & J. M. West (Eds.), *Preliminary review of adaptation options for climate-sensitive ecosystems and resources: A report by the U.S. Climate Change Science Program and the Subcommittee on Climate Change Research* (pp. 4-1–4-68). Washington, DC: U.S. Environmental Protection Agency.

Barten, P. K., Jones, J. A., Achterman, G. L., et al. (2008). *Hydrologic effects of a changing forest landscape* (180pp). Washington, DC: National Academies Press.

Birdsey, R., Pregitzer, K., & Lucier, A. (2006). Forest carbon management in the United States: 1600–2100. *Journal of Environmental Quality, 35*, 1461–1469.

Brown, T. J., Hall, B. L., & Westerling, A. L. (2004). The impact of twenty-first century climate change on wildland fire danger in the Western United States: An applications perspective. *Climatic Change, 62*, 365–388.

Dale, V. H., Joyce, L. A., McNulty, S., et al. (2001). Climate change and forest disturbances. *BioScience, 51*, 723–734.

Davison, J. E., Coe, S., Finch, D., et al. (2011). Bringing indices of species vulnerability to climate change into geographic space: An assessment across the Coronado National Forest. *Biodiversity and Conservation*. doi:10.1007/s10531-011-0175-0.

Dessai, S., & Wilby, R. (2011). *How can developing country decision makers incorporate uncertainty about climate risks into existing planning and policymaking processes?* World Resources Report Uncertainty Series. Washington, DC: World Resources Institute. http://www.worldresourcesreport.org/decision-making-in-depth/managing-uncertainty. 15 Sep 2012.

Enquist, C. A. F. (2002). Predicted regional impacts of climate change on the geographical distribution and diversity of tropical forest in Costa Rica. *Journal of Biogeography, 29*, 519–564.

Fagre, D. B., Charles, C. W., Allen, C. D., et al. (2009). *Thresholds of climate change in ecosystems: A report by the U.S. Climate Change Science Program and the Subcommittee on Global Change Research* (70pp). Reston: U.S. Department of the Interior, Geological Survey.

Halofsky, J. E., Peterson, D. L., O'Halloran, K. A., & Hawkins Hoffman, C. (2011). *Adapting to climate change at Olympic National Forest and Olympic National Park* (Gen. Tech. Rep. PNW-GTR-844, 130pp). Portland: U.S. Department of Agriculture, Forest Service, Pacific Northwest Research Station.

Hamon, W. R. (1961). Estimating potential evapotranspiration. *Journal of the Hydraulic Division of the Proceedings of the American Society of Civil Engineering, 87*, 107–120.

Hardy, C. C. (2005). Wildland fire hazard and risk: Problems, definitions, and context. *Forest Ecology and Management, 211*, 73–82.

Heath, L. S., Smith, J., Skog, K., et al. (2011). Managed forest carbon estimates for the U.S. greenhouse gas inventory, 1990–2008. *Journal of Forestry, 109*, 167–173.

Hobson, K. A., & Schieck, J. (1999). Change in bird communities in boreal mixedwood forest: Harvest and wildfire effects over 30 years. *Ecological Applications, 9*, 849–863.

Howe, R. W., & Roberts, L. J. (2005). Sixteen years of habitat-based bird monitoring in the Nicolet National Forest. In C. J. Ralph & T. D. Rich (Eds.), *Bird conservation implementation and integration in the Americas: Proceedings of the third international partners in flight conference* (Gen. Tech. Rep. PSW-GTR-191, pp. 963–973). Albany: U.S. Department of Agriculture, Forest Service, Pacific Southwest Research Station.

Iverson, L. R., Mathews, S. N., Prasad, A. M., et al. (2012). Development of risk matrices for evaluating climatic change responses of forested habitats. *Climatic Change, 114*, 231–243.

Iverson, L., Prasad, A. M., Matthews, S., & Peters, M. (2008). Estimating potential habitat for 134 eastern US tree species under six climate scenarios. *Forest Ecology and Management, 254*, 390–406.

Iverson, L., Prasad, A. M., Matthews, S., & Peters, M. (2011). Lessons learned while integrating habitat, dispersal, disturbance, and life-history traits into species habitat models under climate change. *Ecosystems, 14*, 1005–1020.

Jackson, R. B., Jobbagy, E. G., & Nosetto, M. D. (2009). Ecohydrology in a human dominated landscape. *Ecohydrology, 2*, 383–389.

Johnson, M. C., Kennedy, M. C., & Peterson, D. L. (2011). Simulating fuel treatment effects in dry forests of the Western United States: Testing the principles of a fire-safe forest. *Canadian Journal of Forest Research, 41*, 1018–1030.

Joyce, L. A., Blate, G. M., Littell, J. S., et al. (2008). National forests. In S. H. Julius & J. M. West (Eds.), *Preliminary review of adaptation options for climate-sensitive ecosystems and resources: A report by the U.S. Climate Change Science Program and the Subcommittee on Climate Change Research* (pp. 3-1–3-127). Washington, DC: U.S. Environmental Protection Agency.

Karl, T. R., Melillo, J. M., Peterson, T. C., & Hassol, S. J. (Eds.) (2009). *Global climate change impacts in the United States*. A report of the U.S. Global Change Research Program (192pp). Cambridge: Cambridge University Press.

Kurz, W. A., Stinson, G., Rampley, G. J. D., et al. (2008). Risk of natural disturbances makes future contribution of Canada's forests to the global carbon cycle highly uncertain. *Proceedings of the National Academy of Sciences USA, 105*, 1551–1555.

Lenihan, J. M., Drapek, R., Bachelet, D., & Neilson, R. P. (2003). Climate change effects on vegetation distribution, carbon, and fire in California. *Ecological Applications, 13*, 1667–1681.

Littell, J. S., & Gwozdz, R. B. (2011). Chapter 5: Climatic water balance and regional fire years in the Pacific Northwest, USA: Linking regional climate and fire at landscape scales. In D. McKenzie, C. Miller, & D. A. Falk (Eds.), *The landscape ecology of fire* (pp. 117–139). New York: Springer.

Littell, J. S. (n.d.). *Relationships between area burned and climate in the western United States: Vegetation-specific historical and future fire*. Manuscript in preparation. Anchorage: U.S. Geological Survey, Climate Science Center.

Littell, J. S., McKenzie, D., Peterson, D. L., & Westerling, A. L. (2009). Climate and wildfire area burned in western U.S. ecoprovinces, 1916-2003. *Ecological Applications, 19*, 1003–1021.

Matthews, S. N., Iverson, L. R., Prasad, A. M., et al. (2011a). Modifying climate change habitat models using tree species-specific assessments of model uncertainty and life history factors. *Forest Ecology and Management, 262*, 1460–1472.

Matthews, S. N., Iverson, L. R., Prasad, A. P., & Peters, M. P. (2011b). Changes in potential habitat of 147 North American breeding bird species in response to redistribution of trees and climate following predicted climate change. *Ecography*. doi:10.1111/j.1600-0587.2010.06803.x.

McDonald, R. I., Green, P., Balk, D., et al. (2011). Urban growth, climate change, and freshwater availability. *Proceedings of the National Academy of Sciences USA, 108*, 6312–6317.

McInerney, D., & Keller, K. (2008). Economically optimal risk reduction strategies in the face of uncertain climate thresholds. *Climatic Change, 91*, 29–41.

McKenzie, D., Gedalof, Z., Peterson, D. L., & Mote, P. (2004). Climatic change, wildfire, and conservation. *Conservation Biology, 18*, 890–902.

Miller, J. D., Safford, H. D., Crimmins, M., & Thode, A. E. (2008). Quantitative evidence for increasing forest fire severity in the Sierra Nevada and southern Cascade Mountains, California and Nevada, USA. *Ecosystems, 12*, 16–32.

Milly, P. C. D., Betancourt, J., Falkenmark, M., et al. (2008). Stationarity is dead: Whither water management? *Science, 319*, 573–574.

Pan, Y., Birdsey, R. A., Fang, J., et al. (2011). A large and persistent carbon sink in the world's forests. *Science, 333*, 988–993.

Peterson, D. L., Halofsky, J., & Johnson, M. C. (2011a). Managing and adapting to changing fire regimes in a warmer climate. In D. McKenzie, C. Miller, & D. Falk (Eds.), *The landscape ecology of fire* (pp. 249–267). New York: Springer.

Peterson, D. L., Millar, C. I., Joyce, L. A., et al. (2011b). *Responding to climate change on national forests: A guidebook for developing adaptation options* (Gen. Tech. Rep. PNW-GTR-855, 109pp). Portland: U.S. Department of Agriculture, Forest Service, Pacific Northwest Research Station.

Prasad, A., Iverson, L., Matthews, S., & Peters, M. (2009). Atlases of tree and bird species habitats for current and future climates. *Ecological Restoration, 27*, 260–263.

Prasad, A., Iverson, L., Peters, M., et al. (2010). Modeling the invasive emerald ash borer risk of spread using a spatially explicit cellular model. *Landscape Ecology, 25*, 353–369.

Price, C., & Rind, D. (1994). Possible implications of global climate change on global lightning distributions and frequencies. *Journal of Geophysical Research, 99*, 10823–10831.

Ranger, N., & Garbett-Shiels, S.-L. (2011). *How can decision-makers in developing countries incorporate uncertainty about future climate risks into existing planning and policy-making processes?* World Resources Report Uncertainty Series. Washington, DC: World Resources Institute. http://www.worldresourcesreport.org/decision-making-in-depth/managing-uncertainty. 15 Sep 2012.

Reeve, N., & Toumi, R. (1999). Lightning activity as an indicator of climate change. *Quarterly Journal of the Royal Meteorological Society, 125*, 893–903.

Reich, P. B. (2011). Taking stock of forest carbon. *Nature Climate Change, 1*, 346–347.

Rodenhouse, N. L. (1992). Potential effects of climatic change on migrant landbirds. *Conservation Biology, 6*, 263–272.

Rodewald, A. D., & Shustack, D. P. (2008). Consumer resource matching in urbanizing landscapes: Are synanthropic species over-matching? *Ecology, 89*, 515–521.

Rogers, B. M., Neilson, R. P., Drapek, R., Lenihan, J. M., et al. (2011). Impacts of climate change on fire regimes and carbon stocks of the U.S. Pacific Northwest. *Journal of Geophysical Research, 116*, G03037. doi:10.1029/2011JG001695.

Ryan, M. G., & Archer, S. R. (2008). Land resources: Forests and arid lands. In P. Backlund, A. Janetos, J. Hatfield, et al. (Eds.), *The effects of climate change on agriculture, land resources, water resources, and biodiversity. A report by the U.S. Climate Change Science Program and the Subcommittee on Global Change Research* (pp. 75–120). Washington, DC: U.S. Environmental Protection Agency.

Solomon, S., Quin, D., Manning, M., et al. (Eds.) (2007). *Climate change 2007: The physical science basis. Contribution of working group I to the fourth assessment report of the Intergovernmental Panel on Climate Change* (996pp). Cambridge: Cambridge University Press.

Turetsky, M. R., Kane, E. S., Harden, J. W., et al. (2010). Recent acceleration of biomass burning and carbon losses in Alaskan forests and peatlands. *Nature Geoscience, 4*, 27–31.

U.S. Department of Agriculture, Forest Service [USDA FS]. (2011). *Climate change atlas—Northern Research Station*. http://www.nrs.fs.fed.us/atlas

U.S. Department of Agriculture, Forest Service [USDA FS]. (2012). *Future of America's forest and rangelands: Forest Service 2010 Resources Planning Act assessment* (Gen. Tech. Rep. WO-87, 198pp). Washington, DC: U.S. Department of Agriculture, Forest Service.

U.S. Environmental Protection Agency [USEPA]. (2011a). Land use, land-use change, and forestry. Annex 3.12. Methodology for estimating net carbon stock changes in forest land remaining forest lands. In: Inventory of U.S. greenhouse gas emissions and sinks: 1990–2009. Publication #430-R-11-005. Washington, DC: U.S. Environmental Protection Agency.

U.S. Environmental Protection Agency [USEPA]. (2011b). Methodological descriptions for additional source or sink categories. Annex 3, section 3.12. In: Inventory of U.S. greenhouse gas emissions and sinks: 1990–2009 (pp. A254–A284). Washington, DC: U.S. Environmental Protection Agency.

Vörösmarty, C. J., McIntyre, P. B., Gessner, M. O., et al. (2010). Global threats to human water security and river biodiversity. *Nature, 467*, 555–561.

Woodall, C. W., Domke, G. M., Riley, K. L., et al. (n.d.). Developing a framework for assessing climate change risks to U.S. forest carbon stocks across large temporal and spatial scales. Unpublished manuscript. St. Paul: U.S. Forest Service, Northern Research Station.

Woodall, C. W., Skog, K., Smith, J. E., & Perry, C. H. (2011). Maintenance of forest contribution to global carbon cycles. In: G. Robertson, P. Gaulke, R. McWilliams (Eds.), National report on sustainable forests–2010. FS-979. Washington, DC: U.S. Department of Agriculture, Forest Service: II59–II65. Criterion 5.

Yohe, G. (2010). Risk assessment and risk management for infrastructure planning and investment. *The Bridge, 40*(3), 14–21.

Yohe, G., & Leichenko, R. (2010). Adopting a risk-based approach. *Annals of the New York Academy of Sciences, 1196*, 29–40.

Part IV
Scientific Issues and Priorities

Chapter 10
Research and Assessment in the Twenty-First Century

Toral Patel-Weynand, David L. Peterson, and James M. Vose

10.1 Improving the Accuracy and Certainty of Climate Change Science

We have heard more than one natural resource manager remark that keeping up with scientific information on climate change is like drinking from a fire hose! The sheer volume of scientific literature makes it challenging to sort through and evaluate evolving concepts and interpretations of climate change effects, as suggested by the fire hose simile. This proliferation of scientific information is providing a foundation for quantifying forest-climate relationships and projecting the effects of continued warming on a wide range of forest resources and ecosystem services. Certainty about climate change effects and understanding of risk to biosocial values has increased as more evidence has accrued.

The recent expansion in scientific analysis of the effects of climate on ecological disturbance has provided empirical data on how wildfire, insects, and other disturbances respond to warmer climatic periods. However, more information is needed on the interaction of ecological disturbances and other environmental stressors, especially for large spatial and temporal scales. Thresholds for climatic triggers of

T. Patel-Weynand
National Research and Development, U.S. Forest Service, Arlington, VA, USA
e-mail: tpatelweynand@fs.fed.us

D.L. Peterson (✉)
Pacific Northwest Research Station, U.S. Forest Service, Seattle, WA, USA
e-mail: peterson@fs.fed.us

J.M. Vose
Southern Research Station, U.S. Forest Service, Raleigh, NC, USA
e-mail: jvose@fs.fed.us

environmental change are poorly understood, and although simulation modeling can suggest how and when thresholds might be exceeded, additional empirical data are needed to confirm thresholds, and research is needed to improve the accuracy of process modeling at different spatial scales. Our understanding of stress complexes in forest ecosystems needs to be expanded to additional ecosystems, including better quantitative descriptions of stressor interactions.

Despite a century of ecological research on human-altered landscapes, our ability to interpret ecological change in the context of human land use and social values is incomplete. Documenting the effects of land use at small spatial and temporal scales is relatively straightforward, but we need to improve our ability to quantify the effects of land use on climate-ecosystem relationships at large spatial and temporal scales. Inferences about climate change effects will be more relevant if various land uses, including evaluation of future alternatives, are considered in a context that incorporates humans, rather than excluding them or considering their actions to be "unnatural" or negative. A framework for quantifying ecosystem services is needed that can be transported across different organizations and that includes a wide range of biosocial values.

Several general scientific issues also need additional focus. First, the value and appropriate interpretations of empirical (statistical) models versus process (mechanistic) models for projecting climate change effects warrant a rich discussion within the scientific community. Conceived from different first principles (e.g., assumed equilibrium [empirical] vs. dynamic [process] climate-species relationships), output from these types of models often differs considerably or is difficult to reconcile because of different assumptions, spatial resolution, and hierarchical levels (e.g., species vs. life form). Resource managers and other users of model information cannot be expected to understand the workings of complex simulation models. Therefore, it is incumbent on the scientific community to do a better job of stating model assumptions, sensitivities, and uncertainties, and to clearly indicate the appropriate contexts for interpreting and using model output.

Second, the direct effects of elevated carbon dioxide (CO_2) on forest ecosystems need to be clarified. Most existing inferences are based on experimental treatments on seedlings and small trees, and on output from simulation models that assume certain types of growth responses. Including CO_2 stimulation (or not) can drive the output of vegetation effects models to such an extent that this factor alone determines the direction of simulated response to climate. A unified effort by scientists to resolve the significant challenges in scaling and interpreting data on direct CO_2 effects (especially in mature forests) is needed to quantify future vegetation productivity and competition among plant species.

Third, effects models that can project multi-centennial patterns of vegetation distribution, disturbance, and biogeochemical cycling dynamics would provide longer term scenarios for planning and policy decisions. Most output from climate change effects models extends to 2100, the limit of projections for most global climate models. This may be sufficient for short-rotation (25–50 years) production forestry, but only scratches the surface for forest ecosystems in which trees can survive for hundreds of years.

We recommend that future research:

- Develop and implement new approaches to understand the effects of elevated CO_2 in mature and diverse forests. The knowledge gained from free-air CO_2 enrichment (FACE) experiments provided a solid understanding of short-term CO_2 responses in young forest stands for a limited number of species (Norby and Zak 2011). However, additional information is needed, including at least some evaluation of whole mature forest stands and physiological measurements of individual trees within stands. Studies in different forest ecosystems are needed to provide a broad perspective on how elevated CO_2 will affect forest productivity and other factors.
- Develop a standard approach for tracking carbon (C) dynamics in different forest ecosystems over space and time. This will improve ecological knowledge and provide consistent input to C accounting systems. It will be important to ensure that C measurements and C accounting can be used in a straightforward manner by resource managers.
- Identify appropriate uses and limitations of remote sensing imagery for detecting the effects of climatic variability and change in forest ecosystems. Remote sensing data from a variety of platforms are now more accessible than in the past, although these data can generally be analyzed and interpreted by only a few specialists. If tools to access, analyze, and help interpret the most reliable and relevant remote sensing data were easier to use, resource managers could obtain timely feedback on forest stress on a routine basis.
- Determine which ongoing and long-term forest measurements are useful or could be modified for tracking the effects of climate change. This may be a small subset of the monitoring data currently being collected on biophysical characteristics of forest ecosystems. Building on existing infrastructure for monitoring will be more efficient than developing new monitoring programs, thus extending time series of measurements taken with established protocols.
- Identify standard approaches for evaluating uncertainty and risk in vulnerability assessments and adaptation planning. Straightforward qualitative and quantitative frameworks will advance the decision making process on both public and private lands.
- Evaluate recently developed processes and tools for vulnerability assessment and adaptation planning to identify which ones are most effective for "climate smart" management on public and private lands. The availability of straightforward social and logistic protocols for eliciting and reviewing scientific information and stakeholder input will make climate change engagement more effective and timely.

It will be especially important to frame the above topics at the appropriate spatial and temporal scales in order to provide relevant input for different climate change issues. In addition, climatic data at different spatial scales need to be matched with applications at different spatial scales to be relevant for climate smart management (Wiens and Bachelet 2010). Despite the value of downscaled climatic and effects data, it should be recognized that the appropriate grain and extent of these data differ by resource (hydrology vs. vegetation vs. wildlife) and resource use (timber management vs. water supply vs. access for recreation). Sharing of information and experiences within and among organizations involved in climate change activities will facilitate incorporation of robust methods and applications across any particular landscape.

10.2 Toward an Ongoing National Assessment

Are we prepared to confront and respond to climate-related forest changes within the context of forest management? The answer lies in our ability to recognize potential loss, quantify risk, examine options, identify tradeoffs, anticipate rare but high-consequence events, and invest commensurate with risk. The challenge before us will require new tools, information, and technology, as well as the experience of resource managers.

As noted above, knowledge gaps exist in our ability to project how forest ecosystems will respond to the direct and indirect effects of climate change. Although ongoing research is addressing some of these knowledge gaps, the complexity of some scientific issues means that many management and policy decisions will continue to be based on imperfect information.

A long-term, consistent process to evaluate climatic risks and opportunities is needed to provide information that supports decision making at various levels. To that end, one objective of the 2014 U.S. National Climate Assessment (NCA) is to improve climate assessment capabilities in an integrated fashion. The current NCA approach is more focused than past climate assessments in supporting adaptation and mitigation, and in evaluating current scientific knowledge relative to climate effects and trends. The U.S. Global Change Research Program and NCA are working toward establishing a permanent national assessment capacity.

Natural resource assessment will be more powerful if the work of stakeholders and scientists across the United States is integrated in an ongoing and continuous process. It will be especially important to track specific climatic stressors, observe and project effects of climate change within regions and sectors, and rapidly deliver data and Web-based products that are relevant for decision making. Ongoing assessment of the effectiveness of mitigation and adaptation practices will also be needed. This is no small task. A truly successful national assessment process will require participation by federal, state, and local government agencies, nongovernmental organizations, academia, and tribal and private interests.

Several emerging issues identified in previous chapters urgently need to be incorporated in national and regional monitoring systems. Ecological disturbance, invasive species, urban forests, forest conversion to other uses, fragmentation of forest habitat, and C cycling are all dynamic entities for which timely monitoring is needed to inform effective adaptation options. Biosocial monitoring is also needed to track the effects of climate change on ecosystem services, human health, water and watersheds, energy and bioenergy, and forest industry.

Identifying areas where forests are most vulnerable to change (i.e., have low resistance and resilience) and where the effects of change on ecosystem services will be greatest is a significant challenge for resource managers. One would expect forest ecosystems and species near the limits of their biophysical requirements to be vulnerable, but the complexities of fragmented landscapes and multiple stressors are likely to alter response thresholds. Under these conditions, management approaches that anticipate and respond to change by guiding development and adaptation of forest ecosystem structure and function will be needed to sustain desired ecosystem services across large landscapes.

Some periodic assessments of forest resources have been implemented in the United States. For example, sustained efforts such as the Forest and Rangeland Renewable Resources Planning Act (RPA) assessment and periodic efforts such as the National Sustainability Report (USDA FS 2011) provide integrated national-scale information. The U.S. Forest Service Forest Inventory and Analysis (FIA) program measures forest growth and related parameters using consistent protocols at 10-year intervals across the nation. FIA is primarily a large-scale inventory, and climate change would need to have a significant effect on net forest growth for FIA data to detect it. However, these data can be used to calculate forest C flux approximated as a change in forest stocks over time. Accounting for C and managing ecosystems raises significant questions because of uncertainty about how C pools will change with climate. Coordination of remotely sensed data with on-the-ground data from inventory and monitoring is a powerful approach for quantifying climate-related trends in resource conditions—inferences are more convincing when multiple lines of evidence are available.

The most recent RPA assessment (USDA FS 2012) summarizes current conditions, trends, and forecasts for the next 50 years. Recent changes to the assessment include (1) presenting conditions, trends, and forecasts in a global context, (2) utilizing three global climate models and multiple emission scenarios (A1B, A2, and B2), and (3) integrating the analysis with socioeconomic factors (e.g., wood product markets and the price of timber). The RPA assessment indicates that forest area in the United States peaked at 253 million ha in 2010 and will decline through 2060 to between 243 and 247 million ha. Product markets, population, income, and climate all interact to determine future forest area, biomass, and forest C. Climate will influence the outcomes, and although significant variation exists across potential climate futures, it is still small relative to human factors in the short run.

10.3 Improving Risk Assessment

Many organizations are working to identify potential climate change vulnerabilities and effects, along with adaptation options to address them, but disparate analyses and interventions need to be incorporated in the context of risk assessment. A risk-based framework (see Chap. 9) needs to be further developed and agreed upon as a standard means for evaluating the consequences and likelihood of climate change effects. The NCA provides a simple set of guidelines for risk assessment (Yohe and Leichenko 2010), based on the risk and uncertainty framework developed by the Intergovernmental Panel on Climate Change (Moss and Schneider 2000). Risk assessment is now being incorporated in several national and state climate change management efforts. For example, all four National Research Council panel reports of "America's Climate Choices" incorporate this framework, as does the draft Adaptation Plan for the United States.

Although risk management frameworks have been used (often informally) in natural resource management for many years, it is a new approach for projecting climate change effects, and some time may be needed for scientists and resource managers to feel comfortable with it. Risk assessment should generally be specific to a particular region and time period, modified by an estimate of confidence in projections of climate change effects. Refining and expanding existing risk management frameworks will provide a consistent approach for addressing climate change vulnerabilities, so that risk can be evaluated iteratively over time as scientific information is updated.

References

Moss, R. H., & Schneider, S. H. (2000). Uncertainties in the IPCC TAR: Recommendations to lead authors for more consistent assessment and reporting. In R. Pachauri, T. Taniguchi, & K. Tanaka (Eds.), *Guidance papers on the cross cutting issues of the Third Assessment Report of the IPCC* (pp. 31–51). Geneva: World Meteorological Organization.

Norby, R. J., & Zak, D. R. (2011). Ecological lessons from free-air CO_2 enrichment (FACE) experiments. *Annual Review of Ecology and Systematics, 42*, 181–203.

U.S. Department of Agriculture, Forest Service [USDA FS]. (2011). *National report on sustainable forests—2010* (FS-979). Washington, DC: U.S. Department of Agriculture.

U.S. Department of Agriculture, Forest Service [USDA FS]. (2012). *Future of America's forest and rangelands: Forest Service 2010 Resources Planning Act assessment* (Gen. Tech. Rep. WO-87, 198pp). Washington, DC: U.S. Department of Agriculture.

Wiens, J. A., & Bachelet, D. (2010). Multiple scales of climate change. *Conservation Biology, 24*, 51–62.

Yohe, G., Leichenko, R. (2010). Adopting a risk-based approach. In *Climate change adaptation in New York City: Building a risk management response* (pp. 29–40). New York: New York City Climate Change Adaptation Task Force, Annals of the New York Academy of Sciences.

Index

A
Abies balsamea, 130
Accounting framework, 156–158
Acer
  A. rubrum, 39, 132
  A. saccharum, 37, 130
Acidic deposition, 130, 134
Adaptation/adaptive
  capacity, 106, 107, 192, 200
  framework, 188, 190
  learning, 186
  options, 21, 107, 136, 190, 193, 204, 206, 208, 213, 216, 250, 251
  planning, 184, 194, 207, 214, 249
  seasonality, 78, 122, 123
  strategies, 44, 46, 184–187, 189–194, 198, 201–204, 206, 212, 213, 216, 237, 238
  tactics, 189–194
Adelges tsugae, 31, 63, 136
Afforestation, 159–163, 168, 171
Agriculture, 100, 113, 120, 128, 131, 136, 158, 160, 163, 174
Agrilus planipennis, 127, 238
Ailanthus altissima, 67, 71
Air pollution, 30, 76, 78, 101, 138, 208
Alaska, 4, 5, 14, 44, 65, 78–80, 105, 115–117, 121, 137, 203, 232, 233
Alaska cedar, 117
Alder, 64
Allelopathy, 67
American basswood, 131
American beech, 131, 132
American Carbon Registry, 171
Aquatic systems, 166
Archips pinus, 163
Arctic Oscillation, 56

Area burned, 5–8, 57, 58, 115, 122, 124, 162, 232–236
Armillaria, 64
Artemisia tridentata, 77
Aspen leaf miner, 65
Assisted migration, 41–42, 190
Atlantic Multidecadal Oscillation, 56
Atmospheric chemicals, 102

B
Balsam fir, 130, 132, 134, 138
Balsam woolly adelgid, 63
Bark beetle, 6, 26, 58–63, 65, 66, 76–79, 122, 124, 125
Berberis thunbergii, 67, 71
Betula
  B. alleghaniensis, 132
  B. papyrifera, 130
Biochar, 165
Bioclimatic envelope, 42
Biocontrol, 73, 121
Biodiversity, 63, 65, 66, 102, 138, 153, 160, 166, 204–207, 212, 213
Bioenergy, 34, 35, 99, 100, 108, 250
Biogeochemical cycling, 26, 34, 69, 248
Biomass, 35, 44, 67, 75, 99, 101, 102, 122, 124, 128, 153, 154, 156, 159, 161, 164, 165, 167–170, 203, 230–233, 251
  energy, 153, 159, 161, 164, 165, 167–170
Biosocial systems, 224, 225
Biosocial values, 247, 248
Biotic interactions, 37, 73, 239
Birds, 35, 118, 134, 135, 138, 226, 239–241
Black ash, 237, 238
Black spruce, 39, 115, 130, 132
Boreal forest, 28, 63, 115, 129, 231, 233, 241

Bristlecone pine, 60
British Columbia, 5, 6
Broadleaf deciduous forest, 129
Bromus tectorum, 67, 69, 71
Buffelgrass, 77
Building material, 103, 158
Buildings, 67, 101, 102, 167
Bureau of Land Management, 44, 188, 200

C

Callitropsis nootkatensis, 117
Canada thistle, 67, 71
Carbon
  accounting, 156, 158, 168
  cycle, 28, 117, 152, 153
  dynamics, 7, 82, 155, 164, 170, 236, 249
  emission, 46, 137, 152, 157, 165, 167–171, 230, 231, 402
  management, 152, 153, 155–161, 163–169, 173, 189, 190
  mitigation, 159–163, 171–174
  neutrality, 170
  offset, 151–153, 156, 166, 169, 171
  sequestration, 34, 35, 99, 101, 106, 108, 129, 134, 137, 153–155, 203
  sink, 29, 44, 117, 159, 162, 230
  source, 29, 117, 230
  stock, 26, 30, 44, 46, 152–156, 159, 162–169, 172, 199, 230–232
  storage, 26–30, 44, 56, 64, 99, 102, 125, 152, 153, 155–162, 164, 165, 167, 172, 173
Carbon dioxide (CO_2), 4, 25, 59, 72, 99, 118, 151, 233, 248
  emissions, 4, 26, 101, 151, 159, 164, 170
  enrichment, 27, 72, 249
Carbon:nitrogen (C:N) ratio, 30
Cardinalis cardinalis, 239
Carolina hemlock, 63
CCRF. *See* Climate change response framework (CCRF)
Cenchrus ciliaris, 77
Centaurea, 67, 69, 71
  C. melitensis, 71
  C. solstitialis, 67
Cheatgrass, 67, 69–71, 77
Chequamegon-Nicolet National Forest, 210
Chinese privet, 68
Choristoneura
  C. fumiferana, 63
  C. occidentalis, 62
Cirsium arvense, 67, 71

Clean Air Act, 189
Clean Water Act of 1977, 189
Climate Action Reserve, 171
Climate Change Resource Center, 195, 196
Climate Change Response Framework (CCRF), 209, 210
Climatic envelope, 42, 133, 135
Climatic extreme, 114, 124, 134
Climatic variability, 30, 76, 100, 124, 183, 191, 203, 225, 233, 249
Co-benefits, 159–164
Cogongrass, 68, 69
Collaboration, 126, 185, 188, 196, 201, 202, 206, 207, 214–217
Common persimmon, 41
Competition, 40, 41, 72, 134, 160, 165, 248
Concrete, 167, 184
Conservation, 37, 44, 96–99, 121, 127, 170, 172, 173, 188, 195, 196, 198, 200, 203–205, 212, 228, 229
Conservation Reserve Program, 172, 173
Council on Environmental Quality, 173
C3 plants, 72
C4 plants, 72
Crown fire, 69, 126, 137, 153, 164, 233, 236

D

Dead wood, 26, 152, 153, 164, 174, 230–232
Decision making, 94, 153, 186, 191, 196, 200, 207, 216, 226, 249, 250
Decomposition, 27, 28, 30, 102, 115, 117, 134, 137
Defoliating insects, 62–64
Deforestation, 105, 153, 159–163, 171
Demographic area, 35, 44
Dendroctonus
  D. frontalis, 45, 61
  D. mexicanus, 62
  D. ponderosae, 5, 60, 122
  D. rufipennis, 5, 116, 123
Denitrification, 134
Department of the Interior (DOI), 198–200
Devils Postpile National Monument (DEPO), 207
Diospyros virginiana, 41
DISTRIB system, 238
Disturbance
  ecological, 5, 73, 76, 114, 241, 247, 250
  multiple, 7
  regime, 7, 8, 31, 43, 44, 46, 55–82, 115, 116, 133, 225, 227
Dothistroma needle blight, 65

Index

Dothistroma
  D. *pini*, 65
  D. *septosporum*, 65
Douglas-fir, 121, 123, 125, 128
Downscaling, 21, 118
Drought, tolerance, 171
Dynamic global vegetation model (DGVM), 39

E

Early detection and rapid response, 73
Eastern Broadleaf Forest, 129, 130
Eastern cottonwood, 127
Eastern hemlock, 30, 63, 73, 132, 136, 138
Eastern redcedar, 127
Ecological disturbance, 5, 73, 76, 114, 241, 247, 250
Ecological restoration, 8, 186
Economic effects, 58, 66
Economic policy, 94, 108
Ecoprovince, 233, 234
Ecosystem services, 9, 42–46, 61, 66, 93, 94, 97–99, 101, 104, 108, 115, 123, 152, 170, 184, 188, 190, 192, 205, 206, 208, 211, 215, 216, 226, 227, 229, 247, 248, 250, 251
  nonmarket, 99
Eddy covariance, 26–27
Education, 105, 107, 120, 184–186, 191, 196, 199, 203, 206–208, 212–215, 217
El Niño-Southern Oscillation (ENSO), 21, 56, 119, 120
Emerald ash borer, 66, 127, 128, 238
Emission scenario, 9, 13, 14, 17–21, 27, 37, 39, 57, 122, 125, 134, 233, 238, 251
Endangered Species Act, 187
Enhanced sequestration, 171
ENSO. *See* El Niño-Southern Oscillation (ENSO)
Environmental policy(ies), 173, 189
  and regulations, 8
Erosion, 31, 34, 42, 68, 74–75, 100, 120, 125, 128, 138, 139, 160, 162, 172, 208
Eucalyptus, 34
European privet, 131
Evapotranspiration, 31, 33, 118, 227, 228
Exposure, 74, 107, 108, 211, 225, 237, 239, 240

F

Fagus grandifolia, 131
Farm Service Agency, 172, 173

Fertilizer
  nitrogen, 162
  phosphorus, 165
FIA. *See* Forest Inventory and Analysis program (FIA)
Fire
  exclusion, 56, 79, 126–128, 233
  frequency, 34, 67, 68, 120, 232, 233, 236
  hazard, 68, 127, 128, 131, 153, 161, 164, 186, 232
  intensity, 56, 58, 68, 126, 164, 233, 236
  management, 8, 201, 205, 235
  regime, 56, 57, 69, 70, 115, 116, 118, 119, 122, 208, 235, 236
  risk, 57, 58, 67, 74, 75, 103, 161, 208, 232, 235, 236
  severity, 5, 8, 30, 57, 124, 125, 233, 235
  suppression, 8, 56–58, 122, 131, 235
Fish, 117, 121, 138, 172, 188, 198, 204, 206
Flood, 17, 31, 34, 43, 45, 74, 139, 186, 201, 213, 228, 229, 238
Forest
  biomass, 44, 156
  boreal, 28, 63, 115, 129, 231, 233, 241
  conversion, 250
  disturbance, 44, 45, 58, 59, 65, 76, 103, 212
  floor, 230, 231
  fragmentation, 35, 96, 134
  growth, 21, 28, 99, 102, 134, 158, 160, 161, 165–166, 169, 203, 251
  habitat, 119, 235–239, 241, 250
  management, 44, 94, 97–99, 105, 108, 136, 151, 153, 156–158, 161, 163, 164, 170–172, 174, 185, 187, 193, 207, 208, 213, 215, 216, 224, 225, 250
  plantation, 34, 69, 136, 159, 169
  productivity, 27, 28, 42, 45, 56, 58, 76, 100, 125, 132, 134, 137, 204, 249
  products, 24, 26, 28, 42, 45, 56, 58, 76, 100, 106, 125, 127, 132, 134, 137, 151, 153, 157–159, 161, 165–167, 169, 172, 201, 249
  structure, 5, 8, 31, 56, 68, 76, 103, 124, 155, 207, 216, 236
  temperate, 29, 115
Forest and Rangeland Renewable Resources Planning Act assessment, 251
Forest growth, 21, 28, 99, 102, 134, 158, 160, 161, 165–166, 169, 203, 251
Forest Inventory and Analysis (FIA)
  program, 26–27, 35, 37, 39, 169, 174, 231, 251

Forest management, 44, 94, 97–99, 105, 108, 136, 151, 156, 158, 161, 163, 164, 171–172, 174, 185, 187, 193, 207, 208, 213, 215–217, 224, 225, 250
Forest measurements, 249
Forest ownership
  family, 96
  fuelwood, 98
  private, 95
  public, 99
  rural, 97
Fossil fuel, 4, 8, 14, 34, 99, 102, 155–168, 173
Fragmentation, 35, 96, 131, 133, 134, 170, 208, 250
Fraxinus nigra, 237
Fraxinus pennsylvanica, 127
Fuel
  loading, 124, 127, 234–236
  treatment, 8, 46, 58, 72, 126, 138, 161, 164–165, 186, 188, 235, 236
Fungi, 60

G
GCMs. *See* Global climate models (GCMs)
Genetic diversity, 194
Genotypes, 42, 165
GHG. *See* Greenhouse gas (GHG)
Giant sequoia, 203
Glaciers, 5, 76, 117, 206
Global climate models (GCMs), 13–15, 17, 18, 21, 37, 39, 57, 122, 134, 238, 248, 251
Grass, 67, 68, 118, 227
Great Plains, 16, 126–129
Green ash, 127, 128
Greenhouse gas (GHG), 8, 13, 14, 18, 21, 29, 37, 59, 99, 107, 115, 153, 156, 157, 162, 163, 166, 169, 171, 173, 174, 212, 230, 231
  emissions, 8, 14, 37, 156
  emission scenarios, 13, 14, 21
Groundwater, 31, 34, 74, 120, 228
Gypsy moths, 63

H
Habitat, 7, 8, 35, 37, 41, 42, 56, 65, 66, 69, 71, 95, 101–103, 117, 119, 123, 127, 128, 130, 133, 134, 138, 160, 162, 172, 173, 183, 188, 192, 200, 201, 208, 227, 233, 235–241, 250
Hawaii, 4, 117–121
Heat wave, 16

Hemlock woolly adelgid, 30, 45, 63, 64, 136
Herbicide, 73
Herbivory, 57, 72
Historic range of variation, 8
Human communities, 107, 108, 129, 131, 199
Human values, 94
Hurricane, 3, 7, 80–82, 104, 137
Hurricane Sandy, 3
Hydrologic cycle, 17, 34, 117
Hydrology, 5, 9, 29, 33, 115, 117, 132, 134, 206, 250

I
Ice storm, 56, 136
Ignition, 56, 58, 232, 233
Imperata cylindrica [L.], 68, 69
Incentive programs, 158, 161, 170–173
Insect outbreak(s), 5, 26, 33, 55, 77, 78, 100, 103, 121, 124, 128, 129, 137, 187, 206, 208
Insects, 5, 21, 26, 30, 33–35, 39, 44, 45, 55, 56, 58–66, 74, 77–80, 100, 102, 103, 116, 120–122, 124, 125, 129, 136–138, 161, 165, 166, 169, 173, 187, 206–208, 213, 225, 232, 238, 247
Intergovernmental Panel on Climate Change, 13, 105, 174, 230, 252
Invasive plants, 63, 66–73, 103, 116, 118, 119, 134
Invasive species, 5, 55, 56, 69, 72, 73, 79, 116, 118–120, 124, 131, 133, 201, 203, 205, 208, 213, 250
Inyo National Forest (INF), 207
Ips confusus, 5, 60, 123

J
Jack pine budworm, 63
Janet's looper, 65
Japanese barberry, 67, 71
Juniper, 66, 75, 105, 127
Juniperus, 66, 127
  J. virginiana, 127

K
Kudzu, 68, 71, 131
Kyoto Protocol, 8

L
Landfill, 166–168
Landslide, 31, 34, 45, 72, 74–75

Land use, 9, 26, 31, 43–44, 56, 57, 94, 97–100, 102, 103, 105, 108, 127, 129, 131, 133, 136, 139, 153, 156–163, 170, 173, 202, 215, 224, 233, 248
Land-use change, 94, 99–100, 129
Laurentian Mixed Forest, 129
LCA. *See* Life cycle assessment (LCA)
Leaf area, 27, 32–34
Leakage, 157–161, 163
Life cycle assessment (LCA), 158, 170
　attributional, 158
　consequential, 158
Lightning, 80, 213, 232, 233
Ligustrum
　L. sinense, 68, 131
　L. vulgare, 131
Limber pine, 60
Liquidambar styraciflua, 41
Liriodendron tulipifera, 41
Loblolly pine, 41, 72, 137, 138
Lodgepole pine, 5, 78, 122, 125, 128
Longleaf pine, 69–70, 72, 81, 137, 138, 208
Lumber, 44, 158

M
Managed Forest Law Program, 172
Management
　adaptive, 185, 186, 189, 194, 200, 210, 212, 215, 216
　ecosystem-based, 187
　forest, 44, 94, 97–99, 105, 108, 136, 151, 153, 156–159, 161, 163, 164, 170–172, 174, 185, 187, 193, 207, 208, 213, 215–217, 224, 225, 250
　resource, 94, 103, 121, 185, 186, 196, 213, 239, 252
Markets, 34, 44, 98, 99, 102, 108, 153, 166, 170, 251
Marshall Islands, 117, 120
Mauna Loa, 4
MC1 model, 39, 122, 234
Methane, 28, 29, 59, 167, 230
Mexican pine beetle, 62
Microclimate, 98, 102
Micurapteryx salicifoliella, 65
Midwest, 15, 16, 18, 56, 129–132, 208
Mitigation, 44, 46, 97–99, 107, 108, 120, 129, 156, 157, 159–164, 170–174, 187, 192, 196, 199, 203, 212, 228, 250
Mixed models, 186–187
Model(ing)
　empirical, 73
　ensemble, 13, 14, 133

global climate, 13, 14, 37, 39, 57, 122, 134, 248, 251
　mechanistic, 41, 233, 248
　simulation, 41, 118, 125, 234, 248
Monitoring, 4, 61, 128, 155–158, 173, 174, 184, 186, 191, 194, 196, 198–200, 202, 212, 215, 216, 231, 232, 237, 249–251
Mountain pine beetle, 5, 60, 61, 66, 77, 122, 125, 127, 128
Mourning warbler, 240, 241
Mutualism, 60

N
National Climate Assessment, 58, 61, 113, 114, 250
National Environmental Policy Act, 173, 189
National forests, 44, 45, 121, 188, 197, 206–212, 214, 216
National Greenhouse Gas Inventory (NGHGI), 230, 231
National Oceanographic and Atmospheric Administration, 200
National Park Service (NPS), 200
National Sustainability Report, 251
Native Americans, 94, 97, 204
Natural resource-based communities, 104–105
Natural Resources Conservation Service, 172, 174
Nepytia janetae, 65
Net primary production (NPP), 27, 29
New Jersey pinelands, 61
NGHGI. *See* National Greenhouse Gas Inventory (NGHGI)
Nitrate, 30, 67, 134
Nitrification, 134
Nitrogen
　deposition, 25, 28–30, 70, 133, 134, 136
　mineralization, 30, 134
Non-equilibrium system, 8
Nongovernmental organizations, 97, 99, 204–205, 212, 250
Northeast, 4, 16, 18, 29, 32, 56, 63, 132–136, 208, 229
Northern cardinal, 239–241
Northern hardwoods, 132, 134
Northern whitecedar, 130
Northwest, 4, 5, 16, 18, 56, 57, 64, 105, 115, 121–123, 164, 204, 229, 232, 234
NPP. *See* Net primary production (NPP)
Nutrient
　cycling, 26–27, 30, 67, 136
　mineralization, 28

O

Oak-hickory, 37, 129, 131, 132
Olympic National Forest/Olympic National Park (ONFP), 206–207
Oporornis philadelphia, 240
Oxydendrum arboreum, 41

P

Pacific Decadal Oscillation (PDO), 5–7, 20, 21, 56, 233
Pacific Islands, 117–121
Paleoecological record, 8
Palmer Drought Severity Index (PDI), 19, 233
Paper
  birch, 130, 132, 134
  products, 98, 167
Parcelization, 96, 131, 132
Pathogens, 26, 40, 42, 45, 55, 56, 58–66, 72, 74, 78, 118, 120, 129, 133, 134, 136–138, 165, 166, 201
PDO. *See* Pacific Decadal Oscillation (PDO)
Peatland, 78, 130
Permafrost, 29, 78, 80, 115
Phaeocryptopus gaeumannii, 64
Phenology, 33, 59, 64, 239
Photosynthesis, 27–29, 78
Phyllocnistis populiella, 65
Physiology, 40, 59, 239
Picea
  P. mariana, 39, 115
  P. rubens, 132
Pinus
  P. albicaulis, 5, 60
  P. aristata, 60
  P. contorta var. *latifolia*, 5, 122
  P. edulis, 5, 33, 60, 123
  P. flexilis, 60
  P. palustris, 69, 72, 137, 208
  P. ponderosa, 72, 125, 127
  P. rigida, 41
  P. taeda, 72, 137
Pinyon ips, 5, 60, 66, 123
Pinyon pine, 5, 33, 60, 75, 77, 123–125
Pitch pine, 41
Planning rule, 198, 214
Plant
  association, 8
  pathogen, 64
Plantation, 34, 35, 42, 44, 69, 99, 136, 157–159, 165, 169, 170, 203, 225
Policy, 8, 9, 45, 58, 94, 97–100, 103, 107, 108, 122, 131, 153, 157, 158, 161, 173, 189, 196, 198, 200, 203, 205, 214–215, 248, 250
Pollution, 30, 76, 78, 98, 101, 134, 138, 208
Ponderosa pine, 72, 75, 125, 127, 128
Populus, 34,
  P. deltoides, 127
  P. tremuloides, 27, 67, 127
Potential evapotranspiration (PET), 227–229
Potential vegetation type, 8
Prairie Parklands, 129, 131
Precipitation
  extreme, 17, 18, 32, 74, 137, 227
  variability, 32, 45, 74–75
Prescribed fire, 56, 58, 72, 80, 81, 136–138
Process-based model, 35, 73
Productivity, 5, 7, 8, 27–29, 42, 44, 45, 55, 56, 58, 67, 70, 71, 76, 100, 108, 115, 124, 125, 129, 132, 136–138, 153, 165, 204, 248, 249
Propagule pressure, 71–73
Pseudotsuga menziesii, 121, 164
Public land management, 8, 44, 184
Public lands, 127, 136, 173–174, 184, 189, 200, 207, 208
Pueraria
  P. lobata, 131
  P. montana, 68, 71
Punctuated equilibrium, 7

Q

Quaking aspen, 27, 65, 127, 134
Quercus
  Q. alba, 237
  Q. coccinea, 131
  Q. falcata, 41
  Q. rubra, 131
  Q. velutina, 131

R

Radiative forcing, 14
Rainfall intensity, 138
Real estate investment trusts (REITs), 97
Realignment, 189, 193
Recreation, 44, 73, 96–98, 104–106, 108, 127, 139, 160, 162, 201, 206–208, 250
Red maple, 39, 132
Red oak, 41, 131
Red spruce, 132, 138
Reference condition, 8, 191
Reforestation, 99, 153, 159, 171

Index 259

Regional Greenhouse Gas Initiative (RGGI), 171
Registries, 171
Regulations, 8, 9, 42, 43, 99, 103, 108, 153, 161, 170–173, 191, 197, 214–215, 217
Remote sensing, 249
Representative concentration pathways (RCPs), 14
Reproductive rate, 136
Research, 9, 13, 15, 20, 37, 39, 73, 76, 113, 121, 128, 129, 136, 185, 196, 197, 200, 203, 205, 216, 226, 230, 232, 247–252
Reservations, 104–106, 204
Resilience, 8–9, 97, 104, 107, 118, 129, 131, 153, 186, 189, 190, 193, 196, 198, 200, 201, 203–205, 214, 216, 226, 236, 239, 251
Resistance, 46, 62, 63, 71–74, 165, 189, 190, 193, 194, 201, 208, 216, 238, 251
Resource management, 94, 103, 121, 185, 186, 196, 213, 239, 252
Response, 27–33, 39, 42–46, 56, 63, 71–77, 82, 98–99, 103, 107, 108, 117, 122, 123, 132, 134, 139, 153, 158, 183, 185–189, 193–196, 199–204, 206–217, 228, 229, 231, 232, 238–240, 248, 249, 251
Restoration, 8, 57, 72, 94, 120, 161, 172, 184, 186, 191, 193, 204, 205, 208, 213, 214
Riparian systems, 128
Risk
  analysis, 28, 185
  assessment, 73, 190, 223–241, 251–252
  framework, 225, 226, 230, 232
  management, 224, 226, 252
  matrix, 231, 235, 237–239
River discharge, 31
Roads, 72, 96, 186, 206, 214, 229
Root pathogen, 64
Rotation, 34, 44, 164, 169, 225
Roundwood, 98, 169,
Runoff, 20, 34, 67, 74, 75, 139, 227
Rural, 4, 44, 94, 97, 98, 100, 105, 108, 115, 198, 205

S
Sagebrush, 77
Sapwood, 33
Scale
  spatial, 9, 21, 43, 66, 123, 154, 155, 187, 188, 195, 196, 205, 248, 249
  temporal, 31, 186, 187, 192, 214, 247–249
Science-management partnership, 185, 206, 207, 209, 212, 215, 216

Scientific
  information, 113, 121, 201, 247, 249, 252
  literature, 195, 227, 232, 247
Sea level rise, 7, 20–21, 80, 82, 118–120
Sedimentation, 34, 67, 139, 208
Seed dispersal, 41
Sensitivity, 39, 58, 78, 81, 107, 123, 192, 195, 211, 239
Sequoiadendron giganteum, 203
Shelterbelt, 127, 128
Shoshone National Forest, 207
Sierra Nevada, 39, 78, 79, 125, 203, 208
Ski industry, 206
Snow, 4, 33, 34, 60, 71, 74–76, 117, 132, 139, 206
Snowmelt, 5, 28, 34, 70, 75
Snow water equivalent (SWE), 4, 5
Social effects, 137
Social interactions, 94, 103–107
Social system, 8, 44, 93, 114, 224–226
Socioeconomic, 58, 66, 94–99, 104, 107, 226–228, 235, 251
Socioeconomic systems, 229
Soil
  carbon, 27, 28, 164
  erosion, 34, 68, 128, 138, 139, 208
  fauna, 34
  organic carbon, 115, 152, 153, 230, 231
  organic matter, 137
  productivity, 165
  warming, 30
  water, 27, 31, 33, 66, 71, 75
Sourwood, 41
Southeast, 4, 14, 16, 17, 45, 57, 58, 69, 78, 80, 82, 136–139, 162
Southern hardwoods, 132
Southern pine, 164
Southern pine beetle, 5, 45, 60, 61
Southern red oak, 41
Species
  composition, 7, 29–33, 39, 55, 56, 65, 66, 68, 74, 80, 81, 101, 102, 115, 129, 132, 134, 216
  distribution model, 35, 37, 73, 238, 239
  migration, 121
Spillover, 158
Spotted knapweed, 67, 71
Spruce beetle, 5, 60, 116, 123
Spruce budworm, 62, 63
Spruce-fir forest, 65, 133
Statistical model, 37, 57, 121, 234, 241, 248
Steel, 167
Stomatal conductance, 31–33

Storm, 26, 31, 55, 56, 74, 119, 120, 134, 136, 137, 227, 229
Storm intensity, 7, 34, 119
Streamflow, 31, 33–35, 128, 132, 160, 162, 228
Stress complex, 78–82, 125, 129, 138, 248
Stressor, 7, 43, 46, 55–82, 107, 129, 136, 153, 165, 185, 194, 198, 201, 211, 212, 214, 225, 247, 248, 250, 251
Substitution, 108, 156–158, 161, 164–168
Sudden oak death, 64
Sugar maple, 37, 39, 130, 132
Surface fire, 57, 69, 164
Sustainability, 103, 132, 136, 173, 184, 187, 188, 198, 203, 204, 212, 213, 215, 251
Sweetgum, 41
Swiss needle cast, 64
System boundaries, 155–158, 166
System for Assessing Vulnerability of Species (SAVS), 196, 239, 240

T
Tamarix, 69
Tax incentives, 171
Temperature, maximum, 3, 4, 16, 17, 125
Tent caterpillar, 63
Thinning, 46, 61, 72, 74, 164, 165, 169, 173, 214, 235
Threshold, 9, 76–78, 81, 121, 124, 191, 194, 211, 225, 226, 234, 241, 247, 248, 251
Thuja occidentalis L., 130
Tilia americana, 131
Timber
  harvest, 8, 9, 56, 69, 97, 156, 166, 171, 214, 240
  industry, 127, 137, 138, 171
  production, 44, 45, 68, 95–99, 106
  products, 93, 108, 166
Timber investment management organizations (TIMOs), 97
Toolkit, 186
Transpiration, 31–34, 39, 40, 102, 163, 227
Tree
  breeding, 265
  cover, 44, 101, 102
  growth, 25, 28, 65, 66, 121, 123, 125, 128
  of heaven, 65, 71
  migration, 35–36, 131
  mortality, 5, 30, 33, 45, 58, 59, 61, 63, 74, 116, 125, 131, 164, 230
  regeneration, 5, 30, 68, 232
Treeline, 76
Tribal forests, 105–106, 188
Tropospheric ozone, 29, 133, 136

Tsuga
  T. canadensis, 30, 63, 73, 132
  T. caroliniana, 63
Tussock moth, 63

U
Uncertainty, 9, 21, 42, 43, 45, 63, 73, 80–83, 105, 118, 149, 155, 158, 160, 161, 166, 184–187, 217, 224, 226, 248, 251, 252
United Nations Framework Convention on Climate Change, 157, 174, 230
Urban
  forest, 101, 102, 108, 128, 159–163, 171, 250
  heat island, 45, 101
Urbanization, 97, 100, 127, 208
Use Value Appraisal Program, 172
U.S. Forest Service (USFS), 7, 8, 26, 41, 58, 61, 69, 81, 95, 117, 119, 123, 126, 129, 132, 136, 172, 174, 195–199, 204, 208, 209, 214, 223, 227, 230, 232, 235, 239, 251
U.S. Global Change Research Program (USGCRP), 113, 114, 192, 250
Uwharrie National Forest (UNF), 208

V
Vapor pressure deficit (VPD), 32, 122
Vegetation, 7, 8, 27, 35, 39, 55–57, 60, 65, 67–69, 74, 75, 101–103, 117, 118, 121–123, 125, 126, 128, 133, 162, 165, 188, 195, 206, 207, 227, 233, 248, 250
Verification, 173
Verified Carbon Standard, 171
Vulnerability
  assessment, 107, 189, 191, 192, 196, 197, 200–202, 205–212, 215, 216, 226, 249
  contextual, 107
  outcome, 107
  social, 94, 105–107

W
Wallow Fire, 124, 126
Water
  balance model, 227
  flow, 45, 227
  quality, 44, 68, 98, 108, 134, 136, 138, 160, 162, 172, 214, 227
  storm, 227, 229

Index

supply, 75, 120, 123, 139, 162, 206, 227, 228, 250
Watershed, 188, 190, 191, 198, 201, 204–206, 211, 213, 214, 227, 250
Watershed Vulnerability Assessment, 211–212
Weather pattern, 4, 233
Western Governors' Association (WGA), 201
Western Mountain Initiative (WMI), 76
Western spruce budworm, 62
Wetlands, 27, 28, 78, 98, 115, 117, 121, 130
Whitebark pine, 5, 60
White oak, 131, 134, 237, 238
White pine blister rust, 60
Wildfire
    management, 57, 58
    severity, 57
    suppression, 56–58, 131
Wildland-urban interface (WUI), 57, 58, 72, 94, 98, 100, 102–103, 108, 122, 235

Wildlife, 42, 56, 60, 63, 65, 67–69, 95, 102, 123, 127, 134, 160, 162, 172, 173, 185, 188, 192, 198, 202, 204, 206, 208, 250
Willow leafblotch miner, 67
Windbreak, 127, 128
Wisconsin, 172, 210, 235–241
Wood products, 44, 66, 98, 99, 127, 151–153, 155–160, 164–169, 238, 251
WUI. *See* Wildland-urban interface (WUI)

X
Xylem, 33

Y
Yellow birch, 132
Yellow poplar, 41, 237, 238
Yellow star-thistle, 67, 71
Yellowstone, 56, 57, 128, 207

Printed by Publishers' Graphics LLC
FMRO140506.23.35.43